Lecture Notes in Computer Science 11737

More information about this series at http://www.springer.com/series/7410

Cristina Pérez-Solà · Guillermo Navarro-Arribas ·
Alex Biryukov · Joaquin Garcia-Alfaro (Eds.)

Data Privacy Management, Cryptocurrencies and Blockchain Technology

ESORICS 2019 International Workshops, DPM 2019
and CBT 2019, Luxembourg, September 26–27, 2019
Proceedings

Springer

Editors
Cristina Pérez-Solà (ID)
Universitat Oberta de Catalunya
Barcelona, Spain

Guillermo Navarro-Arribas (ID)
Universitat Autonoma de Barcelona
Bellaterra, Spain

Alex Biryukov
University of Luxembourg
Esch-sur-Alzette, Luxembourg

Joaquin Garcia-Alfaro (ID)
Institut Mines-Télécom
Evry, France

ISSN 0302-9743   ISSN 1611-3349   (electronic)
Lecture Notes in Computer Science
ISBN 978-3-030-31499-6   ISBN 978-3-030-31500-9   (eBook)
https://doi.org/10.1007/978-3-030-31500-9

LNCS Sublibrary: SL4 – Security and Cryptology

This Springer imprint is published by the registered company Springer Nature Switzerland AG
The registered company address is: Gewerbestrasse 11, 6330 Cham, Switzerland

Foreword from the DPM 2019 Program Chairs

This volume contains the proceedings of the 14th Data Privacy Management International Workshop (DPM 2017), held in Luxembourg, on September 26, 2019. The workshop, as previous editions, was organized as part of the 24nd European Symposium on Research in Computer Security (ESORICS) 2019. The DPM series started in 2005 when the first workshop took place in Tokyo (Japan). Since then, the event has been held in different venues: Atlanta, USA (2006); Istanbul, Turkey (2007); Saint Malo, France (2009); Athens, Greece (2010); Leuven, Belgium (2011); Pisa, Italy (2012); Egham, UK (2013); Wroclaw, Poland (2014); Vienna, Austria (2015); Crete, Greece (2016); Oslo, Norway (2017); and Barcelona, Spain (2018).

The aim of DPM is to promote and stimulate international collaboration and research exchange on areas related to the management of privacy-sensitive information. This is a very critical and important issue for organizations and end-users. It poses several challenging problems, such as translation of high-level business goals into system-level privacy policies, administration of sensitive identifiers, data integration and privacy engineering, among others.

In this workshop edition we received 26 submission, and each one was evaluated on the basis of significance, novelty, and technical quality. The Program Committee performed an excellent job and all submissions went through a careful review process. In the end, eight full papers, and two short/position papers were accepted for publication and presentation at the event.

We would like to thank everyone who helped at organizing the event, including all the members of the Organizing Committee of both ESORICS and DPM 2019. Our gratitude also goes to Peter Ryan, to the ESORICS 2019 general chair, Peter Roenne and Magali Martin, ESORICS 2019 local organization chairs, and Joaquin Garcia-Alfaro, the workshops chair of ESORICS 2019. Last, but by no means least, we thank all the DPM 2019 Program Committee members, all the additional reviewers, all the authors who submitted papers, and all the workshop attendees.

Finally, we want to acknowledge the support received from the sponsors of the workshop: Universitat Autonoma de Barcelona (UAB), Internet Interdisciplinary Institute (IN3) from the Universitat Oberta de Catalunya (UOC), UNESCO Chair in Data Privacy, Institut Mines-Telecom (Telecom SudParis), CNRS Samovar UMR 5157 (R3S team), and projects TIN2017-87211-R and RTI2018-095094-B-C22 "CONSENT" from the Spanish MINECO.

August 2019

Cristina Pérez-Solà
Guillermo Navarro-Arribas

Organization

14th International Workshop on Data Privacy Management – DPM 2019

Program Chairs

Cristina Pérez-Solà   Universitat Oberta de Catalunya, Spain
Guillermo Navarro-Arribas   Universitat Autonoma de Barcelona, Spain

Program Committee

Archita Agarwal   Brown University, USA
Jordi Casas-Roma   Universitat Oberta de Catalunya, Spain
Jordi Castellà-Roca   Universitat Rovira i Virgili, Spain
Mauro Conti   University of Padua, Italy
Frédéric Cuppens   TELECOM Bretagne, France
Nora Cuppens-Boulahia   IMT Atlantique, France
Sabrina De Capitani di   University of Milan, Italy
  Vimercati
Jose Maria de Fuentes   Universidad Carlos III de Madrid, Spain
Josep Domingo-Ferrer   Universitat Rovira i Virgili, Spain
Christian Duncan   Quinnipiac University, USA
Sebastien Gambs   Université du Québec à Montréal, Canada
Joaquin Garcia-Alfaro   Telecom SudParis, France
Marit Hansen   Unabhängiges Landeszentrum für Datenschutz
                                 Schleswig-Holstein, Germany
Jordi Herrera-Joancomartí   Universitat Autònoma de Barcelona, Spain
Marc Juarez   Katholieke Universiteit Leuven, Belgium
Christos Kalloniatis   University of the Aegean, Greece
Florian Kammueller   Middlesex University London and TU Berlin,
                                 UK/Germany
Sokratis Katsikas   Center for Cyber and Information Security, NTNU,
                                 Norway
Hiroaki Kikuchi   Meiji University, Japan
Evangelos Kranakis   Carleton University, Canada
Alptekin Küpçü   Koç University, Turkey
Costas Lambrinoudakis   University of Piraeus, Greece
Maryline Laurent   Institut Mines-Telecom, France
Giovanni Livraga   University of Milan, Italy

Brad Malin	Vanderbilt University, USA
Fabio Martinelli	IIT-CNR, Italy
Chris Mitchell	Royal Holloway, University of London, UK
Anna Monreale	University of Pisa, Italy
Jordi Nin	ESADE, Universitat Ramon Llull, Spain
Melek Önen	EURECOM, France
Javier Parra-Arnau	Universitat Rovira i Virgili, Spain
Silvio Ranise	FBK-Irst, Italy
Kai Rannenberg	Goethe University Frankfurt, Germany
Ruben Rios	University of Malaga, Spain
Pierangela Samarati	University of Milan, Italy
David Sanchez	Universitat Rovira i Virgili, Spain
Claudio Soriente	NEC Laboratories Europe, Spain
Iraklis Symeonidis	SnT/APSIA, Luxembourg
Vicenç Torra	Maynooth University, Ireland
Yasuyuki Tsukada	Kanto Gakuin University, Japan
Alexandre Viejo	Universitat Rovira i Virgili, Spain
Isabel Wagner	De Montfort University, UK
Lena Wiese	Georg-August Universität Göttingen, Germany
Nicola Zannone	Eindhoven University of Technology, The Netherlands

Additional Reviewers

Osman Biçer
Alberto Blanco-Justicia
Majid Hatamian
Eleni Laskarina Makri
Sergio Martinez
Francesca Pratesi
Sanaz Taheri Boshrooyeh
Federico Turrin
Panagiotis Zagouras

Foreword from the CBT 2019 Program Chairs

This volume contains the proceedings of the Third International Workshop on Cryptocurrencies and Blockchain Technology (CBT 2019) held in Luxembourg, during September 26–27, 2019, in conjunction with the 24th European Symposium on Research in Computer Security (ESORICS) 2019.

Cryptocurrencies and blockchain technology is an area of research which is currently going through rapid development combining progress in IT and security technologies such as novel cryptogaphic techniques with economic insight and societal needs. As technology matures one can see a move from building blocks and proofs of concept to higher levels and concrete applications. To that end, the CBT workshop aims to provide a forum where researchers in this area can carefully analyze current systems and propose new ones in order to create a scientific background for the solid development of this new field.

In response to the call for papers, we received 32 submissions that were carefully reviewed by the Program Committee of 25 members as well as additional reviewers. Most of the submission received three reviews. The Program Committee selected ten full papers and five short papers for presentation at the workshop. The selected papers cover aspects of smart contracts, second layer and off-chain transactions, economic incentives, privacy, and applications.

Furthermore, the workshop was enhanced with keynote talks sponsored by the Research Institute (cf. https://researchinstitute.io/), BART (Blockchain Advanced Research & Technologies), Inria Saclay, Institut Mines-Télécom, and SAMOVAR (URM 5157 of CNRS).

Special thanks to all the authors who submitted papers to CBT 2019, the Program Committee and additional reviewers, who worked hard to review the submissions and discussed the final program. Last but not least, we would like to thank the ESORICS 2019 general chair, Peter Ryan, and the ESORICS 2019 local organization chairs, Peter Roenne and Magali Martin, for all their help and support.

August 2019

Alex Biryukov
Joaquin Garcia-Alfaro

Organization

Third International Workshop on Cryptocurrencies and Blockchain Technology – CBT 2019

Program Chairs

Alex Biryukov   University of Luxembourg, Luxembourg
Joaquin Garcia-Alfaro   Institut Mines-Telecom, TELECOM SudParis, France

Program Committee

Daniel Augot   Inria Saclay, France
Jean-Philippe Aumasson   Kudelski, Switzerland
George Bissias   University of Massachusetts at Amherst, USA
Joseph Bonneau   NYU, USA
Rainer Böhme   Universität Innsbruck, Austria
Christian Decker   Blockstream, Switzerland
Sergi Delgado-Segura   UCL, UK
Arthur Gervais   Imperial College London, UK
Hannes Hartenstein   KIT, Germany
Jordi Herrera-Joancomarti   UAB, Spain
Man Ho Au   The Hong Kong Polytechnic University, SAR China
Ghassan Karame   NEC Research, Germany
Aniket Kate   Purdue University, USA
Eleftherios Kokoris-Kogias   EPFL, Switzerland
Patrick McCorry   UCL, UK
Shin'ichiro Matsuo   Georgetown University, USA
Pedro Moreno-Sanchez   TU Wien, Autria
Guillermo Navarro-Arribas   UAB, Spain
Cristina Pérez-Solá   UOC, Spain
Bart Preneel   Katholieke Universiteit Leuven, Belgium
Tim Ruffing   Blockstream, Switzerland
Fatemeh Shirazi   Web3 Foundation, Switzerland
Ewa Syta   Trinity College, USA
Khalifa Toumi   SystemX, France
Edgar Weippl   SBA Research, Austria

Additional Reviewers

Sébastien Andreina
Daniel Feher
Michael Fröwis
Jan Grashoefer
Marc Leinweber
Wenting Li
Donghang Lu
Philipp Schindler

Clara Schneidewind
Brian Shaft
Oliver Stengele
Nicholas Stifter
Giuseppe Vitto
Karl Wuest
Alexei Zamyatin
Ren Zhang

Steering Committee

Rainer Böhme   Universität Innsbruck, Austria
Joaquin Garcia-Alfaro   Institut Mines-Telecom, France
Hannes Hartenstein   Karlsruher Institut für Technologie, Germany
Jordi Herrera-Joancomartí   Universitat Autònoma de Barcelona, Spain

Contents

CBT Workshop: Lightning Networks and Level 2

CBT Workshop: Smart Contracts and Applications

CBT Workshop: Payment Systems, Privacy and Mining

DPM Workshop: Privacy Preserving Data Analysis

PINFER: Privacy-Preserving Inference

Logistic Regression, Support Vector Machines, and More, over Encrypted Data

Marc Joye[✉] and Fabien Petitcolas

OneSpan, Brussels, Belgium
{marc.joye,fabien.petitcolas}@onespan.com

Abstract. The foreseen growing role of outsourced machine learning services is raising concerns about the privacy of user data. This paper proposes a variety of protocols for privacy-preserving regression and classification that (i) only require additively homomorphic encryption algorithms, (ii) limit interactions to a mere request and response, and (iii) that can be used directly for important machine-learning algorithms such as logistic regression and SVM classification. The basic protocols are then extended and applied to simple feed-forward neural networks.

Keywords: Machine learning as a service · Linear regression · Logistic regression · Support vector machines · Feed-forward neural networks · Data privacy · Additively homomorphic encryption

1 Introduction

The popularity and hype around machine learning, combined with the explosive growth of user-generated data, is pushing the development of *machine learning as a service* (MLaaS). A typical application scenario of MLaaS is shown in Fig. 1. It involves a client sending data to a service provider (server) owning and running a trained machine learning model for a given task (e.g., medical diagnosis). Both the input data and the model should be kept private: for obvious privacy reasons on the client's side and to protect intellectual property on the server's side.

In this paper we look at various protocols allowing the realisation of such scenario in a minimum number of message exchanges between both parties. Our assumption is that both the client and the server are honest but curious, that is, they both follow the protocol but may record information all along with the aim, respectively, to learn the model and to breach the client's privacy. Our design is guided by the following *ideal* requirements, in decreasing importance:

1. **Input confidentiality**—The server does not learn anything about the input data x provided by the client;
2. **Output confidentiality**—The server does not learn the outcome $\hat{y}$ of the calculation;

© Springer Nature Switzerland AG 2019
C. Pérez-Solà et al. (Eds.): DPM 2019/CBT 2019, LNCS 11737, pp. 3–21, 2019.
https://doi.org/10.1007/978-3-030-31500-9_1

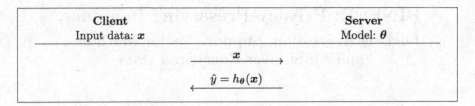

Fig. 1. A server offering MLaaS owns a model defined by its parameters θ. A client needs the prediction $h_\theta(x)$ of this model for a new input data x. This prediction is a function of the model and of the data.

3. **Minimal model leakage**—The client does not learn any other information about the model beyond what is revealed by the successive outputs.

With respect to the issue of model leakage, it is noted that the client gets access to the outcome, i.e., the value of $h_\theta(x)$, which may leak information about θ, violating Requirement 3. This is unavoidable and not considered as an attack within our framework. Possible countermeasures to limit the leakage on the model include rounding the output or adding some noise to it [19].

Related Work. Earliest works for private machine learning evaluation [2,17] were concerned with training models in a privacy-preserving manner. More recent implementations for linear regression, logistic regression, as well as neural networks are offered by SecureML [18]. The case of support vector machines (SVM) is, for example, covered in [21]. On the contrary, this paper deals with the problem of privately *evaluating* a linear machine-learning model, including linear/logistic regression and SVM classification. In [4], Bos et al. suggest to evaluate a logistic regression model by replacing the sigmoid function with its Taylor series expansion. They then apply fully homomorphic encryption so as to get the output result through a series of multiplications and additions over encrypted data. They observe that using terms up to degree 7 the Taylor expansion gives roughly two digits of accuracy to the right decimal. Kim et al. [16] argue that such an expansion does not provide enough accuracy on real-world data sets and propose another polynomial approximation. For SVM classification, Zhang et al. [21, Protocol 2] propose to return an encryption of the raw output. The client decrypts it and applies the discriminating function to obtain the corresponding class. Unfortunately, this leaks more information than necessary on the model. A similar path is taken by Barni et al. in [3] for feed-forward neural networks. Extracting the model (even partially) is nevertheless more difficult in their case because of the inherent complexity of the model. Moreover, to further obfuscate it (and thereby limit the potential leakage), the authors suggest to randomly permute computing units (neurons) sharing the same activation function or to add dummy ones. For classification, the approach put forward by Bost et al. [5] is closest to ours. They construct three classification protocols fulfilling our design criteria (Requirements 1–3): hyperplane decision, naïve Bayes, and decision trees. An approach orthogonal to ours that introduces privacy in regression or classification is differential privacy [8]. Crucially, it can be combined with secure computation, in our case by incorporating noise in the input vectors or in the model parameters. Differential privacy can thus be used to enhance the privacy properties of our protocols.

Our Contributions. Our paper follows the line of work by Bost et al., making use only of *additively* homomorphic encryption (i.e., homomorphic encryption supporting additions). We devise new privacy-preserving protocols for a variety of important prediction tasks. The protocols we propose either improve on [5] or address machine-learning models not covered in [5]. In particular, we aim at minimising the number of message exchanges to a mere request and response. This is important when latency is critical. An application of [5, Protocol 4] to binary SVM classification adds a round-trip to the comparison protocol whereas our implementation optimally only needs a *single* round-trip, all included. Likewise, a single round-trip is needed in our private logistic regression protocol, as in [4,21]. But contrary to [4,21], the resulting prediction is *exact* in our case (i.e., there is no loss of accuracy) and does not require the power of fully homomorphic encryption. With respect to neural networks, we adapt our protocols to binarised networks and to networks relying of the popular ReLU activation; see Sect. 4.2. As far as we know, this results in the first privacy-preserving implementation of the non-linear ReLU function from additively homomorphic encryption.

2 Preliminaries

2.1 Linear Models and Beyond

Owing to their simplicity, linear models (see, e.g., [1, Chapter 3] or [13, Chapters 3 and 4]) should not be overlooked: They are powerful tools for a variety of machine learning tasks and find numerous applications.

Problem Setup. In a nutshell, machine learning works as follows. Each particular problem instance is characterised by a set of d features which may be viewed as a vector $(x_1, \ldots, x_d)^\intercal$ of $\mathbb{R}^d$. For practical reasons, a fixed coordinate $x_0 = 1$ is added. We let $\mathcal{X} \subseteq \{1\} \times \mathbb{R}^d$ denote the input space and $\mathcal{Y}$ the output space.
There are two phases:

- The *learning phase* consists in approximating a target function $f \colon \mathcal{X} \to \mathcal{Y}$ from $\mathcal{D} = \left\{ (\boldsymbol{x}_i, y_i) \in \mathcal{X} \times \mathcal{Y} \mid y_i = f(\boldsymbol{x}_i) \right\}_{1 \leqslant i \leqslant n}$, a training set of n pairs of elements. Note that the target function can be noisy. The output of the learning phase is a function $h_\theta \colon \mathcal{X} \to \mathcal{Y}$ drawn from some hypothesis set of functions.
- In the *testing phase*, when a new data point $\boldsymbol{x} \in \mathcal{X}$ comes in, it is evaluated on h_θ as $\hat{y} = h_\theta(\boldsymbol{x})$. The hat on y indicates that it is a predicted value.

Since h_θ was chosen in a way to "best match" f on the training set $\mathcal{D}$, it is expected that it will provide a good approximation on a new data point. Namely, we have $h_\theta(x_i) \approx y_i$ for all $(x_i, y_i) \in \mathcal{D}$ and we should have $h_\theta(x) \approx f(x)$ for $(x, \cdot) \notin \mathcal{D}$. Of course, this highly depends on the problem under consideration, the data points, and the hypothesis set of functions.

In particular, linear models for machine learning use a hypothesis set of functions of the form $h_\theta(x) = g(\theta^\mathsf{T} x)$ where $\theta = (\theta_0, \theta_1, \ldots, \theta_d)^\mathsf{T} \in \mathbb{R}^{d+1}$ are the model parameters and $g \colon \mathbb{R} \to \mathcal{Y}$ is a function mapping the linear calculation to the output space. Specific choices for g are discussed hereafter.

Linear Regression. A *linear regression* model assumes that the real-valued target function f is linear—or more generally affine—in the input variables. In other words, it is based on the premise that f is well approximated by an affine map; i.e., g is the identity map: $f(x_i) \approx g(\theta^\mathsf{T} x_i) = \theta^\mathsf{T} x_i$, $1 \leqslant i \leqslant n$, for training data $x_i \in \mathcal{X}$ and weight vector $\theta \in \mathbb{R}^{d+1}$.

The linear regression algorithm relies on the least squares method to find the coefficients of θ. Once θ has been computed, it can be used to produce estimates on new data points $x \in \mathcal{X}$ as $\hat{y} = \theta^\mathsf{T} x$.

Support Vector Machines. Another important problem is to classify data into different classes. This corresponds to a target function f whose range $\mathcal{Y}$ is discrete. Of particular interest is the case of two classes, say $+1$ and -1, in which case $\mathcal{Y} = \{+1, -1\}$.

In dimension d, an hyperplane Π is given by an equation of the form $\theta_0 + \theta_1 X_1 + \theta_2 X_2 + \cdots + \theta_d X_d = 0$ where $\theta' = (\theta_1, \ldots, \theta_d)^\mathsf{T}$ is the normal vector to Π and $\theta_0 / \|\theta'\|$ indicates the offset from the origin. When the training data are *linearly separable*, there is some hyperplane Π such that for each $(x_i, y_i) \in \mathcal{D}$, one has $y_i \, \theta^\mathsf{T} x_i \geqslant 1$, $(1 \leqslant i \leqslant n)$. The training data points x_i satisfying $y_i \, \theta^\mathsf{T} x_i = 1$ are called *support vectors*. When the training data are not linearly separable, it is not possible to satisfy the previous hard constraint $y_i \, \theta^\mathsf{T} x_i \geqslant 1$, $(1 \leqslant i \leqslant n)$. So-called "slack variables" $\xi_i = \max(0, 1 - y_i \, \theta^\mathsf{T} x_i)$ are generally introduced in the optimisation problem. They tell how large a violation of the hard constraint there is on each training point.

There are many possible choices for θ. For better classification, the separating hyperplane Π is chosen so as to maximise the *margin*; namely, the minimal distance between any training data point and Π. Now, from the resulting model θ, when a new data point x comes in, its class is estimated as the sign of the discriminating function $\theta^\mathsf{T} x$; i.e., $\hat{y} = \mathrm{sign}(\theta^\mathsf{T} x)$.

When there are more than two classes, the optimisation problem returns several vectors θ_k, each defining a boundary between a particular class and all the others. The classification problem becomes an iteration to find out which θ_k maximises $\theta_k^\mathsf{T} x$ for a given test point x.

Logistic Regression. *Logistic regression* is widely used in predictive analysis to output a probability of occurrence. The logistic function is defined by the sigmoid function $\sigma \colon \mathbb{R} \to [0,1]$, $t \mapsto \sigma(t) = \frac{1}{1+e^{-t}}$. The logistic regression model returns $h_\theta(\boldsymbol{x}) = \sigma(\boldsymbol{\theta}^\mathsf{T}\boldsymbol{x}) \in [0,1]$, which can be interpreted as the probability that $\boldsymbol{x}$ belongs to the class $y = +1$. The SVM classifier thresholds the value of $\boldsymbol{\theta}^\mathsf{T}\boldsymbol{x}$ around 0, assigning to $\boldsymbol{x}$ the class $y = +1$ if $\boldsymbol{\theta}^\mathsf{T}\boldsymbol{x} > 0$ and the class $y = -1$ if $\boldsymbol{\theta}^\mathsf{T}\boldsymbol{x} < 0$. In this respect, the logistic function is seen as a soft threshold as opposed to the hard threshold, $+1$ or -1, offered by SVM. Other threshold functions are possible. Another popular soft threshold relies on tanh, the hyperbolic tangent function, whose output range is $[-1,1]$.

Remark 1. Because the logistic regression algorithm predicts probabilities rather than just classes, we fit it through likelihood optimisation. Specifically, given the training set $\mathcal{D}$, we learn the model by maximising $\prod_{y_i=+1} p_i \cdot \prod_{y_i=-1}(1 - p_i)$ where $p_i = \sigma(\boldsymbol{\theta}^\mathsf{T}\boldsymbol{x}_i)$. This deviates from the general description of our problem setup, where the learning is directly done on the pairs $(\boldsymbol{x}_i, y_i)$. However, the testing phase is unchanged: the outcome is expressed as $h_\theta(\boldsymbol{x}) = \sigma(\boldsymbol{\theta}^\mathsf{T}\boldsymbol{x})$. It therefore fits our framework for private inference, that is, the private evaluation of $h_\theta(\boldsymbol{x}) = g(\boldsymbol{\theta}^\mathsf{T}\boldsymbol{x})$ for a certain function g; the sigmoid function σ in this case.

2.2 Cryptographic Tools

Representing Real Numbers. So far, we have discussed a number of machine learning models using real numbers, but the cryptographic tools we intend to use require working on integers. In order to operate over encrypted data, we need to accurately represent real numbers as elements of the finite message set $\mathcal{M} \subset \mathbb{Z}$.

To do that, since all input variables of machine learning models are typically rescaled in the range $[-1,1]$, one could use a fixed point representation. A real number x with a fractional part of at most P bits uniquely corresponds to signed integer $z = x \cdot 2^P$. Hence, with a fixed-point representation, a real number x is represented by $z = \lfloor x \cdot 2^P \rfloor$, where integer P is called the bit-precision. The sum of $x_1, x_2 \in \mathbb{R}$ is performed as $z_1 + z_2$ and their multiplication as $\lfloor (z_1 \cdot z_2)/2^P \rfloor$.

Additively Homomorphic Encryption. It is useful to introduce some notation. We let $[\![\cdot]\!]$ and $]\!]\cdot[\![$ denote the encryption and decryption algorithms, respectively. The message space is an additive group $\mathcal{M} \cong \mathbb{Z}/M\mathbb{Z}$. It consists of integers modulo M and we view it as $\mathcal{M} = \{-\lfloor M/2 \rfloor, \ldots, \lceil M/2 \rceil - 1\}$ in order to keep track of the sign. Ciphertexts are noted with Gothic letters. If $\boldsymbol{m} = (m_1, \ldots, m_d) \in \mathcal{M}^d$ is a vector, we write $\mathfrak{m} = [\![\boldsymbol{m}]\!]$ as a shorthand for $(\mathfrak{m}_1, \ldots, \mathfrak{m}_d) = ([\![m_1]\!], \ldots, [\![m_d]\!])$. The minimal security notion that we require is semantic security [12]; in particular, encryption is probabilistic.

Algorithm $[\![\cdot]\!]$ being *additively* homomorphic (over $\mathcal{M}$) means that given any two plaintext messages m_1 and m_2 and their corresponding ciphertexts $\mathfrak{m}_1 = [\![m_1]\!]$ and $\mathfrak{m}_2 = [\![m_2]\!]$, we have $\mathfrak{m}_1 \boxplus \mathfrak{m}_2 = [\![m_1 + m_2]\!]$ and $\mathfrak{m}_1 \boxminus \mathfrak{m}_2 = [\![m_1 - m_2]\!]$ for some publicly known operations $\boxplus$ and $\boxminus$ on ciphertexts. By induction, for a given integer scalar $r \in \mathbb{Z}$, we also have $[\![r \cdot m_1]\!] := r \odot \mathfrak{m}_1$.

It is worth noting here that the decryption of $(\mathfrak{m}_1 \boxplus \mathfrak{m}_2)$ gives $(m_1 + m_2)$ as an element of $\mathcal{M}$; that is, $]\!] \, \mathfrak{m}_1 \boxplus \mathfrak{m}_2 \, [\!] \equiv m_1 + m_2 \pmod{M}$. Similarly, we also have $]\!] \, \mathfrak{m}_1 \boxminus \mathfrak{m}_2 \, [\!] \equiv m_1 - m_2 \pmod{M}$ and $]\!] \, r \odot \mathfrak{m}_1 \, [\!] \equiv r \cdot m_1 \pmod{M}$.

Private Comparison Protocol. In [6,7], Damgård et al. present a protocol for comparing private values. It was later extended and improved in [9] and [15,20]. The protocol makes use of an additively homomorphic encryption scheme. It compares two non-negative ℓ-bit integers. The message space is $\mathcal{M} \cong \mathbb{Z}/M\mathbb{Z}$ with $M \geqslant 2^\ell$ and is supposed to behave like an integral domain

DGK+ protocol. A client possesses a private ℓ-bit value $\mu = \sum_{i=0}^{\ell-1} \mu_i \, 2^i$ and a server possesses a private ℓ-bit value $\eta = \sum_{i=0}^{\ell-1} \eta_i \, 2^i$. They seek to respectively obtain bits δ_C and δ_S such that $\delta_C \oplus \delta_S = [\mu \leqslant \eta]$ (where $\oplus$ represents the exclusive OR operator, and $[\mathsf{Pred}] = 1$ if predicate Pred is true, and 0 otherwise). Following [15, Fig. 1], the DGK+ protocol proceeds in four steps:

1. The client encrypts each bit μ_i of μ under its public key and sends $[\![\mu_i]\!]$, $0 \leqslant i \leqslant \ell - 1$, to the server.
2. The server chooses uniformly at random a bit δ_S and defines $s = 1 - 2\delta_S$. It also selects $\ell + 1$ random non-zero scalars $r_i \in \mathcal{M}$, $-1 \leqslant i \leqslant \ell - 1$.
3. Next, the server computes[1]

$$
\begin{cases}
[\![h_i^*]\!] = r_i \odot \big([\![1]\!] \boxplus [\![s \cdot \mu_i]\!] \boxminus [\![s \cdot \eta_i]\!] \boxplus (\boxplus_{j=i+1}^{\ell-1} [\![\mu_j \oplus \eta_j]\!])\big) \\
\hspace{7.5cm} \text{for } \ell - 1 \geqslant i \geqslant 0, \quad (1) \\
[\![h_{-1}^*]\!] = r_{-1} \odot \big([\![\delta_S]\!] \boxplus \boxplus_{j=0}^{\ell-1} [\![\mu_j \oplus \eta_j]\!]\big)
\end{cases}
$$

and sends the $\ell + 1$ ciphertexts $[\![h_i^*]\!]$ in a *random order* to the client.
4. Using its private key, the client decrypts the received $[\![h_i^*]\!]$'s. If one is decrypted to zero, the client sets $\delta_C = 1$. Otherwise, it sets $\delta_C = 0$.

Remark 2. At this point, neither the client, nor the server, knows whether $\mu \leqslant \eta$ holds. One of them (or both) needs to reveal its share of δ ($= \delta_C \oplus \delta_S$) so that the other can find out. Following the original DGK protocol [6], this modified comparison protocol is secure in the semi-honest model (i.e., against honest but curious adversaries).

3   Basic Protocols of Privacy-Preserving Inference

In this section, we present three families of protocols for private inference. They aim to satisfy the ideal requirements given in the introduction while keeping the number of exchanges to a bare minimum. Interestingly, they only make use of additively homomorphic encryption.

We keep the general model presented in the introduction, but now work with integers only. The client holds $\boldsymbol{x} = (1, x_1, \ldots, x_d)^{\mathsf{T}} \in \mathcal{M}^{d+1}$, a private feature

---

[1] Given $[\![\mu_i]\!]$, the server obtains $[\![\eta_i \oplus \mu_i]\!]$ as $[\![\mu_i]\!]$ if $\eta_i = 0$, and as $[\![1]\!] \boxminus [\![\mu_i]\!]$ if $\eta_i = 1$.

vector, and the server possesses a trained machine-learning model given by its parameter vector $\boldsymbol{\theta} = (\theta_0, \ldots, \theta_d)^\mathsf{T} \in \mathcal{M}^{d+1}$ or, in the case of feed-forward neural networks a set of matrices made of such vectors. At the end of protocol, the client obtains the value of $g(\boldsymbol{\theta}^\mathsf{T}\boldsymbol{x})$ for some function g and learns nothing else; the server learns nothing. To make the protocols easier to read, for a real-valued function g, we abuse notation and write $g(t)$ for an integer t assuming g also includes the conversion to real values; see Sect. 2.2. We also make the distinction between the encryption algorithm $[\![\cdot]\!]$ using the client's public key and the encryption algorithm $\{\!\!\{\cdot\}\!\!\}_s$ using the server's public key and stress that, not only keys are different, but the algorithm could also be different. We use $]\!\!]\cdot[\![$ and $\}\!\!\}\cdot\{\!\!\{_s$ for the respective corresponding decryption algorithms.

3.1   Private Linear/Logistic Regression

Evaluating a linear regression model over encrypted data is straightforward and requires just one round of communication; see Appendix A.1. Things get more complicated for logistic regression. At first sight, it seems counter-intuitive that additively homomorphic encryption could suffice to evaluate a logistic regression model over encrypted data. After all, the sigmoid function, $\sigma(t)$, is non-linear. The key observation is that the sigmoid function is *injective*. So the client does not learn more about the model $\boldsymbol{\theta}$ from $t := \boldsymbol{\theta}^\mathsf{T}\boldsymbol{x}$ than it can learn from $\hat{y} := \sigma(t)$ since the value of t can be recovered from $\hat{y}$ using $t = \sigma^{-1}(\hat{y}) = \ln(\hat{y}/(1-\hat{y}))$.

Our Core Protocol. The protocol we propose for privacy-preserving linear or logistic regression is detailed in Fig. 2. Let (pk_C, sk_C) denote the client's matching pair of public encryption key/private decryption key for an additively homomorphic encryption scheme $[\![\cdot]\!]$. We use the notation of Sect. 2.2. If B is an upper bound on the inner product (in absolute value), the message space $\mathcal{M} = \{-\lfloor M/2 \rfloor, \ldots, \lceil M/2 \rceil - 1\}$ should be such that $M \geqslant 2B + 1$.

1. In a first step, the client encrypts its feature vector $\boldsymbol{x} \in \mathcal{M}^{d+1}$ under its public key pk_C and gets $[\![\boldsymbol{x}]\!] = ([\![x_0]\!], [\![x_1]\!], \ldots, [\![x_d]\!])$. The ciphertext $[\![\boldsymbol{x}]\!]$ along with the client's public key are sent to the server.
2. In a second step, from its model $\boldsymbol{\theta}$, the server computes an encryption of the inner product over encrypted data as: $\mathsf{t} = [\![\boldsymbol{\theta}^\mathsf{T}\boldsymbol{x}]\!] = [\![\theta_0]\!] \boxplus \boxplus_{j=1}^{d} \theta_j \odot [\![x_j]\!]$. The server returns t to the client.
3. In a third step, the client uses its private decryption key sk_C to decrypt t, and gets the inner product $t = \boldsymbol{\theta}^\mathsf{T}\boldsymbol{x}$ as a signed integer of $\mathcal{M}$.
4. In a final step, the client applies the g function to obtain the prediction $\hat{y}$ corresponding to input vector $\boldsymbol{x}$.

Dual Approach. The previous protocol encrypts with the client's public key pk_C. In the dual approach, the server's public key is used for encryption.

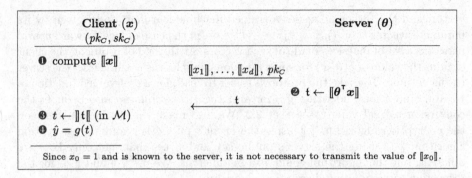

Fig. 2. Privacy-preserving regression. Encryption is done using the client's public key and noted $[\![\cdot]\!]$. The server learns nothing. Function g is the identity map for linear regression and the sigmoid function for logistic regression.

Let (pk_S, sk_S) denote the public/private key pair of the server for some additively homomorphic encryption scheme $(\{\!|\cdot|\!\}_s, \}\!\cdot\!\{_s)$. The message space $\mathcal{M}$ is unchanged.

In this case, the server needs to publish an encrypted version $\{\!|\boldsymbol{\theta}|\!\}_s$ of its model. The client must therefore get a copy of $\{\!|\boldsymbol{\theta}|\!\}_s$ once, but can then engage in the protocol as many times as it wishes. One could also suppose that each client receives a different encryption of $\boldsymbol{\theta}$ using a server's encryption key specific to the client, or that a key rotation is performed on a regular basis. This protocol uses a mask μ which is chosen uniformly at random in $\mathcal{M}$. Consequently, it is important to see that t^* ($\equiv \boldsymbol{\theta}^{\mathsf{T}}\boldsymbol{x}+\mu \pmod{M}$) is also uniformly distributed over $\mathcal{M}$. Thus, the server gains no bit of information from t^*. The different steps are summarised in Fig. 3.

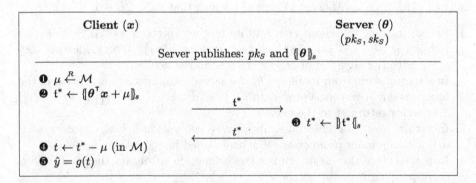

Fig. 3. Dual approach for privacy-preserving regression. Here, encryption is done using the server's public key pk_S and noted $\{\!|\cdot|\!\}_s$. Function g is the identity map for linear regression and the sigmoid function for logistic regression.

Extensions. The proposed methods are not limited to the identity map or the sigmoid function but generalise to any injective function g. This includes the hyperbolic tangent function, the arctangent function, the softsign function, the softplus function, the leaky ReLU function, and more. For any injective function g, there is no more information leakage in returning $\boldsymbol{\theta}^\mathsf{T}\boldsymbol{x}$ than $g(\boldsymbol{\theta}^\mathsf{T}\boldsymbol{x})$.

3.2   Private SVM Classification

As discussed in Sect. 2.1, SVM inference can be abridged to the evaluation of the sign of an inner product. However, the sign function is clearly not injective. Our idea is to make use of a privacy-preserving comparison protocol. For concreteness, we consider the DGK+ protocol; but any privacy-preserving comparison protocol could be adapted.

Our Core Protocol. As explained in Appendix A.2, a naïve implementation for private SVM has important issues. To address them we select the message space much larger than the upper bound B on the inner product.

Specifically, if $\boldsymbol{\theta}^\mathsf{T}\boldsymbol{x} \in [-B, B]$ then, letting ℓ be the bit-length of B, the message space $\mathcal{M} = \{-\lfloor M/2 \rfloor, \ldots, \lceil M/2 \rceil - 1\}$ should be dimensioned such that $M \geqslant 2^\ell(2^\kappa + 1) - 1$ for some security parameter κ. Let μ be an $(\ell + \kappa)$-bit integer that is chosen such that $\mu \geqslant B$. By construction we will then have $0 \leqslant \boldsymbol{\theta}^\mathsf{T}\boldsymbol{x} + \mu < M$ so that the decrypted value modulo M corresponds to the actual integer value. As will become apparent, this presents the further advantage of optimising the bandwidth requirements: the number of exchanged ciphertexts depends on the length of B and not on the length of M (notice that $M = \#\mathcal{M}$).

Our resulting core protocol for private SVM classification of a feature vector $\boldsymbol{x}$ is illustrated in Fig. 4 and includes the following steps:

0. The server publishes pk_s and $\{\!\{\boldsymbol{\theta}\}\!\}_s$.
1. Let κ be a security parameter. The client starts by picking uniformly at random in $[2^\ell - 1, 2^{\ell+\kappa})$ an integer $\mu = \sum_{i=0}^{\ell+\kappa-1} \mu_i 2^i$.
2. In a second step, the client computes, over encrypted data, the inner product $\boldsymbol{\theta}^\mathsf{T}\boldsymbol{x}$ and masks the result with μ to get $\mathsf{t}^* = \{\!\{t^*\}\!\}_s$ with $t^* = \boldsymbol{\theta}^\mathsf{T}\boldsymbol{x} + \mu$.
3. Next, the client individually encrypts the first ℓ bits of μ with its own encryption key to get $[\![\mu_i]\!]$, for $0 \leqslant i \leqslant \ell - 1$, and sends t^* and the $[\![\mu_i]\!]$'s to the server.
4. Upon reception, the server decrypts t^* to get $t^* := \}\!\{t^*\{\!\}_s \bmod M = \boldsymbol{\theta}^\mathsf{T}\boldsymbol{x} + \mu$ and defines the ℓ-bit integer $\eta := t^* \bmod 2^\ell$.
5. The DGK+ protocol is now applied to two ℓ-bit values $\underline{\mu} := \mu \bmod 2^\ell = \sum_{i=0}^{\ell-1} \mu_i 2^i$ and $\eta = \sum_{i=0}^{\ell-1} \eta_i 2^i$. The server selects the $(\ell+1)$-th bit of t^* for δ_S (i.e., $\delta_S = \lfloor t^*/2^\ell \rfloor \bmod 2$), defines $s = 1 - 2\delta_S$, and forms the $[\![h_i^*]\!]$'s (with $-1 \leqslant i \leqslant \ell - 1$) as defined by Eq. (1). The server permutes randomly the $[\![h_i^*]\!]$'s and sends them to the client.
6. The client decrypts the $[\![h_i^*]\!]$'s and gets the h_i^*'s. If one of them is zero, it sets $\delta_C = 1$; otherwise it sets $\delta_C = 0$.

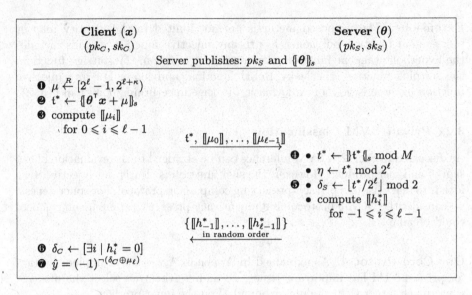

Fig. 4. Privacy-preserving SVM classification. Note that some data is encrypted using the client's public key pk_C, while other, is encrypted using the server's public key pk_S. They are noted $[\![\cdot]\!]$ and $\{\!\!\{\cdot\}\!\!\}_s$ respectively.

7. As a final step, the client obtains the predicted class as $\hat{y} = (-1)^{\neg(\delta_C \oplus \mu_\ell)}$, where μ_ℓ denotes bit number ℓ of μ.

Again, the proposed protocol keeps the number of interactions between the client and the server to a minimum: a request and a response.

Correctness. Remember that, by construction, $\boldsymbol{\theta}^\mathsf{T}\boldsymbol{x} \in [-B, B]$ with $B = 2^\ell - 1$, that $\mu \in [2^\ell - 1, 2^{\ell+\kappa})$, and by definition that $t^* := \{\!\!\{t^*\}\!\!\}_s \bmod M$ with $t^* = \{\!\!\{\boldsymbol{\theta}^\mathsf{T}\boldsymbol{x}+\mu\}\!\!\}_s$. Hence, in Step 4, the server gets $t^* = \boldsymbol{\theta}^\mathsf{T}\boldsymbol{x}+\mu \bmod M = \boldsymbol{\theta}^\mathsf{T}\boldsymbol{x}+\mu$ (over $\mathbb{Z}$) since $0 \leqslant \boldsymbol{\theta}^\mathsf{T}\boldsymbol{x} + \mu \leqslant 2^\ell - 1 + 2^{\ell+\kappa} - 1 < M$. Let $\delta := \delta_C \oplus \delta_s = [\underline{\mu} \leqslant \eta]$ (with $\underline{\mu} := \mu \bmod 2^\ell$ and $\eta := t^* \bmod 2^\ell$) denote the result of the private comparison in Steps 5 and 6 with the DGK+ protocol.

Either of those two conditions holds true

$$\begin{cases} 0 \leqslant \boldsymbol{\theta}^\mathsf{T}\boldsymbol{x} < 2^\ell \iff 1 \leqslant \frac{\boldsymbol{\theta}^\mathsf{T}\boldsymbol{x}+2^\ell}{2^\ell} < 2 \iff \left\lfloor \frac{\boldsymbol{\theta}^\mathsf{T}\boldsymbol{x}+2^\ell}{2^\ell} \right\rfloor = 1 \\ -2^\ell < \boldsymbol{\theta}^\mathsf{T}\boldsymbol{x} < 0 \iff 0 < \frac{\boldsymbol{\theta}^\mathsf{T}\boldsymbol{x}+2^\ell}{2^\ell} < 1 \iff \left\lfloor \frac{\boldsymbol{\theta}^\mathsf{T}\boldsymbol{x}+2^\ell}{2^\ell} \right\rfloor = 0 \end{cases},$$

and so, since $t^* = \boldsymbol{\theta}^\mathsf{T}\boldsymbol{x} + \mu$, $[\boldsymbol{\theta}^\mathsf{T}\boldsymbol{x} \geqslant 0] \in \{0, 1\}$, and $\delta_s = \lfloor t^*/2^\ell \rfloor \bmod 2$:

$$\begin{aligned} [\boldsymbol{\theta}^\mathsf{T}\boldsymbol{x} \geqslant 0] &= \left\lfloor \tfrac{\boldsymbol{\theta}^\mathsf{T}\boldsymbol{x}+2^\ell}{2^\ell} \right\rfloor = \left\lfloor \tfrac{t^*-\mu}{2^\ell} \right\rfloor + 1 = \left\lfloor \tfrac{t^*}{2^\ell} \right\rfloor - \left\lfloor \tfrac{\mu}{2^\ell} \right\rfloor + \left\lfloor \tfrac{\eta-\underline{\mu}}{2^\ell} \right\rfloor + 1 \\ &= \left\lfloor \tfrac{t^*}{2^\ell} \right\rfloor - \left\lfloor \tfrac{\mu}{2^\ell} \right\rfloor + \delta = \left(\left\lfloor \tfrac{t^*}{2^\ell} \right\rfloor - \left\lfloor \tfrac{\mu}{2^\ell} \right\rfloor + \delta \right) \bmod 2 \\ &= \left(\left\lfloor \tfrac{\mu}{2^\ell} \right\rfloor + \delta_C \right) \bmod 2 = \mu_\ell \oplus \delta_C. \end{aligned}$$

Now, noting $\mathrm{sign}(\boldsymbol{\theta}^\mathsf{T}\boldsymbol{x}) = (-1)^{\neg[\boldsymbol{\theta}^\mathsf{T}\boldsymbol{x} \geqslant 0]}$, we get the desired result.

Security. The security of the protocol of Fig. 4 follows from the fact that the inner product $\boldsymbol{\theta}^\mathsf{T}\boldsymbol{x}$ is statistically masked by the random value μ. Security parameter κ guarantees that the probability of an information leak due to a carry is negligible. The security also depends on the security of the DGK+ comparison protocol, which is provably secure (cf. Remark 2).

A Heuristic Protocol. The previous protocol, thanks to the use of the DGK+ algorithm offers provable security guarantees but incurs the exchange of $2(\ell+1)$ ciphertexts. Here we aim to reduce the number of ciphertexts and introduce a new heuristic protocol. This protocol requires the introduction of a signed factor λ, such that $|\lambda| > |\mu|$, and we now use both μ and λ to mask the model. To ensure that $\lambda\boldsymbol{\theta}^\mathsf{T}\boldsymbol{x} + \mu$ remains within the message space, we pick λ in $\mathcal{B}$ where $\mathcal{B} := \left[- \left\lceil \frac{\lceil M/2 \rceil}{B+1} \right\rceil, \left\lfloor \frac{\lceil M/2 \rceil}{B+1} \right\rfloor \right]$. Furthermore, to ensure the effectiveness of the masking, $\mathcal{B}$ should be sufficiently large; namely, $\#\mathcal{B} > 2^\kappa$ for a security parameter κ, hence $M > 2^\ell(2^\kappa - 1)$.

The protocol, which is illustrated in Fig. 5, runs as follows:

1. The client encrypts its input data $\boldsymbol{x}$ using its public key, and sends its key and the encrypted data to the server.
2. The server draws at random a signed scaling factor $\lambda \in \mathcal{B}$, $\lambda \neq 0$, and an offset factor $\mu \in \mathcal{B}$ such that $|\mu| < |\lambda|$ and $\mathrm{sign}(\mu) = \mathrm{sign}(\lambda)$. The server then defines the bit δ_s such that $\mathrm{sign}(\lambda) = (-1)^{\delta_s}$ and computes an encryption t^* of the shifted and scaled inner product $t^* = (-1)^{\delta_s} \cdot (\lambda\boldsymbol{\theta}^\mathsf{T}\boldsymbol{x} + \mu)$ and sends t^* to the client.
3. In the final step, the client decrypts t^* using its private key, recovers t^* as a signed integer of $\mathcal{M}$, and deduces the class of the input data as $\hat{y} = \mathrm{sign}(t^*)$.

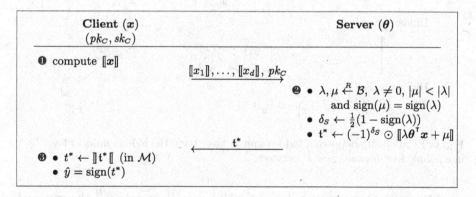

Fig. 5. Heuristic protocol for privacy-preserving SVM classification.

Correctness. The constraint $|\mu| < |\lambda|$ with $\lambda \neq 0$ ensures that $\hat{y} = \mathrm{sign}(\boldsymbol{\theta}^\mathsf{T}\boldsymbol{x})$. Indeed, as $(-1)^{\delta_s} = \mathrm{sign}(\lambda) = \mathrm{sign}(\mu)$, we have $t^* = (-1)^{\delta_s}(\lambda\boldsymbol{\theta}^\mathsf{T}\boldsymbol{x} + \mu) =$

$|\lambda||\boldsymbol{\theta}^{\mathsf{T}}\boldsymbol{x} + |\mu| = |\lambda|(\boldsymbol{\theta}^{\mathsf{T}}\boldsymbol{x} + \epsilon)$ with $\epsilon := |\mu|/|\lambda|$. Hence, whenever $\boldsymbol{\theta}^{\mathsf{T}}\boldsymbol{x} \neq 0$, we get $\hat{y} = \mathrm{sign}(t^*) = \mathrm{sign}(\boldsymbol{\theta}^{\mathsf{T}}\boldsymbol{x} + \epsilon) = \mathrm{sign}(\boldsymbol{\theta}^{\mathsf{T}}\boldsymbol{x})$ since $|\boldsymbol{\theta}^{\mathsf{T}}\boldsymbol{x}| \geqslant 1$ and $|\epsilon| = |\mu|/|\lambda| < 1$. If $\boldsymbol{\theta}^{\mathsf{T}}\boldsymbol{x} = 0$ then $\hat{y} = \mathrm{sign}(\epsilon) = 1$.

Security. We stress that the private comparison protocol we use in Fig. 5 does not come with formal security guarantees. In particular, the client learns the value of $t^* = \lambda\boldsymbol{\theta}^{\mathsf{T}}\boldsymbol{x} + \mu$ with $\lambda, \mu \in \mathcal{B}$ and $|\mu| < |\lambda|$. Some information on $t := \boldsymbol{\theta}^{\mathsf{T}}\boldsymbol{x}$ may be leaking from t^* and, in turn, on $\boldsymbol{\theta}$ since $\boldsymbol{x}$ is known to the client. The reason resides in the constraint $|\mu| < |\lambda|$. So, from $t^* = \lambda\boldsymbol{\theta}^{\mathsf{T}}\boldsymbol{x} + \mu$, we deduce $\log|t^*| \leqslant \log|\lambda| \cdot \log(|t| + 1)$. For example, when t has two possible very different "types" of values (say, very large and very small), the quantity $\log|t^*|$ can be enough to discriminate with non-negligible probability the type of t. This may possibly leak information on $\boldsymbol{\theta}$. That does not mean that the protocol is necessarily insecure but it should be used with care.

4 Application to Neural Networks

Typical *feed-forward neural networks* are represented as large graphs. Each node on the graph is often called a *unit*, and these units are organised into layers. At the very bottom is the input layer with a unit for each of the coordinates $x_j^{(0)}$ of the input vector $\boldsymbol{x}^{(0)} := \boldsymbol{x} \in \mathcal{X}$. Then various computations are done in a bottom-to-top pass and the output $\hat{y} \in \mathcal{Y}$ comes out all the way at the very top of the graph. Between the input and output layers, a number of *hidden layers* are evaluated. We index the layers with a superscript (l), where $l = 0$ for the input layer and $1 \leqslant l < L$ for the hidden layers. Layer L corresponds to the output. Each unit of each layer has directed connections to the units of the layer below; see Fig. 6.

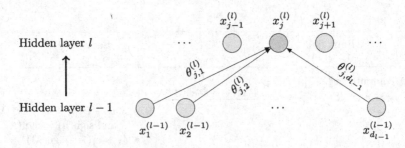

Fig. 6. Relationship between a hidden unit in layer l and the hidden units of layer $l-1$ in a simple feed-forward neural network.

We keep the convention $x_0^{(l)} := 1$ for all layers. If we note $\boldsymbol{\theta}_j^{(l)}$ the vector of weight coefficients $\theta_{j,k}^{(l)}$, $0 \leqslant k \leqslant d_{l-1}$, where d_l is the number of units in layer l, then $x_j^{(l)}$ can be expressed as:

$$x_j^{(l)} = g_j^{(l)}\left(\left(\boldsymbol{\theta}_j^{(l)}\right)^{\mathsf{T}}\boldsymbol{x}^{(l-1)}\right), \quad 1 \leqslant j \leqslant d_l \ . \tag{2}$$

Functions $g_j^{(l)}$ are non-linear functions such as the sign function or the Rectified Linear Unit (ReLU) function (see Sect. 4.2). Those functions are known as *activation functions*. The weight coefficients characterise the model and are known only to the owner of the model. Each hidden layer depends on the layer below, and ultimately on the input data $x^{(0)}$, known solely to the client.

Generic Solution. A generic solution can easily be devised from Eq. (2): for each inner product computation, and therefore for each unit of each hidden layer, the server computes the encrypted inner product and the client computes the output of the activation function in the clear. In more detail, the evaluation of a neural network can go as follows.

0. The client starts by encrypting its input data and sends it to the server.
1. Then, as illustrated in Fig. 7, for each hidden layer l, $1 \leqslant l < L$:
   (a) The server computes d_l encrypted inner products t_j corresponding to each unit j of the layer and sends those to the client.
   (b) The client decrypts the inner products, applies the required activation function $g_j^{(l)}$, re-encrypts, and sends back d_l encrypted values.
2. During the last round ($l = L$), the client simply decrypts the t_j values and applies the corresponding activation function $g_j^{(L)}$ to each unit j of the output layer. This is the required result.

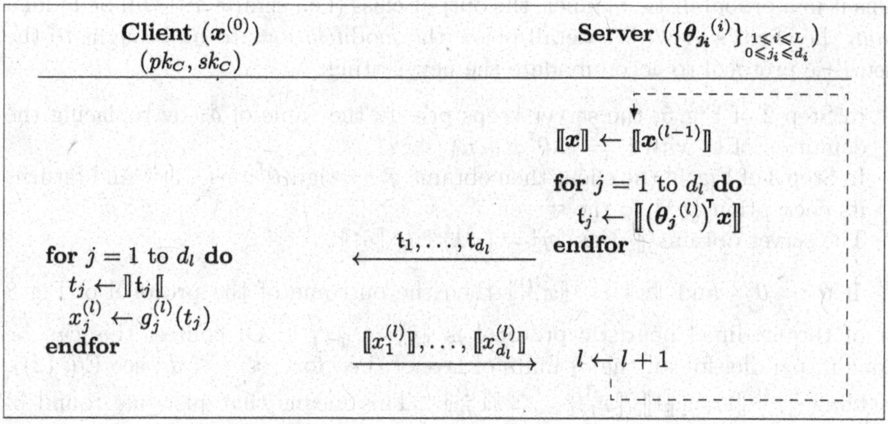

Fig. 7. Generic solution for privacy-preserving evaluation of feed-forward neural networks. Evaluation of hidden layer l.

For each hidden layer l, exactly two messages (each comprising d_l encrypted values) are exchanged. The input and output layers only involve one exchange; from the client to the server for the input layer and from the server back to the client for the output layer.

In the following two sections, we improve this generic solution for two popular activation functions: the sign and the ReLU functions. In the new proposed implementations, everything is kept encrypted—from start to end. The raw signals are hidden from the client's view in all intermediate computations.

4.1 Sign Activation

Binarised neural networks implement the sign function as activation function. This is very advantageous from a hardware perspective [14].

Section 3.2 describes two protocols for the client to get the sign of $\boldsymbol{\theta}^{\mathsf{T}}\boldsymbol{x}$. In order to use them for binarised neural networks in a setting similar to the generic solution, the server needs to get an encryption of $\mathrm{sign}(\boldsymbol{\theta}^{\mathsf{T}}\boldsymbol{x})$ for each computing unit j in layer l under the client's key from $[\![\boldsymbol{x}]\!]$, where $[\![\boldsymbol{x}]\!] := [\![\boldsymbol{x}^{(l-1)}]\!]$ is the encrypted output of layer $l-1$ and $\boldsymbol{\theta} := \boldsymbol{\theta}_j^{(l)}$ is the parameter vector for unit j in layer l.

We start with the core protocol of Fig. 4. It runs in dual mode and therefore uses the server's encryption. Exchanging the roles of the client and the server almost gives rise to the sought-after protocol. The sole extra change is to ensure that the server gets the classification result encrypted. This can be achieved by masking the value of δ_C with a random bit b and sending an encryption of $(-1)^b$. The resulting protocol is depicted in Fig. 8.

In the heuristic protocol (cf. Fig. 5), the server already gets an encryption of $[\![\boldsymbol{x}]\!]$ as an input. It however fixes the sign of t^* to that of $\boldsymbol{\theta}^{\mathsf{T}}\boldsymbol{x}$. If now the server flips it in a probabilistic manner, the output class (i.e., $\mathrm{sign}(\boldsymbol{\theta}^{\mathsf{T}}\boldsymbol{x})$) will be hidden from the client's view. We detail below the modifications to be brought to the heuristic protocol to accommodate the new setting:

- In Step 2 of Fig. 5, the server keeps private the value of δ_S by replacing the definition of t^* with $t^* = [\![\lambda\boldsymbol{\theta}^{\mathsf{T}}\boldsymbol{x} + \mu]\!]$.
- In Step 3 of Fig. 5, the client then obtains $\hat{y}^* := \mathrm{sign}(\boldsymbol{\theta}^{\mathsf{T}}\boldsymbol{x})\cdot(-1)^{\delta_S}$ and returns its encryption $[\![\hat{y}^*]\!]$ to the server.
- The server obtains $[\![\hat{y}]\!]$ as $[\![\hat{y}]\!] = (-1)^{\delta_S} \odot [\![\hat{y}^*]\!]$.

If $\boldsymbol{\theta} := \boldsymbol{\theta}_j^{(l)}$ and $[\![\boldsymbol{x}]\!] := [\![\boldsymbol{x}^{(l)}]\!]$ then the outcome of the protocol of Fig. 8 or of the modified heuristic protocol is $[\![\hat{y}]\!] = [\![x_j^{(l)}]\!]$. Of course, this can be done in parallel for all the d_l units of layer l (i.e., for $1 \leqslant j \leqslant d_l$; see Eq. (2)), yielding $[\![\boldsymbol{x}^{(l)}]\!] = ([\![1]\!], [\![x_1^{(l)}]\!], \dots, [\![x_{d_l}^{(l)}]\!])$. This means that just one round of communication between the server and the client suffices per hidden layer.

4.2 ReLU Activation

A widely used activation function is the ReLU function. It allows a network to easily obtain sparse representations and features cheaper computations as there is no need for computing the exponential function [10].

The ReLU function can be expressed from the sign function as

$$\mathrm{ReLU}(t) = \tfrac{1}{2}(1 + \mathrm{sign}(t)) \cdot t \ . \tag{3}$$

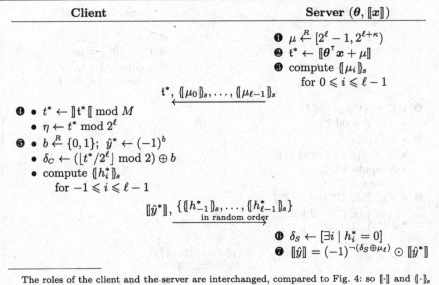

The roles of the client and the server are interchanged, compared to Fig. 4: so $\llbracket \cdot \rrbracket$ and $\{\!\!\{ \cdot \}\!\!\}_s$ are interchanged and so are δ_C and δ_S; $s = 1 - 2\delta_C$.

In Step 7, we abuse the $\hat{y}$ notation to mean either the input to the next layer or the final output.

Fig. 8. Privacy-preserving binary classification with inputs and outputs encrypted under the client's public key. This serves as a building block for the evaluation over encrypted data of the sign activation function in a neural network and shows the computations and message exchanges for one unit in one hidden layer.

Back to our setting, the problem is for the server to obtain $\llbracket \mathrm{ReLU}(t) \rrbracket$ from $\llbracket t \rrbracket$, where $t = \boldsymbol{\theta}^{\mathsf{T}} \boldsymbol{x}$ with $\boldsymbol{x} := \boldsymbol{x}^{(l-1)}$ and $\boldsymbol{\theta} := \boldsymbol{\theta}_j^{(l)}$, in just one round of communication per hidden layer. We saw in the previous section how to do it for the sign function. The ReLU function is more complex to apprehend. If we use Eq. (3), the difficulty is to let the server evaluate a *product* over encrypted data. To get around that, the server super-encrypts $\llbracket \boldsymbol{\theta}^{\mathsf{T}} \boldsymbol{x} \rrbracket$, gets $\{\!\!\{ \llbracket \boldsymbol{\theta}^{\mathsf{T}} \boldsymbol{x} \rrbracket \}\!\!\}_s$, and sends it the client. According to its secret share the client sends back the pair $\left(\{\!\!\{ \llbracket 0 \rrbracket \}\!\!\}_s, \{\!\!\{ \llbracket \boldsymbol{\theta}^{\mathsf{T}} \boldsymbol{x} \rrbracket \}\!\!\}_s \right)$ or $\left(\{\!\!\{ \llbracket \boldsymbol{\theta}^{\mathsf{T}} \boldsymbol{x} \rrbracket \}\!\!\}_s, \{\!\!\{ \llbracket 0 \rrbracket \}\!\!\}_s \right)$. The server then uses its secret share to select the correct item in the received pair, decrypts it, and obtains $\llbracket \mathrm{ReLU}(\boldsymbol{\theta}^{\mathsf{T}} \boldsymbol{x}) \rrbracket$. For this to work, it is important that the client re-randomises $\{\!\!\{ \llbracket \boldsymbol{\theta}^{\mathsf{T}} \boldsymbol{x} \rrbracket \}\!\!\}_s$ as otherwise the server could distinguish it from $\{\!\!\{ \llbracket 0 \rrbracket \}\!\!\}_s$.

Actually, a simple one-time pad suffices to implement the above solution. To do so, the server chooses a random mask $\mu \in \mathcal{M}$ and "super-encrypts" $\llbracket \boldsymbol{\theta}^{\mathsf{T}} \boldsymbol{x} \rrbracket$ as $\llbracket \boldsymbol{\theta}^{\mathsf{T}} \boldsymbol{x} + \mu \rrbracket$. The client re-randomises it as $\mathsf{t}^{**} := \llbracket \boldsymbol{\theta}^{\mathsf{T}} \boldsymbol{x} + \mu \rrbracket \boxplus \llbracket 0 \rrbracket$, computes $\mathsf{o} := \llbracket 0 \rrbracket$, and returns the pair $(\mathsf{o}, \mathsf{t}^{**})$ or $(\mathsf{t}^{**}, \mathsf{o})$, depending on its secret share. The server uses its secret share to select the correct item and "decrypts" it. If the server (obliviously) picked o, it already has the result in the right form; i.e., $\llbracket 0 \rrbracket$. Otherwise the server has to remove the mask μ so as to get $\llbracket \boldsymbol{\theta}^{\mathsf{T}} \boldsymbol{x} \rrbracket \leftarrow \mathsf{t}^{**} \boxminus \llbracket \mu \rrbracket$.

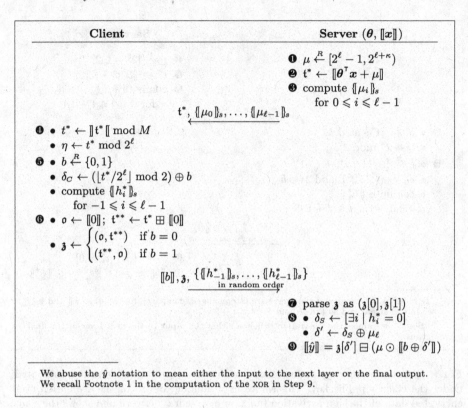

We abuse the $\hat{y}$ notation to mean either the input to the next layer or the final output.
We recall Footnote 1 in the computation of the XOR in Step 9.

Fig. 9. Privacy-preserving ReLU evaluation with inputs and outputs encrypted under the client's public key. The first five steps are as in Fig. 8. This building block is directed to neural networks using the ReLU activation and shows the computations and messages exchanges for one unit in one hidden layer.

In order to allow the server to (obliviously) remove or not the mask, the client also sends an encryption of the pair index; e.g., 0 for the pair $(\mathfrak{o}, \mathfrak{t}^{**})$ and 1 for the pair $(\mathfrak{t}^{**}, \mathfrak{o})$.

Figure 9 details an implementation of this with the DGK+ comparison protocol. Note that to save on bandwidth the same mask μ is used for the comparison protocol and to "super-encrypt" $[\![\boldsymbol{\theta}^{\mathsf{T}}\boldsymbol{x}]\!]$. The heuristic protocol can be adapted in a similar way.

Remark 3. It is interesting to note that the new protocols readily extend to any piece-wise linear function, such as the clip function $\mathrm{clip}(t) = \max(0, \min(1, \frac{t+1}{2}))$ (a.k.a. hard-sigmoid function).

5   Conclusion

In this work, we presented several protocols for privacy-preserving regression and classification. Those protocols only require additively homomorphic encryption and limit interactions to a mere request and response. They are secure against semi-honest adversaries. They can be used as-is in generalised linear models (including logistic regression and SVM classification) or applied to other machine-learning algorithms. As an illustration, we showed how they nicely adapt to binarised neural networks or to feed-forward neural networks with the ReLU activation function.

A   A More Private Protocols

A.1   Linear Regression

As seen in Sect. 2.1, linear regression produces estimates using the identity map for g: $\hat{y} = \boldsymbol{\theta}^{\mathsf{T}}\boldsymbol{x}$. Since $\boldsymbol{\theta}^{\mathsf{T}}\boldsymbol{x}$ is linear, given an encryption $[\![\boldsymbol{x}]\!]$ of $\boldsymbol{x}$, the value of $[\![\boldsymbol{\theta}^{\mathsf{T}}\boldsymbol{x}]\!]$ can be homomorphically evaluated, in a provably secure way [11].

Therefore, the client encrypts its feature vector $\boldsymbol{x}$ under its public key and sends $[\![\boldsymbol{x}]\!]$ to the server. Using $\boldsymbol{\theta}$, the server then computes $[\![\boldsymbol{\theta}^{\mathsf{T}}\boldsymbol{x}]\!]$ and returns it the client. Finally, the client decrypts $[\![\boldsymbol{\theta}^{\mathsf{T}}\boldsymbol{x}]\!]$ and gets the output $\hat{y} = \boldsymbol{\theta}^{\mathsf{T}}\boldsymbol{x}$. This is straightforward and only requires one round of communication.

A.2   SVM Classification

A Naïve Protocol. A client holding a private feature vector $\boldsymbol{x}$ wishes to evaluate $\mathrm{sign}(\boldsymbol{\theta}^{\mathsf{T}}\boldsymbol{x})$ where $\boldsymbol{\theta}$ parametrises an SVM classification model. In the primal approach, the client can encrypt $\boldsymbol{x}$ and send $[\![\boldsymbol{x}]\!]$ to the server. Next, the server computes $[\![\eta]\!] = [\![\boldsymbol{\theta}^{\mathsf{T}}\boldsymbol{x} + \mu]\!]$ for some random mask μ and sends $[\![\eta]\!]$ to the client. The client decrypts $[\![\eta]\!]$ and recovers η. Finally, the client and the server engage in a private comparison protocol with respective inputs η and μ, and the client deduces the sign of $\boldsymbol{\theta}^{\mathsf{T}}\boldsymbol{x}$ from the resulting comparison bit $[\mu \leqslant \eta]$.

There are two issues. If we use the DGK+ protocol for the private comparison, at least one extra exchange from the server to the client is needed for the client to get $[\mu \leqslant \eta]$. This can be fixed by considering the dual approach. A second, more problematic, issue is that the decryption of $[\![\eta]\!] := [\![\boldsymbol{\theta}^{\mathsf{T}}\boldsymbol{x} + \mu]\!]$ yields η as an element of $\mathcal{M} \cong \mathbb{Z}/M\mathbb{Z}$, which is not necessarily equivalent to the *integer* $\boldsymbol{\theta}^{\mathsf{T}}\boldsymbol{x} + \mu$. Note that if the inner product $\boldsymbol{\theta}^{\mathsf{T}}\boldsymbol{x}$ can take any value in $\mathcal{M}$, selecting a smaller value for $\mu \in \mathcal{M}$ to prevent the modular reduction does not solve the issue because the value of η may then leak information on $\boldsymbol{\theta}^{\mathsf{T}}\boldsymbol{x}$.

A Heuristic Protocol (Dual Approach). The bandwidth usage with the heuristic comparison protocol (cf. Fig. 5) could be even reduced to one ciphertext and a single bit with the dual approach. From the published encrypted model $\{\![\boldsymbol{\theta}]\!\}_s$, the client could homomorphically compute and send to the server $\mathsf{t}^* = \{\![\lambda\boldsymbol{\theta}^{\mathsf{T}}\boldsymbol{x} + \mu]\!\}_s$ for random $\lambda, \mu \in \mathcal{B}$ with $|\mu| < |\lambda|$. The server would then decrypt t^*, obtain t^*, compute $\delta_s = \frac{1}{2}(1 - \text{sign}(t^*))$, and return δ_s to the client. Analogously to the primal approach, the output class $\hat{y} = \text{sign}(\boldsymbol{\theta}^\mathsf{T}\boldsymbol{x})$ is obtained by the client as $\hat{y} = (-1)^{\delta_s} \cdot \text{sign}(\lambda)$. However, and contrarily to the primal approach, the potential information leakage resulting from t^*—in this case on $\boldsymbol{x}$—is now on the server's side, which is in contradiction with our Requirement 1 (input confidentiality). We do not further discuss this variant.

References

1. Abu-Mostafa, Y.S., Magdon-Ismail, M., Lin, H.T.: Learning From Data: A Short Course. AMLbook.com, New York (2012). http://amlbook.com
2. Agrawal, R., Srikant, R.: Privacy-preserving data mining. ACM SIGMOD Rec. **29**(2), 439–450 (2000). https://doi.org/10.1145/335191.335438
3. Barni, M., Orlandi, C., Piva, A.: A privacy-preserving protocol for neural-network-based computation. In: MM&Sec 2006, pp. 146–151. ACM (2006). https://doi.org/10.1145/1161366.1161393
4. Bos, J.W., Lauter, K., Naehrig, M.: Private predictive analysis on encrypted medical data. J. Biomed. Inf. **50**, 234–243 (2014). https://doi.org/10.1016/j.jbi.2014.04.003
5. Bost, R., Popa, R.A., Tu, S., Goldwasser, S.: Machine learning classification over encrypted data. In: NDSS 2015. The Internet Society (2015). https://doi.org/10.14722/ndss.2015.23241
6. Damgård, I., Geisler, M., Krøigaard, M.: Homomorphic encryption and secure comparison. Int. J. Appl. Cryptogr. **1**(1), 22–31 (2008). https://doi.org/10.1504/IJACT.2008.017048
7. Damgård, I., Geisler, M., Krøigaard, M.: A correction to 'efficient and secure comparison for on-line auctions'. Int. J. Appl. Cryptogr. **1**(4), 323–324 (2009). https://doi.org/10.1504/IJACT.2009.028031
8. Dwork, C., Feldman, V.: Privacy-preserving prediction. In: COLT 2018. PMLR, vol. 75, pp. 1693–1702. PMLR (2018). http://proceedings.mlr.press/v75/dwork18a/dwork18a.pdf
9. Erkin, Z., Franz, M., Guajardo, J., Katzenbeisser, S., Lagendijk, I., Toft, T.: Privacy-preserving face recognition. In: Goldberg, I., Atallah, M.J. (eds.) PETS 2009. LNCS, vol. 5672, pp. 235–253. Springer, Heidelberg (2009). https://doi.org/10.1007/978-3-642-03168-7_14
10. Glorot, X., Bordes, A., Bengjio, Y.: Deep sparse rectifier neural networks. In: AISTAT 2011. PMLR, vol. 15, pp. 315–323. PMLR (2011). http://proceedings.mlr.press/v15/glorot11a/glorot11a.pdf
11. Goethals, B., Laur, S., Lipmaa, H., Mielikäinen, T.: On private scalar product computation for privacy-preserving data mining. In: Park, C., Chee, S. (eds.) ICISC 2004. LNCS, vol. 3506, pp. 104–120. Springer, Heidelberg (2005). https://doi.org/10.1007/11496618_9
12. Goldwasser, S., Micali, S.: Probabilistic encryption. J. Comput. Syst. Sci. **28**(2), 270–299 (1984). https://doi.org/10.1016/0022-0000(84)90070-9
13. Hastie, T., Tibshirani, R., Friedman, J.: The Elements of Statistical Learning. Springer Series in Statistics, 2nd edn. Springer, New York (2009). https://doi.org/10.1007/978-0-387-84858-7

14. Hubara, I., Courbariaux, M., Soudry, D., El-Yaniv, R., Bengio, Y.: Binarized neural networks. In: NISP 2016, pp. 4107–4115. Curran Associates, Inc. http://papers. nips.cc/paper/6573-binarized-neural-networks.pdf

15. Joye, M., Salehi, F.: Private yet efficient decision tree evaluation. In: Kerschbaum, F., Paraboschi, S. (eds.) DBSec 2018. LNCS, vol. 10980, pp. 243–259. Springer, Cham (2018). https://doi.org/10.1007/978-3-319-95729-6_16

16. Kim, M., Song, Y., Wang, S., Xia, Y., Jiang, X.: Secure logistic regression based on homomorphic encryption: design and evaluation. JMIR Med. Inform. 6(2), e19 (2018). https://doi.org/10.2196/medinform.8805

17. Lindell, Y., Pinkas, B.: Privacy preserving data mining. In: Bellare, M. (ed.) CRYPTO 2000. LNCS, vol. 1880, pp. 36–54. Springer, Heidelberg (2000). https:// doi.org/10.1007/3-540-44598-6_3

18. Mohassel, P., Zhang, Y.: SecureML: A system for scalable privacy-preserving machine learning. In: IEEE S&P 2017, pp. 19–38. IEEE Computer Society (2017). https://doi.org/10.1109/SP.2017.12

19. Tramèr, F., Zhang, F., Juels, A., Reiter, M.K., Ristenpart, T.: Stealing machine learning models via prediction APIs. In: USENIX Security 2016, pp. 601–618. USENIX Association (2016). https://www.usenix.org/system/files/conference/ usenixsecurity16/sec16_paper_tramer.pdf

20. Veugen, T.: Improving the DGK comparison protocol. In: WIFS 2012, pp. 49–54. IEEE (2012). https://doi.org/10.1109/WIFS.2012.6412624

21. Zhang, J., Wang, X., Yiu, S.M., Jiang, Z.L., Li, J.: Secure dot product of outsourced encrypted vectors and its application to SVM. In: SCC@AsiaCCS 2017, pp. 75–82. ACM (2017). https://doi.org/10.1145/3055259.3055270

Integral Privacy Compliant Statistics Computation

Navoda Senavirathne[1,2(✉)] and Vicenç Torra[1,2]

[1] School of Informatics, University of Skövde, Skövde, Sweden
navoda.senavirathne@his.se, vtorra@ieee.org
[2] Hamilton Institute, Maynooth University, Maynooth, Ireland

Abstract. Data analysis is expected to provide accurate descriptions of the data. However, this is in opposition to privacy requirements when working with sensitive data. In this case, there is a need to ensure that no disclosure of sensitive information takes place by releasing the data analysis results. Therefore, privacy-preserving data analysis has become significant. Enforcing strict privacy guarantees can significantly distort data or the results of the data analysis, thus limiting their analytical utility (i.e., differential privacy). In an attempt to address this issue, in this paper we discuss how "integral privacy"; a re-sampling based privacy model; can be used to compute descriptive statistics of a given dataset with high utility. In integral privacy, privacy is achieved through the notion of stability, which leads to release of the least susceptible data analysis result towards the changes in the input dataset. Here, stability is explained by the relative frequency of different generators (re-samples of data) that lead to the same data analysis results. In this work, we compare the results of integrally private statistics with respect to different theoretical data distributions and real world data with differing parameters. Moreover, the results are compared with statistics obtained through differential privacy. Finally, through empirical analysis, it is shown that the integral privacy based approach has high utility and robustness compared to differential privacy. Due to the computational complexity of the method we propose that integral privacy to be more suitable towards small datasets where differential privacy performs poorly. However, adopting an efficient re-sampling mechanism can further improve the computational efficiency in terms of integral privacy.

Keywords: Privacy-preserving statistics ·
Privacy-preseving data analysis · Descriptive statistics

1 Introduction

Privacy preserving data analysis has become a strong requirement with the use of sensitive data in data analysis. The privacy requirement remains such that no analysis done on sensitive data should lead to any disclosure of sensitive information. Several definitions of what privacy means have been introduced

© The Author(s) 2019
C. Pérez-Solà et al. (Eds.): DPM 2019/CBT 2019, LNCS 11737, pp. 22–38, 2019.
https://doi.org/10.1007/978-3-030-31500-9_2

in the literature. They are computational definitions that permit us to build algorithms to provide solutions satisfying these privacy guarantees. Examples of such definitions include k-anonymity and differential privacy.

In [1], the concept of *integral privacy* (IP) was introduced with respect to machine and statistical learning models, which focuses on how the models are affected as the underlying data changes. In a real world scenario, the collected data we use for analysis may update over time. And this brings up the requirement to regenerate the data analysis results. An adversary who has access to previous and new (regenerated) data analysis results should not be able to infer any sensitive information despite of having access to auxiliary information. The privacy model suggests achieving privacy through releasing stable/robust results that are less likely to change due to small perturbation done to training data. The *stability* of the results are defined in terms of how many different combinations of data (generators) can be used to construct the same result.

In this paper, we study how to apply IP in order to compute descriptive statistics. In particular, we will consider mean, median, IQR, standard deviation, variance, count, sum, min and max. We have proposed a method based on data discretization and re-sampling to compute integrally private statistics. Also, we compare the differentially private statistics with the ones obtained with our approach for their robustness (variability) and accuracy.

The structure of the paper is as follows. In Sect. 2 we review the related work followed by Sect. 3 which explains the preliminary concepts. Section 4 describes the methodology. Evaluation and results are presented in Sect. 5. Section 6 contains the discussion and the paper finishes with a section on conclusions and lines for future work.

2   Related Work

Over the years, many different privacy models have been introduced to attain privacy preserving data analysis. Among them differential privacy [2] stands out due to its mathematical rigour. Differential privacy is considered in the context of statistics, mainly with respect to statistical database systems [3]. In an interactive setting, the data curators want to ensure answering queries submitted by the users does not lead to any form of disclosure. Dwork et al. discuss differentially private statistical estimators and how they can be applied to obtain privacy preserving statistics [4]. Also, in another work Dwork et al. explore the relationship between robust statistics and differential privacy [5]. Even though differential privacy provides a very strong, theoretically sound privacy guarantee, there are some practical limitations [6]. Intuitively, differential privacy states that any possible result of an analysis should be almost equally likely regardless of the presence or absence of specific data records. This goal is achieved through controlled random noise addition. This diminishes the utility of the final outputs greatly. Also, differential privacy is being criticized for its complexity in implementing differentially private mechanisms, the difficulty of adopting such mechanisms into other algorithms, deciding on privacy parameter ϵ, difficulty in estimating the sensitivity of an arbitrary function etc. Therefore, a solution is required that is compliant with the "indistinguishability" principle while capable of providing results with high utility. With that goal in mind in this work, we implement integral privacy [1] in the context of statistics in order to compute descriptive statistics. In some previous works, it is shown that the concept of IP can also be applied in the context of machine learning model section where stable models can be selected to achieve privacy [7,8].

3   Preliminaries

Differential privacy (DP) is the most commonly used privacy model in statistical and machine learning domains. The privacy guarantee of DP is such that the existence of any individual record can not be determined by examining the results of a function that was executed on two neighbouring datasets, which are differing from each other based on a single record. In other words, the result of a function does not change too much as a response to an addition or deletion of one record. This is achieved by introducing some uncertainty to the final result. Formally DP is defined as below.

Definition 1. *A randomized algorithm A is said to be ϵ-differentially private, if for all neighbouring data sets X and X', and for all events $E \subseteq Range(A)$,*

$$Pr[A(X) \in E] \ \leq \ e^{\epsilon} \ Pr[A(X') \in E]$$

Laplacian noise addition is one of the most commonly used mechanisms to implement DP in the case of numerical data. The noise is calibrated based on the "sensitivity" or the maximum variation a function can take [2].

Definition 2. *Let A be a real valued function; then, the global sensitivity of A is defined by*

$$\Delta A = \max_{d(X,X')=1} ||A(X) - A(X')||_1.$$

At the end, the noisy result is computed as $A(X) + Lap(\frac{\Delta A}{\epsilon})$ for $\epsilon > 0$. Here, ϵ is the privacy parameter.

The concept of integral privacy (IP) was first introduced in [1] with respect to machine and statistical learning models, that focuses on how the privacy of the models are affected by the changes done to the underlying dataset. It states that by observing the regenerated results (due to the dataset modification), an adversary with some auxiliary information can infer the modifications done to the input data (data addition and deletion) as it is being reflected by the final results. In [7], an adversarial model is explained with respect to machine learning model selection known as "model comparison attacks", that can be avoided by adhering to IP conditions. The idea is that, when the adversary has information on the previous ML model and the new ML model (trained after the changes applied to the input data) along with full/partial access to the training data used to generate the previous ML model; they can be used together in order to determine which input data have specifically resulted in the given ML model, or to derive an idea on how input data have been changed. Therefore, generating robust/stable results that are less likely to be affected by the input data modification is significant for privacy. The goal of IP is to protect from intruders learning about the database and about the set of modifications applied. DP achieves the above mentioned privacy requirement through random noise addition whereas, IP achieves it by releasing the least susceptible result for input modification.

IP is based on the concept of "generators of an output". Let P be the population (or an estimation of this population) in a given domain $\mathcal{D}$. Let A be an algorithm or a function that given a data set $S \subseteq P$ computes an output $A(S)$ that belongs to another domain $\mathcal{G}$. Then for any $G \in \mathcal{G}$ and some previous knowledge S on the generators, the set of possible generators of G is the set defined by $Gen(G,S) = \{S' | S \subseteq S' \subseteq P, A(S') = G\}$.

The following definition formalizes integral privacy. It is to protect inferences by an intruder who (i) has some partial knowledge S on the original database and on S' the database obtained after modification, (ii) has knowledge on the algorithm/function A applied to both databases, and (iii) on the output of this algorithm when applied to the original database (say, G) and the one obtained when applied to the modified database (say G').

Definition 3. *Let $G, G' \in \mathcal{G}$, let A be the algorithm to compute the function, let $S, S' \subseteq P$ be some background knowledge on the data sets used to compute G and G', and let*

$$\mathbb{M} = \cup_{g \in Gen(G,S), g' \in Gen(G',S')} \{g' \ominus g\}.$$

Then integral privacy is satisfied when the set $\mathbb{M}$ is large *and*

$$\cap_{m \in \mathbb{M}} m = \emptyset.$$

The null intersection is to avoid that all generators share record/s. This would imply that there is a minimum set of modifications that can be inferred from G and G'.

4   Methodology

Inferential analysis of aggregated statistics can be used to obtain a variety of sensitive information about the underlying dataset. Compared to DP, IP looks at privacy preservation from a slightly different angle. As explained in the previous section, the main goal here is to select a statistical or a machine learning model that can be represented by multiple generators. In other words, these are different combinations of input data samples with no shared records among them. In this case, it is infeasible to determine exactly what input data has resulted the specific output even though the adversary has access to crucial auxiliary information.

The implementation of IP achieves this through re-sampling and discretization of outputs. When deriving the answer for a given statistical query (e.g., mean), the proposed integral privacy based method selects the most recurrent result which can be generated by unique input data samples with no intersection among them.

In order to implement the above, it is required to construct the distribution of the outputs of a given function $A()$ considering all possible combinations of the input dataset. As generating all possible combinations are computationally expensive, a re-sampling based approximation method is used to build a sampling distribution of the outputs. A t number of re-samples (S_i) are drawn from the original dataset P and then a specific function $A()$ is computed for each of the re-sample as $m_i = A(S_i)$. In the end, the distribution of function outputs (m_i) is built based on the relative frequency of occurrence of each output. Here, t is a user defined parameter.

A user defined parameter k, is used to define the level of recurrence (frequency). In this context, k works as a frequency threshold. All the responses with the frequency of occurrence greater than k are selected as a candidate response. Then the responses (m_i) with no intersection among its generators are filtered out, and the one with the highest frequency of occurrence or the least error can be selected as the final answer. Parameter k can take any value ≥ 2.

However, it becomes challenging when IP needs to be applied on statistical databases, due to the fact that the range of a function A could be such that $A(S_i) \in \mathbb{R}$. This does not guarantee recurrence in output values as most of the outputs can be unique. Our solution to this problem is applying rounding based data discretization on input data as well as to the final result before determining the relative frequencies. By using data discretization, a continuous data set can be mapped to a finite, discrete set. We discuss our solutions for input and output discretization below.

1. Input discretization - We apply microaggregation (MA) a masking technique where the input data are divided into micro-clusters, and then they are replaced by the cluster representatives. Parameter y defines the number of minimum data points required to form a micro-cluster. As the cluster centroid is used to replace the original values that fall into the particular cluster, the uniqueness of data records is concealed, thus preserving the privacy of the released data. The basic idea is to generate homogeneous clusters over the original data in a way the distance between clusters are maximized. As the value of y increases, more distortion is applied to data and vice versa. Application of microaggregation on a numerical dataset transforms the data into a discrete space.

2. Output discretization - The output values of a given function are rounded off in order to limit the number of unique responses. This improves the frequency of occurrence of a given response value with respect to different data re-samples. In this case, the final answer is rounded-off to a decimal number with fewer digits, r (E.g., 2 decimal points).

As explained above, in order to implement IP, it is required to obtain re-samples of data from the original dataset. In this work bootstrapping is used as

Data: P: Data set;
n: number of re-samples;
A: function;
k: minimum frequency threshold for generators;
Result: Integrally private function results

1   $P^D := $ MA(P,y)   ▷ Input discretization using microaggregation (MA)
2   **for** $i = 1$ **to** n **do**
3      $S_i := bootstrapSample(P^D)$
4      $m_i := A(S_i)$ ▷ Compute function A() for given S_i
5      $md_i := round(m_i, r)$ ▷ Output discretization with r number of decimals
6      $E_M := $ add (md_i, S_i) ▷ Add the results to the re-sample function space
7   **end**
8   **for** *each unique* $md_i \in E_M$ **do**
9      $Frequency_i := frequency(md_i)$ ▷ Derive the distribution of function results
       from E_M
10     $DistributionOfResults :=$
       add $(E_M, Frequency_i, generatorList = append(concat(S_i)))$ ▷
       generatorList is upadted by concatenating all items in re-samples S_i
11   **end**
12   **for** *each* $md_i \in DistributionOfResults$ **do**
13   **if** $frequency(md_i) \geq k \wedge intersection(generatorList_i) == \emptyset$ **then**
14 | $CandidateResultList := $ add (md_i)
15   **end**
16   **end**
17   **return** $CandidateResultList$;

Algorithm 1. Integrally private statistics computation.

the re-sampling technique [9]. It consists of drawing s observations with replacement from the original data. Here s denotes the size of the original data. Sampling with replacement causes the replication of some observations while the exclusion of the others. On average, in a bootstrap sample, there are $0.632 * s$ unique observations. In this case, bootstrapping is selected as it draws samples that match the size of the original dataset (due to sampling with replacement). In this way, when IP is used to compute counting queries, it leads to the correct answer. Initially, we also experimented with sub-sampling technique that generates samples without replacement. Results observed in both cases are very similar (bootstrap results are marginally better than sub-sampling), except for the counting queries. Hence, we opted bootstrapping as the re-sampling technique.

To build each sample S_i^*, s instances are selected with replacement from the original data set (P). The samples are same in size as $s = |P|$. This process is repeated n times to generate the bootstrap distribution.

Algorithm 1 summarizes our method. The algorithm returns an empty list when there are no integrally private results for the function A, given the dataset and other user defined parameters. In that case, the discretization parameters (y in microaggregation, r rounding), number of re-samples (n) or the frequency threshold (k) can be adjusted to generate IP results. However, this can result in high computational cost (when increasing n) or high distortion of the results (when increasing y, r).

The above mentioned method is applied to compute IP solutions for some descriptive statistics. We focus our work in the following ones; mean, median, IQR, standard deviation, variance, count, sum, min and max. The experiments obtained using Algorithm 1 are described in the next section.

In IP, privacy is defined by the notion of stability, which leads to release of the least susceptible data analysis results towards the changes in the input data. Here, stability is explained by the relative frequency of different generators (re-samples of data) that lead to the same data analysis results. Highly stable results are recurring with respect to different data re-samples obtained from the original dataset. Integrally private output $f(x)$ can be considered as stable on the dataset x, if the same result appears more than a given frequency threshold k with respect to unique data re-samples drawn from x which does not share any common data instances among them.

5 Results and Evaluation

This section is focused on evaluating the effectiveness of our approach when computing a set of descriptive statistics. Here, we describe the experimental setting (data in Sect. 5.1, evaluation in Sect. 5.3), analysis of the results and comparison with differential privacy (Sect. 5.4) respectively.

5.1 Data

Six synthetic datasets (1-dimensional) and two real world datasets are used to evaluate the results. The parameters used for creating the synthetic data distributions are described along with the dataset dimension in Table 1. Abalone and breast cancer datasets are downloaded from the UCI data repository.

Table 1. Dataset descriptions.

Dataset	Instances × Columns	Description
Norm I	1000 × 1	Normally distributed with $\mu = 1, \sigma = 1$
Norm II	1000 × 1	Normally distributed with $\mu = 1, \sigma = 5$
Exp I	1000 × 1	Exponentially distributed with $\lambda = 1$
Exp II	1000 × 1	Exponentially distributed with $\lambda = 0.2$
Unif I	1000 × 1	Uniformly distributed in range (min $= 0$, max $= 100$)
Unif II	1000 × 1	Uniformly distributed in range (min$= 0$, max $= 1000$)
Abalone Dataset	4177 × 8	UCI data repository
Breast Cancer	683 × 7	UCI data repository

5.2   Experimental Setup

Algorithm 1 is implemented for calculating the descriptive statistics compliant with IP. Nine basic descriptive statistics have been considered. They are mean, median, standard deviation, min, max, interquartile range (IQR), sum and variance. Algorithm 1 is used to calculate the descriptive statistics compliant with IP. In this case, the number of re-samples (n) extracted from the original dataset is set to 1000 for synthetic datasets, 5000 for the Abalone dataset and 700 for the breast cancer dataset which is roughly closer to the number of instances (i). For both IP and DP, before reporting the final statistics, 10 iterations are carried out per each statistic and then the mean values are reported with their standard deviation and mean absolute relative error (ARE) for evaluation purpose.

For calculating DP statistics Laplacian mechanism is used. As explained in the preliminaries section, to calibrate the noise for DP, we need to derive the sensitivity of the functions. To compute the maximum variation a function can take (global sensitivity), it is essential to know the lower and upper bounds for the domain of a given dataset.

The global sensitivity of a function can be very large, causing high distortion to the computed results. Because of that local sensitivity derived from the dataset is used for some functions (i.e., median, max, min, IQR). For normally and exponentially distributed data the minimum and maximum bounds for the datasets lies between $(-20, 20)$ whereas, for the unif I the values range from $(0, 100)$ and for Unif II from $(0, 1000)$. Given that the Abalone and Breast cancer datasets are biological datasets, no strict domain bounds are introduced to pre-process the data as it presents very limited chances of being boundless. Therefore, when computing function's sensitivity min and max values of the respective datasets are used.

When computing differential privacy statistics mechanisms introduced in the literature are used to estimate the global/local sensitivity of the statistical functions as mentioned below. For median, max and min functions techniques introduced in [10] are used with local sensitivity. For mean calculation *noisy average clamping down* algorithm introduced in [11] is used whereas, for sum queries maximum value in the domain (i.e., in this case the max value of the specific dataset) is used. For IQR calculation *Scale* algorithm proposed in [12] is used. For variance and standard deviation the min and max values computed using the above techniques are used to estimate the function's sensitivity. For counting queries the global sensitivity is set to 1.

For IP, 18 different dataset instances are evaluated based on the data distribution type and discretization parameters. Two discretization phases are used as *output discretization (Out Dis:)*, and *both input and output discretization (in/out Dis:)* combined with input discretization levels, low (L) and high (H). In low discretization level parameter y for microaggregation is set to 2 whereas for high discretization parameter y is set to 20. In all the cases, rounding parameter is set to 2 at the output discretization phase. For DP different data distributions are used with differing ϵ values which indicates the amount of privacy.

5.3   Evaluation Criteria

For the evaluation purposes three measures are used as mentioned below. Here, $A()$ indicates the statistic value to be computed (e.g., mean(), median()), P indicates the original dataset, S_i indicates the re-samples, IP{} indicates integrally private value selection, true value indicates the real statistic value computed on the original dataset and private value indicate the mean IP or DP compliant statistic value. When computing absolute relative error (ARE) the distance between the true value and the private value is divided by maximum among 1 or true value to avoid division by zero. A lower ARE indicate less distorted IP/DP results.

$$IP\ Mean\ Statistic\ Value\ With\ SD = \frac{\sum\limits_{j=1}^{10} IP\{A(S_1)\dots A(S_i)\}}{10} \pm SD \quad (1)$$

$$DP\ MeanStatistic\ Value\ With\ SD = \frac{\sum\limits_{j=1}^{10} \{A(P) + Lap(\frac{\Delta A}{\epsilon})\}}{10} \pm SD \quad (2)$$

$$Absolute\ Relative\ Error(ARE) = \frac{|True\ Value - Private\ Value|}{max\{1, True\ Value\}} \quad (3)$$

5.4   Results and Discussion

Variability/Robustness of the Results. Tables 2 and 3 respectively show IP statistics and DP statistics computed on the synthetic dataset. In this case, we wanted to check the variation of the final results among different iterations. The same is illustrated by Fig. 1. By observing the results, few interesting facts can be noted. Relative to DP in IP the variability of the results is low in many instances. This is indicated by the $\pm SD$ values. Further, this indicates that as opposed to adding Laplacian noise to achieve DP, re-sampling based IP provides more stable/robust answers with less variability despite different iterations. However, DP performs better than IP when calculating sum() and mean(). This behaviour is expected as re-sampling does not provide a correct approximation of the total values. Also, in the case of mean() computation DP results have low variability compared to IP.

In the case of uniformly distributed data IP reports many "NA" values. This indicates that at least there has been one iteration where an IP result was not available with respect to the provided set of parameters. Uniformly distributed data contains integers as opposed to the other two distributions. This might require us to increase the discretization level or the number of re-samples and recheck for IP results. For example, in output discretization we have limited the number of decimal points to 2. In the case of integer data, rounding to the nearest integer or a multiple of some value can be used to avoid "no response (NA)" issue. Moreover, it is noted that robustness of the answers and the discretization level in IP or ϵ in DP has no prominent relationship.

Accuracy of the Results. Variability of the results does not indicate the quality of the computed statistics alone. To measure how accurate the results are absolute relative error (ARE) can be used. Tables 4 and 5 contain a detailed picture of the ARE rate for different data instances. Generally speaking, IP reports a lower error rate compared to DP. However, as discussed earlier with respect to uniformly distributed data sum() and variance() functions fails to find IP compliant solutions within the defined set of parameters. Table 6 shows the summation of ARE rate after excluding the sum() and the variance(). As it can be seen clearly, DP reports a very high ARE rate compared to IP in all the cases. Thus, IP can be seen as the preferable solution in both robustness and accuracy wise.

Different Discretization Methods for IP. When computing IP compliant statistics, three discretization methods are used as, (a) output discretization, (b) output discretization with minimum input discretization and (c) output discretization with high input discretization. Based on the results from Table 4, it can be seen that the output discretization is enough to produce IP results with minimum ARE rate. Sum of ARE is reported as 16.47, 22.28 and 38.74 under the discretization scenario (a), (b) and (c) respectively. However, by using both input and output discretization the frequency of occurrence (k) in a given result can be increased. In other words, this provides a high degree of privacy as having a higher number of generators increase the uncertainty of exactly figuring out the set of generators of a given result. When IP is used with integer data, output discretization required to be more carefully selected to avoid "no response (NA)" scenarios. Usually, increasing the rounding base or the number of re-samples can be seen as an answer to this.

Comparison of IP with DP on Real World Datasets. Here, we carried out the same experiment on the Abalone dataset where IP and DP are used to compute the descriptive statistics. As depicted by Fig. 2, IP reports a much less ARE rate compared to DP. Further, for statistics like count, mean, median, SD, and IQR, the ARE rates are negligible. The highest amount of the errors in IP are reported by the variable "V8" (number of rings) which is an integer attribute. As mentioned earlier, by adjusting the rounding parameters or the number of re-samples the error rate can be further reduced. DP statistics are calculated with $\epsilon = 4$ which should provide a very high data utility. However, compared to the IP solution, in the DP case the error is much higher for all the statistics except the count and the sum values.

Moreover, with respect to IP we collected the frequency of occurrence (k) of the selected IP results for descriptive statistics computed over the 8 variables of the Abalone dataset. In other words, out of 5,000 re-samples what was the average rate of occurrence (ARO) of the selected IP statistic over the 8 variables (number of generators). It is respectively, 5000 for count, 3994 for mean, 3793 for median, 4251 for SD, 4652 for min, 4446 for max, 3891 for IQR, 4245 for variance and 9 for sum. For a total of 5,000 re-samples (approximately the size of the

Table 2. IP statistics and their standard deviation computed for synthetic datasets with different discretization methods. 1,000 re-samples are used for the computation.

Dataset	Count	Mean	Median	SD	Min	Max	IQR	Sum	Variance
Norm I Out Dis:	1000 ± 0	1.04 ± 0	1.1 ± 0.01	0.99 ± 0	−1.77 ± 0.05	3.97 ± 0.01	1.33 ± 0.01	668.9 ± 15.77	0.98 ± 0.01
Norm I in/out Dis:(L)	1000 ± 0	1.04 ± 0.01	1.1 ± 0.01	0.98 ± 0.01	−1.77 ± 0	3.96 ± 0	1.32 ± 0.01	678.03 ± 10.17	0.98 ± 0.01
Norm I in/out Dis:(H)	1000 ± 0	1.04 ± 0.01	1.09 ± 0	0.99 ± 0	−1.33 ± 0	3.4 ± 0	1.31 ± 0.05	670.87 ± 13.12	0.98 ± 0.01
Norm II Out Dis:	1000 ± 0	0.99 ± 0.01	1.05 ± 0.1	4.87 ± 0.02	−16.54 ± 2.78	16.52 ± 0.02	6.45 ± 0.06	680.42 ± 69.27	23.72 ± 0.16
Norm II in/out Dis:(L)	1000 ± 0	1 ± 0.04	0.98 ± 0.09	4.87 ± 0.03	−15.39 ± 1.42	16.54 ± 0	6.48 ± 0.1	661.09 ± 72.42	23.86 ± 0.31
Norm II in/out Dis:(H)	1000 ± 0	1 ± 0.03	0.92 ± 0	4.85 ± 0.02	−11.47 ± 0	12.69 ± 0	6.4 ± 0.13	718.88 ± 13.59	23.56 ± 0.29
Exp I Out Dis:	1000 ± 0	1.06 ± 0	0.73 ± 0.01	1.1 ± 0	0 ± 0	8.67 ± 0.71	1.18 ± 0.01	693.78 ± 18.03	1.21 ± 0.02
Exp I in/out Dis:(L)	1000 ± 0	1.06 ± 0.01	0.73 ± 0	1.1 ± 0.01	0 ± 0	8.2 ± 0	1.18 ± 0.03	675.47 ± 13.49	1.17 ± 0.03
Exp I in/out Dis:(H)	1000 ± 0	1.06 ± 0	0.72 ± 0	1.08 ± 0.01	0.01 ± 0	5.33 ± 0	1.15 ± 0	680.61 ± 4.89	1.15 ± 0.02
Exp II Out Dis:	1000 ± 0	5.34 ± 0.05	3.59 ± 0.1	5.39 ± 0.06	0 ± 0	32.89 ± 3.6	5.98 ± 0.05	3470.72 ± 56.4	29.06 ± 0.55
Exp II in/out Dis:(L)	1000 ± 0	5.32 ± 0.03	3.62 ± 0.1	5.41 ± 0.04	0 ± 0	35.3 ± 0	5.98 ± 0.05	3479.54 ± 54.26	29.51 ± 0.64
Exp II in/out Dis:(H)	1000 ± 0	5.35 ± 0.03	3.76 ± 0	5.4 ± 0.02	0.06 ± 0	25.45 ± 0	5.84 ± 0	3446.93 ± 26.87	29.2 ± 0.99
Unif I Out Dis:	1000 ± 0	50.69 ± 0.02	50.62 ± 0.37	28.95 ± 0.13	0.05 ± 0	99.93 ± 0.03	51.73 ± 0.4	NA ± 185.88	834.24 ± 11.7
Unif I in/out Dis:(L)	1000 ± 0	50.57 ± 0.17	50.75 ± 0.76	28.89 ± 0.05	0.06 ± 0	99.93 ± 0	52.05 ± 0.78	NA ± 122.08	844.04 ± 5.18
Unif I in/out Dis:(H)	1000 ± 0	50.75 ± 0.08	51.02 ± 0	28.91 ± 0.21	1.59 ± 0	99.1 ± 0	51.35 ± 1.41	NA ± 139.89	841.39 ± 9.55
Unif II Out Dis:	1000 ± NA	486.36 ± NA	468.79 ± NA	303.57 ± NA	0.62 ± NA	997.81 ± NA	546.37 ± NA	NA ± NA	NA ± NA
Unif II in/out Dis:(L)	1000 ± NA	484.6 ± NA	475.7 ± NA	303.36 ± NA	0.6 ± NA	997.44 ± NA	540.01 ± NA	NA ± NA	NA ± NA
Unif II in/out Dis:(H)	1000 ± NA	488.25 ± NA	483.47 ± NA	304.25 ± NA	5.37 ± NA	988.37 ± NA	552.93 ± NA	NA ± NA	NA ± NA

Table 3. DP statistics and their standard deviation computed for synthetic datasets with different ϵ values.

Dataset	Count.	Mean	Median	SD	Min	Max	IQR	Sum	Variance
Norm I ($\epsilon = 0.01$)	1100.7 ± 2.29	0.59 ± 0	0.61 ± 2.82	19.78 ± 0.78	8.47 ± 1.01	7.86 ± 1.51	7.86 ± 1.51	1435.68 ± 0.79	5.62 ± 2.04
Norm I ($\epsilon = 2$)	1001.11 ± 0.95	0 ± 0	1.51 ± 0.98	1.19 ± 1.35	−2.46 ± 2.57	3.92 ± 1.14	3.92 ± 1.14	1039.42 ± 1.05	1.95 ± 1.54
Norm I ($\epsilon = 4$)	1000.38 ± 0.68	0 ± 0	1.78 ± 1	1.88 ± 2.3	−1.88 ± 0.4	4.91 ± 1.12	4.91 ± 1.12	1038.11 ± 0.75	−0.47 ± 2.25
Norm II ($\epsilon = 0.01$)	1100.02 ± 1.16	3.51 ± 0	1.65 ± 2.33	115.8 ± 1.48	486.82 ± 2.9	208.31 ± 0.64	208.31 ± 0.64	2651.68 ± 1.46	146.96 ± 1.36
Norm II ($\epsilon = 2$)	1000.87 ± 1.14	0.02 ± 0	0.44 ± 0.39	5.41 ± 1.16	−17.04 ± 2.53	17.47 ± 0.79	17.47 ± 0.79	1004.16 ± 1.29	24.73 ± 0.71
Norm II ($\epsilon = 4$)	1000.88 ± 0.7	0.01 ± 0	1.38 ± 0.82	5.46 ± 1.2	−16.99 ± 0.54	16.97 ± 1.45	16.97 ± 1.45	998.8 ± 1.51	24.63 ± 2.6
Exp I ($\epsilon = 0.01$)	1099.28 ± 0.8	0.9 ± 0	0.83 ± 2.43	30.45 ± 2.07	0.08 ± 0.76	167.36 ± 0.37	167.36 ± 0.37	1962.91 ± 1.03	8.29 ± 1.72
Exp I ($\epsilon = 2$)	1001.56 ± 1.6	0.01 ± 0	1.35 ± 0.85	0.85 ± 1.73	0.59 ± 0.95	9.34 ± 0.45	9.34 ± 0.45	1066.92 ± 0.53	−0.24 ± 2.48
Exp I ($\epsilon = 4$)	1000.16 ± 0.81	0 ± 0	1.44 ± 0.44	1.23 ± 1.28	−0.18 ± 1.38	8.9 ± 1.35	8.9 ± 1.35	1065.72 ± 0.44	0.53 ± 1.02
Exp II ($\epsilon = 0.01$)	1100.38 ± 1.45	3.94 ± 0	3.47 ± 1.01	127.57 ± 2.43	0.98 ± 0.76	842.91 ± 0.92	842.91 ± 0.92	9264.07 ± 1.8	184.24 ± 0.8
Exp II ($\epsilon = 2$)	998.98 ± 1.01	0.03 ± 0	3.68 ± 0.47	5.22 ± 1.36	−0.97 ± 0.75	43.32 ± 1.11	43.32 ± 1.11	5352.31 ± 1.79	29.73 ± 1.03
Exp II ($\epsilon = 4$)	1000.85 ± 2.29	0.02 ± 0	4.52 ± 0.92	6.38 ± 0.5	0.66 ± 1.09	42.02 ± 0.38	42.02 ± 0.38	5342.18 ± 1.43	30.81 ± 2.02
Unif I ($\epsilon = 0.01$)	1099.71 ± 1.63	10.05 ± 0	57.54 ± 1.24	345.43 ± 2.39	7.71 ± 0.42	107.37 ± 1.85	107.37 ± 1.85	60624.79 ± 1.33	1838.98 ± 0.8
Unif I ($\epsilon = 2$)	999.31 ± 1.35	0.1 ± 0	49.95 ± 0.77	31.06 ± 1.19	0.02 ± 0.8	100.09 ± 1.78	100.09 ± 1.78	50678.05 ± 1.62	843.86 ± 3.27
Unif I ($\epsilon = 4$)	1000.32 ± 1.57	0.08 ± 0	50.53 ± 1.31	30.16 ± 1.49	0.39 ± 1.18	99.55 ± 0.51	99.55 ± 0.51	50652.64 ± 1.2	842.97 ± 2.93
Unif II ($\epsilon = 0.01$)	1099.71 ± 0.91	100.22 ± 0	593.97 ± 0.66	3457.43 ± 0.98	12.87 ± 1.13	1106.95 ± 1.83	1106.95 ± 1.83	587122.17 ± 0.92	191734.16 ± 1.11
Unif II ($\epsilon = 2$)	1001.13 ± 0.63	0.99 ± 0	472.36 ± 1.14	319.76 ± 0.91	1.15 ± 0.86	999.76 ± 3.29	999.76 ± 3.29	487829.21 ± 2.88	92763.76 ± 0.99
Unif II ($\epsilon = 4$)	999.79 ± 0.56	0.74 ± 0	475.33 ± 2.86	312.99 ± 1.67	0.77 ± 1.48	997.72 ± 1.91	997.72 ± 1.91	487580.65 ± 2.32	92515.39 ± 2.56

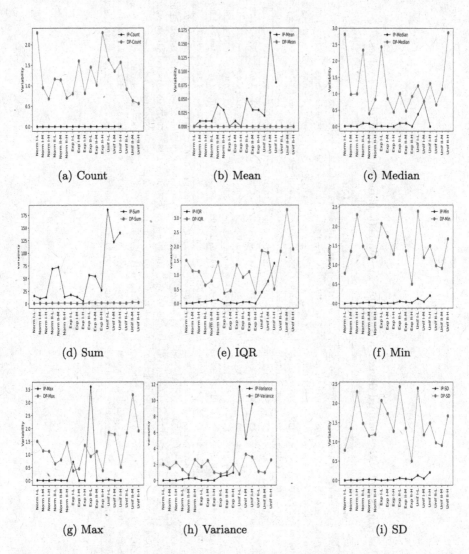

(a) Count   (b) Mean   (c) Median

(d) Sum   (e) IQR   (f) Min

(g) Max   (h) Variance   (i) SD

Fig. 1. Standard deviation of IP and DP statistics over multiple iterations. Each synthetic data distribution is tagged as L, M and H which indicate Low, Medium and High privacy levels. For IP, L indicates output discretization, M indicates low input discretization combined with output discretization and H indicates high input discretization with output discretization. For DP L, M and H respectively indicate ϵ values 0.01,2 and 4.

Table 4. Absolute Relative Error (ARE) of IP statistics computed on synthetic datasets with different discretization methods.

Dataset	Count	Mean	Median	SD	Min	Max	IQR	Sum	Variance
Norm I Out Dis:	0.00	0.00	0.01	0.00	0.05	0.00	0.00	0.35	0.01
Norm I in/out Dis:(L)	0.00	0.00	0.01	0.01	0.05	0.01	0.01	0.35	0.01
Norm I in/out Dis:(H)	0.00	0.00	0.00	0.00	0.28	0.15	0.02	0.35	0.01
Norm II Out Dis:	0.00	0.01	0.03	0.01	1.09	0.01	0.00	0.30	0.04
Norm II in/out Dis:(L)	0.00	0.00	0.04	0.01	1.71	0.01	0.02	0.32	0.10
Norm II in/out Dis:(H)	0.00	0.00	0.09	0.03	3.82	0.97	0.04	0.27	0.21
Exp I Out Dis:	0.00	0.00	0.01	0.01	0.00	0.08	0.00	0.36	0.02
Exp I in/out Dis:(L)	0.00	0.00	0.01	0.01	0.00	0.20	0.00	0.37	0.02
Exp I in/out Dis:(H)	0.00	0.00	0.00	0.01	0.01	0.92	0.02	0.37	0.04
Exp II Out Dis:	0.00	0.01	0.07	0.01	0.00	1.61	0.01	1.79	0.14
Exp II in/out Dis:(L)	0.00	0.01	0.04	0.01	0.00	1.01	0.01	1.79	0.32
Exp II in/out Dis:(H)	0.00	0.02	0.09	0.00	0.03	3.48	0.12	1.82	0.00
Unif I Out Dis:	0.00	0.06	0.05	0.03	0.03	0.01	0.09	NA	5.54
Unif I in/out Dis:(L)	0.00	0.06	0.17	0.09	0.03	0.01	0.15	NA	4.50
Unif I in/out Dis:(H)	0.00	0.12	0.41	0.07	0.86	0.22	0.38	NA	1.79
Unif II Out Dis:	0.00	0.94	3.22	0.18	0.03	0.03	0.23	NA	NA
Unif II in/out Dis:(L)	0.00	2.63	3.11	0.40	0.01	0.12	4.54	NA	NA
Unif II in/out Dis:(H)	0.00	0.89	10.22	0.50	2.58	2.39	5.14	NA	NA

Table 5. Absolute Relative Error (ARE) of DP statistics computed on synthetic datasets with different ϵ values.

Dataset	Count	Mean	Median	SD	Min	Max	IQR	Sum	Variance
Norm I ($\epsilon = 0.01$)	0.10	0.43	0.44	19.03	5.56	0.97	4.89	0.39	4.76
Norm I ($\epsilon = 2$)	0.00	1.00	0.38	0.20	0.33	0.02	1.94	0.00	1.00
Norm I ($\epsilon = 4$)	0.00	1.00	0.63	0.90	0.01	0.23	2.68	0.00	1.48
Norm II ($\epsilon = 0.01$)	0.10	2.42	0.58	112.32	272.23	48.09	151.27	1.60	126.32
Norm II ($\epsilon = 2$)	0.00	0.94	0.53	0.54	0.82	0.23	8.26	0.01	0.99
Norm II ($\epsilon = 4$)	0.00	0.95	0.33	0.59	0.85	0.10	7.88	0.00	0.89
Exp I ($\epsilon = 0.01$)	0.10	0.16	0.10	29.73	0.04	39.72	124.53	0.87	7.28
Exp I ($\epsilon = 2$)	0.00	1.02	0.57	0.24	0.32	0.09	6.12	0.00	1.47
Exp I ($\epsilon = 4$)	0.00	1.03	0.66	0.14	0.10	0.02	5.79	0.00	0.68
Exp II ($\epsilon = 0.01$)	0.10	1.34	0.18	123.70	0.53	201.52	627.16	3.79	158.97
Exp II ($\epsilon = 2$)	0.00	5.11	0.01	0.19	0.52	1.00	27.97	0.02	0.54
Exp II ($\epsilon = 4$)	0.00	5.12	0.78	0.99	0.35	0.68	27.00	0.01	1.65
Unif I ($\epsilon = 0.01$)	0.10	39.13	6.38	320.44	4.15	1.85	41.60	9.64	1024.67
Unif I ($\epsilon = 2$)	0.00	48.73	0.57	2.11	0.01	0.03	36.15	0.05	4.32
Unif I ($\epsilon = 4$)	0.00	48.75	0.04	1.20	0.21	0.11	35.74	0.02	3.41
Unif II ($\epsilon = 0.01$)	0.10	373.32	111.39	3193.41	6.62	27.34	420.31	96.24	101990.65
Unif II ($\epsilon = 2$)	0.00	469.02	0.05	16.21	0.31	0.46	339.99	0.48	510.80
Unif II ($\epsilon = 4$)	0.00	469.26	2.77	9.35	0.11	0.05	338.46	0.24	256.13

dataset) following input and output discretization number of generators seems to be very high showing that the chances to distinguish the exact data records used to compute a given statistic are minimal. However, it is being repeatedly shown that the for summation queries IP might not be an ideal solution.

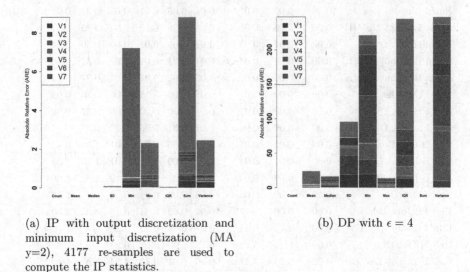

(a) IP with output discretization and minimum input discretization (MA y=2), 4177 re-samples are used to compute the IP statistics.

(b) DP with $\epsilon = 4$

Fig. 2. Absolute relative error (ARE) for descriptive statistics computed over the numerical variables of the Abalone dataset.

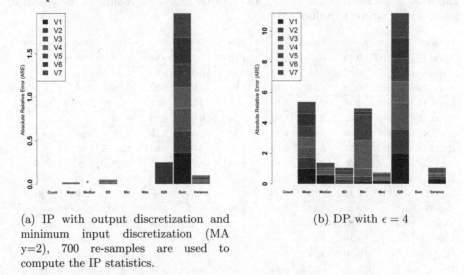

(a) IP with output discretization and minimum input discretization (MA y=2), 700 re-samples are used to compute the IP statistics.

(b) DP with $\epsilon = 4$

Fig. 3. Absolute relative error (ARE) for descriptive statistics computed over the numerical variables of the Breast Cancer dataset.

Figure 3 depicts the computation of IP and DP statistics on UCI breast cancer dataset. To comparatively evaluate the results ARE rates are computed per variable. The results show the same pattern as the Abalone dataset. Compared to DP the ARE rates are low for IP except for calculating the sum. The poor performance of the IP with respect to summation can be attributed to use of re-sampling.

These results shows us that IP results have a high utility value compared to DP in most of the cases. Therefore, this method can be used for releasing aggregated statistics without compromising the privacy of the sensitive data. However, in order to use this in terms of large scale databases, it is required to improve the computational efficiency of the process further.

Table 6. Summation of absolute relative error for different statistics computed using IP and DP

Privacy model	Count	Mean	Median	SD	Min	Max	IQR
IP	0	4.75	17.58	1.38	10.58	11.23	10.78
DP	0.6	1468.73	126.39	3831.29	293.07	322.51	2207.74

6 Conclusion

In this paper, we have discussed how to provide integral privacy for descriptive statistics computation while maintaining the robustness and the utility of the final results. We have proposed a re-sampling and discretization based approach to achieve integral privacy and empirically shown that integral privacy based solution works better than the differential privacy based solution in most of the cases. Especially, it is noted that the proposed solution can easily be used with small datasets where differential privacy usually fails in terms of utility. However, further work is required to minimize the computational cost and to introduce a formal method to derive the minimum number of re-samples required to achieve integral privacy for a given dataset. And also, we hope to develop an inference attack to assess the effectiveness of integral privacy in our future work.

References

1. Torra, V., Navarro-Arribas, G.: Integral privacy. In: Foresti, S., Persiano, G. (eds.) CANS 2016. LNCS, vol. 10052, pp. 661–669. Springer, Cham (2016). https://doi.org/10.1007/978-3-319-48965-0_44
2. Dwork, C., McSherry, F., Nissim, K., Smith, A.: Calibrating noise to sensitivity in private data analysis. In: Halevi, S., Rabin, T. (eds.) TCC 2006. LNCS, vol. 3876, pp. 265–284. Springer, Heidelberg (2006). https://doi.org/10.1007/11681878_14
3. Blum, A., Dwork, C., McSherry, F., Nissim, K.: Practical privacy: the SuLQ framework. In: Proceedings of the Twenty-Fourth ACM SIGMOD-SIGACT-SIGART Symposium on Principles of Database Systems, pp. 128–138. ACM (2005)

4. Dwork, C., Smith, A.: Differential privacy for statistics: what we know and what we want to learn. J. Priv. Confid. $1(2)$ (2010)
5. Dwork, C., Lei, J.: Differential privacy and robust statistics. In: STOC, vol. 9, pp. 371–380 (2009)
6. Clifton, C., Tassa, T.: On syntactic anonymity and differential privacy. In: 2013 IEEE 29th International Conference on Data Engineering Workshops (ICDEW), pp. 88–93. IEEE (2013)
7. Senavirathne, N., Torra, V.: Integrally private model selection for decision trees. Comput. Secur. **83**, 167–181 (2019)
8. Senavirathne, N., Torra, V.: Approximating robust linear regression with an integral privacy guarantee. In: 2018 16th Annual Conference on Privacy, Security and Trust (PST), pp. 1–10. IEEE (2018)
9. Efron, B.: Bootstrap methods: another look at the Jackknife. In: Kotz, S., Johnson, N.L. (eds.) Breakthroughs in Statistics. Springer Series in Statistics (Perspectives in Statistics), pp. 569–593. Springer, New York (1992). https://doi.org/10.1007/978-1-4612-4380-9_41
10. Soria-Comas, J., Domingo-Ferrer, J., Sánchez, D., Megías, D.: Individual differential privacy: a utility-preserving formulation of differential privacy guarantees. IEEE Trans. Inf. Forensics Secur. **12**(6), 1418–1429 (2017)
11. Li, N., Lyu, M., Dong, S., Yang, W.: Differential privacy: from theory to practice. Synth. Lect. Inf. Secur. Priv. Trust. **8**(4), 1–138 (2016)
12. Dwork, C., Lei, J.: Differential privacy and robust statistics. In: Proceedings of the 41st Annual ACM Symposium on Theory of Computing, STOC 2009, Bethesda, MD, USA, 31 May–June 2 2009, pp 371–380 (2009)

Towards Data Anonymization in Data Mining via Meta-heuristic Approaches

Fatemeh Amiri[1,2]([envelope]), Gerald Quirchmayr[1,2], Peter Kieseberg[3], Edgar Weippl[2], and Alessio Bertone[4]

[1] Faculty of Computer Science, University of Vienna, Vienna, Austria
{amirif86,gerald.quirchmayr}@univie.ac.at
[2] SBA Research GmbH, Vienna, Austria
peter.kieseberg@fhstp.ac.at
[3] St. Poelten University of Applied Sciences, St. Poelten, Austria
eweippl@sba-research.org
[4] Radar Cyber Security GmbH, Vienna, Austria
Alessio.bertone@gmail.com

Abstract. In this paper, a meta-heuristics model proposed to protect the confidentiality of data through anonymization. The aim is to minimize information loss as well as the maximization of privacy protection using Genetic algorithms and fuzzy sets. As a case study, Kohonen Maps put in practice through Self Organizing Map (SOM) applied to test the validity of the proposed model. SOM suffers from some privacy gaps and also demands a computationally, highly complex task. The experimental results show an improvement of protection of sensitive data without compromising cluster quality and optimality.

Keywords: Privacy-preserving · Anonymization · Meta heuristics · Genetic algorithm · Fuzzy sets · Vectorization

1 Introduction

Privacy issues become more and more challenging, mainly due to novel regulations like the General Data Protection Regulation (GDPR), or to protect the intrinsic value of the collected data sets. This problem can be summarized as "How handling personal data can be implemented without revealing the confidential data of the owner?" [2]. Data mining as the heart of Big Data, plays a crucial role in this problem, which is addressed by "Privacy-Preserving Data Mining" (PPDM). Most of the existing PPDM approaches use conventional techniques of privacy-preserving like perturbation, generalization, suppression and k-anonymity and new versions (like l-diversity and t-closeness) [3]. A comprehensive study of popular methods in PPDM can be found in [4]. The authors define a broad classification in both categories (soft and traditional ways of privacy-preserving) to reach an innovative idea concerning research gaps and the state of the art of the topic. Results show that using soft computing methods (such

© Springer Nature Switzerland AG 2019
C. Pérez-Solà et al. (Eds.): DPM 2019/CBT 2019, LNCS 11737, pp. 39–48, 2019.
https://doi.org/10.1007/978-3-030-31500-9_3

as machine learning and meta-heuristics) are more promising to find an optimal solution. Recently this methodology was used in PPDM approaches, and first works reveal its usefulness [5]. This paper introduces a meta heuristics model in a specific use case: "instead of anonymizing everything in the database, in case of knowing the sensitive items, we find them in database and we just apply the anonymization just on this portion of data, not all of them!". As a case study, we focus on unsupervised clustering (in this paper SOM were selected) tasks because of their wide application and limited privacy-preserving methods. We use a structural anonymization approach based on meta-heuristics approaches. First, a subset is extracted with a specially designed Genetic Algorithm (GA). The aim is to minimize the selection of sensitive data from the database [6,7]. In our proposed function, we try to minimize a parameter called "hiding failure" that will be explained in detail in Sect. 2. Then, the process of anonymization hides sensitive data. The output of the GA algorithm is used as the input of a fuzzy membership function to anonymize the content of the sensitive subset. Finally, the result of this step is appended to the primary database to be imported for the usual clustering data mining task. Kohonen Map clustering, known as Self Organizing Map (SOM), is used then as the case study, which was selected due to its popularity and its currently existing privacy gaps (See Fig. 1). Despite the similarity with some traditional PPDM methods (e.g. differential privacy and k-anonymity), the novelty of our approach consists in the use of meta-heuristics. We don't apply the anonymizing process on all data, but we try to find the most crucial items and hide them.

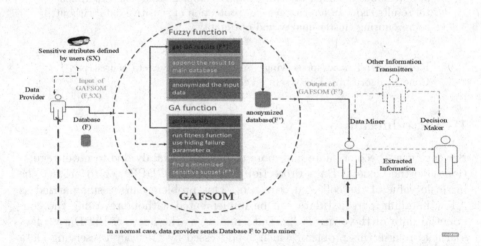

Fig. 1. A simple illustration of the application scenario with the proposed model (GAFSOM). Data provider sends the (F, SX) as the input of GAFSOM (SX is defined by user). The process begins from GA function. Then the process continues by sending the F* to fuzzy function. Anonymized Database F' is the output of GAFSOM which will be sent to data miner

2   Meta-heuristic Approaches for Anonymization

In this paper, a new model based on meta-heuristics is introduced that uses GAs and Fuzzy sets to anonymize selective sensitive items without compromising the utility of a SOM clustering. So that, it is shortened as "GAFSOM", and it is based on the method introduced in [8].

Definition 1 (Input and Output). The input and output of the proposed model GAFSOM are defined as:

Let F be the original database and $F_q(1 \leq q \leq N)$ a record in database with a unique identifier (FID), and a set of sensitive data to be hidden $SX = S_1, SX_2, ..., SX_n$.

SX_i is a field in the current record that should fulfill the user-defined criteria in order to be selected as a sensitive item. Let all of these parameters be input values, and F' be the output of GAFSOM that is an anonymized database.

Definition 2 (Optimization Problem). Given a large database $F = (N, E)$ with N nodes and E attributes, in current use case anonymization apply only on sensitive attributes (SX). An $n \times n$ adjacency matrix is taken as an input. The problem is to find the optimum partition set F^* to achieve an almost minimize Information Loss-IL (F, F^*) value for the given graph F. As an optimization problem we expect:

$$SOM(F) \approx SOM(F^*) \tag{1}$$

That means that the result of SOM clustering on the main database F and the optimal anonymized database extracted from GAFSOM F' is relatively equal. So, privacy and utility factors would be satisfied.

Definition 3 (Fitness Function). Fitness Function is the heart of a GA method and it is used to evaluate the hiding failures of each processed transaction; we assess the hiding failures of each processed transaction in the anonymization process using:

$$\alpha^j(S_x) = \frac{MAX_s x - freq(S_x) + 1}{MAX_s x - \lceil |F| \times MST \rceil + 1} \tag{2}$$

Where $MAX_s x$ is the maximum number of sensitive data of record with $F(S_x)$, $|F|$ is the number of records and $freq(S_x)$ is the frequency of SX in the current record. MST (Minimum Support Threshold) is a pre-defined parameter that limits the number of records to be selected as sensitive records. We use MST as a condition of termination of the GA function as an influencing parameter on the speed and quality of GA function. The overall amount of α per record defined by:

$$\alpha^j = \frac{1}{\sum i = 1^n \alpha^j(s_i) + 1} \tag{3}$$

Definition 4 (Termination of Algorithm). Threshold of termination defines a minimum support threshold ratio MST, as the percentage of the minimum support threshold used in the GA algorithm and plays a crucial role. This factor identifies when the GA method should terminate. In other words, MST is a threshold that identifies when a subset is optimal.

Definition 5 (GA Operation Parameters). The efficiency and convergence rate of the algorithm depends on population size (number of individuals), number of generations, crossover and mutation operations. In the proposed algorithm, we implement different parameters to find an acceptable rate for a real database. The population is the set of individuals in a generation instance. The higher the population size, the higher is the number of individuals present in the mating pool, hence the higher are the chances of individuals with better fitness values, but computational resource consumption is increased. So currently for a 1000 record database, a typical population size is defined with 50 individuals. Table 1 describes the parameters set for this experiment.

Table 1. Parameters used in designed genetic algorithm of GAFSOM.

Parameter	Value
Creation method	Pop Function
Population Size	50
Generations	100
Population Type	bit string
Selection function selection	tournament
Mutation function mutation	uniform, 0.1
Crossover function crossover	arithmetic, 0.8
Elite Count	2
StallGen Limit	100

Crossover recombination is the process of generating off-springs from parents. Crossover is applied in the hope that a new generation will combine good parts of old chromosomes and thus will have better fitness value. However, it is good to leave a few chromosome strings of the old population to survive to the next generation. Arithmetic crossover with the rate of 80% is applied for this method. After crossover, the individuals are subjected to mutation, in order to prevents the algorithm to be confined in some local minima. The mutation scheme for an individual randomly picks two nodes from different sub-nodes and swaps them. It provides diversity to the algorithm. As we do not want to jeopardize pure data, a uniform mutation with 10% rate is applied.

Definition 6 (Fuzzyfing Process). In GAFSOM, the output of the GA method is defined as the input of a fuzzy function. In this paper, the Triangular Membership Function is used to anonymized the sensitive content, using the relation $TriMF$ as:

$$TriMF(D; X, Y, Z) = \begin{cases} 0, & when \quad D < X \\ \frac{D-X}{Y-D} & when \quad XD < Y \\ \frac{Z-D}{Z-Y} & when \quad TD < Z \\ D & D \geq Z \end{cases} \tag{4}$$

Which D is the value in the database. X, Y, Z are the three boundary points. As an example, one could generalize the Age attribute among ((20–30), (30–40), (40–50), (50–70)) and job could be anonymized into the difficulty domain ($high, mediumandlow$).

Definition 7 (Evaluating the Model). For measuring the results of SOM clustering two factors tested:

– Quantization Error (QE): the average distance between current BMU and each data vector [9]:

$$QE = \frac{1}{N} \sum k = 1^N (x_k - m_k) \tag{5}$$

where N is the number of data-vectors and m_k is the best matching prototype of the corresponding x_k data-vector. The smaller is the value of QE, the better is the algorithm.

– Topographic Error (TE): describes how well the SOM preserves the topology of the studied data set:

$$TE = \frac{1}{N} \sum k = 1^N u(x_k) \tag{6}$$

Where N is the number of input samples, and $u(x_k)$ is 1 if the first and second record of x_k are not next to each other, otherwise $u(x_k)$ is 0. A small value of TE is more desirable [10].

3   GAFSOM – A Case Study

In the proposed method, a GA approach is used as a search heuristic to find the optimal sensitive data to anonymize, based on evolutionary principles of genetics to optimization and search problems [11]. The GA framework was chosen because of its flexible formulation and its ability to find solutions given adequate computational resources. Among different techniques of protecting privacy, Anonymization and Generalization are selected which are applied by different membership functions in a fuzzy set. To anonymize the sensitive data, fuzzy sets are used, which relates to classes of objects with unsharp boundaries.

The algorithm starts by deriving the MST and SX that in this version of GAFSOM are user-defined. Next, the GA loop begins to calculate the fitness function and also the crossover and mutation operations. The loop terminates when either a maximum number of generations has been produced, or a satisfactory fitness level MST has been reached for the population. F* is the optimal subset of sensitive data. The next steps anonymize F* using a fuzzy function. Finally, the fuzzyfied(F*) results are imported into the primary database as F', which is the final output and should be relatively similar to the main database F^1. The following sections describe the test cases in detail, the algorithm used as well as a description of the overall process employed to obtain our results.

3.1 Data

A social real database [12] is adapted to evaluate the performance of the proposed algorithm regarding the privacy, execution time, and accuracy of clustering operations. For experiments, we use two real-world datasets named Adult, and Bank Marketing Dataset:

- "Adult": The "Adult" dataset, which is released by the UCI Machine Learning repository for research purpose. There are 14 attributes and 48, 842 records in total.
- "Bank Marketing": This dataset is generated through direct marketing campaigns (phone calls) of a Portuguese banking institution. It contains 45,212 records and 17 attributes.

3.2 Test Cases

We designed two different test cases to evaluate the proposed methods from various perspectives:

Test case 1 (aim: random execution). Looking for clustering results of Adult dataset when the sensitive criteria defined by data miner as "black female who are post graduated and work more than 20 h per week and younger than 30 years old".

Design 1: To find the results, a test case including 1000 records/tuples of the "adult" dataset has been used, which includes age, work-class, gender, education, and race as sensitive attributes. The tuples were selected randomly from main dataset.

Test case 2 (aim: worst case-too many items defined as sensitive). Looking for clustering results of Bank dataset when the sensitive criteria defined by data miner as "any young employee who is married and work at high ranked position like manager".

Design 2: This time the selection of sensitive record was not random. Since we want to check the accuracy of the proposed method in the worst case (more

---

[1] The efficiency and convergence rate of the algorithm depends on the population size (number of individuals), number of generations, crossover and mutation operations. The details of parameters used in GAFSOM is explained in Table 1.

sensitive data to hide), then those tuples with more sensitive data selected for test case. For Bank dataset a bigger test case including 3000 tuples is selected. The selected sensitive attributes are: "age work-class and marital case".

For both tests, we fuzzyfy the selected tuples by the proposed algorithm. The algorithm starts by first selecting tuples that match the sensitivity criteria in order to hide those records optimally. Next sections demonstrate different parameters which prove the usefulness of proposed method.

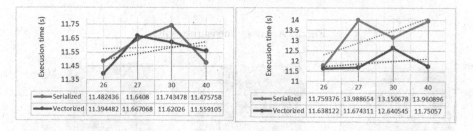

Fig. 2. Execution time for "Adult" and "Bank" dataset with different values for minimum support threshold, MST. Execution time is tested in both serialized and vectorized cases of GA function. Results show that vectorization is faster. Also, the impact of MST parameter on execution times is evaluated. As you can see, with the increasing of MST (horizontal axis) in both datasets, execution time also increases

4   Experiments

In this part, we evaluate the data utility using real databases. Then, privacy protection and information loss of the algorithm in SOM clustering results were tested. The algorithms were implemented in MATLAB [13], and executed on a VM/ Linux Ubuntu platform with four vCPU in Intel(R) Xeon (R) E5-2650 v4 processors and 4 GB memory.

4.1   Execution Time

GAs could be time-consuming, thus significantly impairing the applicability of the process. So, the GA function designed for GAFSOM tested in two cases of serialized and vectorized. In serialized execution in each generation, algorithm fetches the fittest value as the sensitive item. In vectorized execution, the algorithm first fetches all the possible sensitive record as a sub-database and then execute the process. As Fig. 2 shows, the execution time is much less than serial experiments in both databases. We tried to optimize the fitness function by using a hiding failure parameter in designed genetic algorithm. However, with different value, we observed that defining the value near 40 for MST could keep the balance between the utility and privacy, but more precisely the value will be defined in case of testing the method in front of various privacy attacks which currently is beyond the scope of this work.

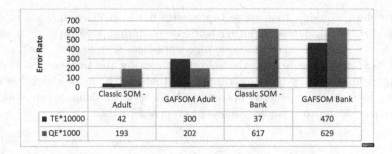

Fig. 3. Accuracy measure results

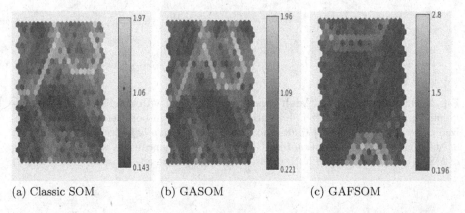

(a) Classic SOM   (b) GASOM   (c) GAFSOM

Fig. 4. Comparison of U-Matrixes between SOM clustering results for main Adult dataset D. As you can see, clustering results of GAFSOM in part c is partially different with classic SOM and GASOM. Thus, despite the usefulness of GAFSOM, such a method should be improved to reduce the difference between clustering results.

4.2   Analysis of the Accuracy and Information Loss of GAFSOM

To validate the proposed model GAFSOM some popular comparative criteria were used, including the average quantization error (QE) in (5) between data vectors and BMUs on the map, as well as topographic error (TE) in (6) counting of errors obtained in the application of the algorithm over the databases. Figure 3 shows the results of these accuracy measures.

Precise measuring of similarity between the raw database and the proposed method is affected by the size of the database and level of anonymization. A new well-promising algorithm with lower anonymization which takes into account the above assumption with less penalty in the similarity of results is being studied, and it is expected to be even more efficient. However, visual comparison of the trained map and U-Matrix between classic SOM and the suggested approach in Fig. 4 and also Fig. 5 show the usefulness of the proposed model for defined aims.

4.3   Topological Analysis of GAFSOM

We carried out the experiments using the SOM toolbox v.2. We set the radius of the lattice to 3/2, the network topology to the hexagonal lattice, and the optimum cluster number to three [1]. To evaluate the precision of the proposed algorithm, results are compared with those of traditional SOM clustering. Figure 4 shows the U-Matrixes of clustering schemes for Adult dataset, and Fig. 5 demonstrates the results for Bank dataset. As Figs. 4 and 5 show, the difference between classic clustering with GASOM is negligible. However, improving the accuracy level of GAFSOM is still necessary to acquire more precision in clustering results. As it can be seen, a considerable topological difference exists in Fig. 5(c). So that, using applied fuzzy function at least with current parameters is not satisfying and should be improved. However, mathematical evaluation formulas prove the usefulness of the proposed method.

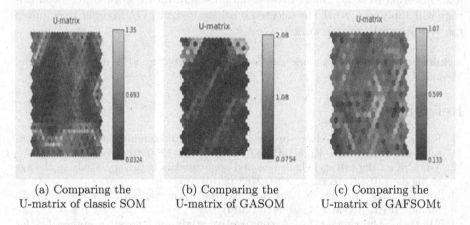

| (a) Comparing the | (b) Comparing the | (c) Comparing the |
| U-matrix of classic SOM | U-matrix of GASOM | U-matrix of GAFSOMt |

Fig. 5. Comparison of U-Matrixes for Bank dataset: (a) presents the clustering result of classic SOM for original dataset. (b) shows clustering applied on the anonymized dataset that is generated by GASOM. As you can see, the trend of clusters is similar to (a). (c) illustrates the experiments of clustering applied to the database retrieved by GAFSOM. This time the similarity between clusters is distorted. Though the accuracy experiments demonstrate the usefulness of GAGSOM, the information loss still needs to be decreased.

4.4   Privacy Attacks

Penetration test of GAFSOM using popular privacy attacks like membership attack is beyond the aim of this paper. Here, we just evaluated the cost and correctness of GAFSOM. We believe that a privacy-preserving approach firstly should prove these aspects and then claim for its power against known breaches and attacks. However, practical techniques of privacy protections meaning suppression and generalization that are implemented in GAFSOM, could show initial protection of sensitive data.

5 Conclusion

PPDM using meta-heuristic techniques brings smarter solutions not only to protect against privacy breaches but also to increase accuracy in the final results of data mining. Consequently, we introduced GAFSOM method in this paper, which uses a combination of genetic algorithm and fuzzy sets for a trade-off between privacy and utility. Our results show that selective deletion of valuable data items is less destructive than general anonymization done by fuzzyfication, so that complying with other similar techniques especially differential privacy is still preferable to taking preemptive steps to de-identify personal information in databases. The overhead of GAFSOM in negligible and using the topological error formulas of clustering its correctness proved. However, using the general fuzzy anonymization should be improved. Our results are highly selective and should be corroborated by applying a wider spectrum of other data mining tasks like clustering to larger, more diverse databases. In future work Differential privacy will apply to perturb the selected sensitive items by GA in order to compare the validity with current work.

Acknowledgments. This work was partially supported by SBA research institute, Vienna, Austria.

References

1. Kaleli, C., Polat, H.: Privacy-preserving SOM-based recommendations on horizontally distributed data. Knowl.-Based Syst. **33**, 124–135 (2012)
2. Yu, S.J.I.a.: Big privacy: challenges and opportunities of privacy, study in the age of big data. IEEE Access **4**, 2751–2763 (2016)
3. Xu, L., Yung, J., Ren, Y.: Information security in big data: privacy and data mining. IEEE Access **2**, 1149–1176 (2014)
4. Amiri, F., Quirchmayr, G.: A comparative study on innovative approaches for privacy-preserving in knowledge discovery. In: ICIME. ACM (2017)
5. Dua, S., Du, X.: Data Mining and Machine Learning in Cybersecurity. Auerbach Publications (2016)
6. Sivanandam, S., Deepa, S.: Genetic algorithm optimization problems. In: Sivanandam, S., Deepa, S. (eds.) Introduction to Genetic Algorithms, pp. 165–209. Springer, Heidelberg (2008). https://doi.org/10.1007/978-3-540-73190-0_7
7. Srinivas, M., Patnaik, L.M.: Genetic algorithms a survey. IEEE Comput. **27**(6), 17–26 (1994)
8. Amiri, F., Quirchmayr, G.: Sensitive data anonymization using genetic algorithms for SOM-based clustering. In: Secureware (2018)
9. Kohonen, T.: Self-Organizing Maps. Springer, Berlin (2001). https://doi.org/10.1007/978-3-642-56927-2
10. Kiviluoto, K.: Topology preservation in self-organizing maps. In: International Conference on Neural Networks, pp. 294–299 (1996)
11. Holland, J.: Adaptation in Artificial and Natural Systems. MIT Press Cambridge, Cambridge (1992)
12. Dheeru, D.K.T.: UCI Machine Learning Repository. University of California: Irvine, School of Information and Computer Sciences (2017)
13. Moler, C.: MATLAB User's Guide. The Mathworks Inc. (1980)

Skiplist Timing Attack Vulnerability

Eyal Nussbaum[✉] and Michael Segal

School of Electrical and Computer Engineering, Communication Systems
Engineering Department, Ben-Gurion University, 84105 Beer-Sheva, Israel
eyalnus@post.bgu.ac.il

Abstract. In this paper we address the structure and behavior of the
probabilistic Skiplist data structure and present an exploit in the form of
a timing attack on the structure. In this exploit, we show how to map the
presumably hidden structure of a Skiplist by timing the return time of
search queries. This data can then be used to perform operations on the
Skiplist which will cause a degradation in its subsequent performance. In
addition, we describe another exploitation of this data to use the Skiplist
as a means of creating a hidden channel between two attackers. Finally,
we propose a new variant of Skiplist we call a Splay Skiplist, which retains
the $O(\log n)$ performance of Skiplist while defending against the stated
exploit.

Keywords: Skiplist · Data structures · Complexity · Security ·
Timing attacks

1 Introduction

The underlying architecture of a database may be comprised of a single or multiple data structures, such as graphs, queues, stacks, trees, etc., each of which
allow access to the data in different ways. The organization of the data structure
holds information regarding the data itself, even though this may have an invol-
untary side-effect that can cause hidden information to be leaked. For example,
knowledge regarding the location of a specific node in a binary search tree discloses information regarding its children and parent nodes, as its position on
a path is related to the values along this path. The run-time of queries is also
dependant on the structure of the database, and in itself can contain information
that should remain hidden.

There are several examples where timing attacks have been used to gather
hidden information from databases, Futoransky et al. [7] describe one such attack
on SQL databases. Additional related analyses of structure vulnerabilities and
timing attacks can be found in [5,9] and [3]. In this paper, we present a possible timing-based vulnerability of the Skiplist data structure introduced by
Pugh [11]. We show how this vulnerability can be used for 2 types of attacks:
one which can adversely affect the search time of the Skiplist, and one which
allows manipulation of the Skiplist to create a hidden channel for data transfer.

© Springer Nature Switzerland AG 2019
C. Pérez-Solà et al. (Eds.): DPM 2019/CBT 2019, LNCS 11737, pp. 49–58, 2019.
https://doi.org/10.1007/978-3-030-31500-9_4

Our threat model is based on a strong adversary, with access to the database and the ability to manipulate it's structure. However, the adversary is limited in that the manipulation must be undetected by someone with access to the database who does not have direct knowledge of the manipulation. Any query performed by a naive user on the database must return the same results as though no action was performed by the attacker. This is similar in behavior to a "man in the middle" attack.

Skiplists themselves have been extensively studied, and several variants have been proposed. The variants found in [2] and [6], for example, are adapted to improve access time for specific use cases. For use in concurrent processes or distributed environments, variants in [10] and [1] have been developed. These however do not address the issue of the timing vulnerability raised in this paper, which may affect existing systems. Skiplist implementations can be seen today in systems such as *MemSQL* [4] and *Redis*, where these vulnerabilities could affect performance. Other implementations included authenticated dictionaries (Goodrich et al. [8]) and privacy preserving authorization revocation systems (Solis and Tsudik [13]). The rest of this paper is organized as follows: in Sect. 2 we outline the data structure's vulnerability and possible avenues of attack, Sect. 3 describes a variation of the Skiplist we name Splay Skiplist which guarantees a defense from the vulnerability stated before. We summarize it in Sect. 4.

2   Skiplist Timing Vulnerability

Skiplists are a probabilistic alternative to balanced trees. Skiplists are balanced by consulting a random number generator and inserting nodes at different levels of the data structure while maintaining an ordered structure. For each new value inserted into the Skiplist, a node is created and placed in its sorted position in the structure. The node is connected to the succeeding and preceding nodes at each level. A full analysis of Skiplist creation and search complexity can be found in [11]. The general algorithm for inserting nodes into a Skiplist is as follows:

- Search for the ordered placement location of the node.
- Insert node into the Skiplist at the lowest level (level 1).
- With 0.5 probability, add the node to the next level.
- Continue adding the node to subsequent levels with 0.5 probability until either the node was not added to the next level, or the max level ($\log n$) has been reached.

In this manner, a Skiplist with n nodes maintains $\log n$ levels, with all nodes existing at least at level 1, and every other level, l, containing approximately $\frac{n}{2^{l-1}}$ nodes (see Fig. 1). Since Skiplist is a randomized structure, any two data structures implemented using the same Skiplist algorithm and having the same set of instructions for insertion/deletion of elements are extremely unlikely to end up with the same structure. Due to this fact, it would be difficult for an attacker with access to the data structure to impose a set of instructions that would force the Skiplist into a "worst case scenario" structure, where all nodes exist only in level 1 (like a regular List data structure), which will harm its run-time. Bethta and Reiter [3] show that the manner in which Skiplists are created, maintained and queried creates a vulnerability with regards to timing attacks in the structure. Their approach is based on a probabilistic analysis of the Skiplist as they attempt to discover the distribution of the structure. We devise a deterministic mapping algorithm based on this vulnerability in the Skiplist which allows for an attacker to cause such a "worst case scenario" for this structure. We will also show this vulnerability can allow an attacker to manipulate the data structure to create a hidden channel between themselves and a collaborator, without this being easily detectable.

2.1 Skiplist Mapping Algorithm

We introduce an algorithm, *SkiplistMap*, that maps the structure of a given Probabilistic Skiplist using the search function on the list. The algorithm is based on the following assumptions:

- The size of the structure, n is known.
- Each node in the structure holds a unique value.
- The range of possible values, X, in the structure is known, and is of size $O(n)$. (this may be achieved when the Skiplist implementation allows for a "first value" and "last value" query, or iteration over the entire list.)
- The runtime of the search algorithm in the structure is consistent, i.e. searching for the same value twice will yield the same runtime.

The *SkiplistMap* algorithm consists of 2 steps:

1. Search Time Mapping, shown in Algorithm 1
2. Skiplist Reconstruction, shown in Algorithm 2

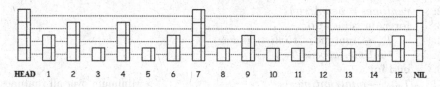

Fig. 1. Skiplist example

Search Time Mapping is done as follows:

- Search for all possible values, x_i, in the Skiplist.
- For each value found, denote its search time T_{x_i}.
- Denote the lowest runtime to be T_{min}.
- Normalize runtimes based on T_{min} such that $T_{min} = 1$.
- Normalized T_{x_i} is the length of the search path to x_i.

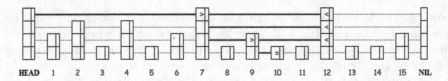

Fig. 2. Search process in Skiplist. Node 10's search path is 6 comparisons.

In our example, we begin by searching for all possible values $x_i \in X, 1 \leq i \leq 15$ in the structure and recording the runtime of each value found. Denote this time T_{x_i}. Figure 2 shows an example of searching for the value 10 in the Skiplist. We can see the highlighted path taken which shows that 6 comparisons were required until the node was found therefore $T_{10} = 6$. Looking at the example in Fig. 1, we can see the runtimes for all nodes are as follows (normalized relative to T_7):

$$T_1 = 3, T_2 = 2, T_3 = 5, T_4 = 3, T_5 = 6, T_6 = 5, T_7 = 1, T_8 = 5, T_9 = 4,$$
$$T_{10} = 6, T_{11} = 7, T_{12} = 2, T_{13} = 6, T_{14} = 7, T_{15} = 5. \tag{1}$$

Once the Skiplist is mapped, the reconstruction can be performed in the following manner:

- Create empty Skiplist with $log(n)$ levels.
- Insert nodes in order of increasing values of x_i, beginning at level 1. Level of insertion chosen by incrementing the level until the mapped search time is found in the newly created Skiplist.

Algorithm 1. Map Skiplist search times

1: **procedure** MAPRUNTIME($skiplist, values$)
2:     $runtimes = newArray$
3:     **for** x_i in $values$ **do**
4:         $T_{x_i} \leftarrow$ runtime of $sliplist.find(x_i)$
5:         $runtimes.append(T_{x_i})$
6:     **end for**
7:     $T_{min} \leftarrow min(runtimes)$ ▷ Minimum over all runtimes
8:     **for** T_{x_i} in $runtimes$ **do**
9:         $T_{x_i} = \frac{T_{x_i}}{T_{min}}$
10:     **end for**
11:     **return** $runtimes$
12: **end procedure**

Figure 3 details the reconstruction of the first 4 nodes our Skiplist as described. Initially, node x_1 is inserted at level 1. Searching for node x_1 reveals a runtime of 1, which does not match the original runtime $T_1 = 3$. Increasing

---

Algorithm 2. Reconstruct Skiplist by search times

1: **procedure** RECONSTRUCTSKIPLIST($values, runtimes, T_{min}$)
2:     $reconstructedList \leftarrow newSkiplist()$
3:     $runtimes \leftarrow sortAscending(runtimes)$ ▷ Sort runtimes from min to max
4:     **for** T_{x_i} in $runtimes$ **do**
5:         $t = 0$ ▷ zero current search time
6:         $L = 1$ ▷ initialize insertion level
7:         **do**
8:             $reconstructedList.insert(x_i, L)$ ▷ Insert x_i at level L
9:             $t \leftarrow$ runtime of $reconstructedList.find(x_i)$
10:            $t = \frac{t}{T_{min}}$ ▷ Normalize runtime
11:            $L = L + 1$ ▷ Increase level for next iteration
12:         **while** $t \neq T_{x_i}$
13:     **end for**
14:     **return** $reconstructedList$
15: **end procedure**

---

the level of x_1 to 2 now yields the correct runtime when searching. Continuing in this manner, we insert x_2 into level 1 and continue increasing its level until the correct runtime $T_2 = 2$ is achieved. This process is repeated for each node until all nodes have been inserted into their respective levels. Note that once a node x_i has been placed at a specific level, inserting any larger node added into the Skiplist (to the right of x_i) does not affect the search time for x_i.

Once the process is complete and the Skiplist structure has been determined, we can perform our attacks. The runtime attack will cause a deterioration in performance of the Skiplist's search function, while the hidden channel attack will allow an attacker to covertly transfer information to a collaborator without altering the data in the structure.

2.2 Runtime Attack

We restructure the Skiplist in a way that will cause it to perform near its worst case scenario search item. To do so, we use the following method:

- Remove items from the data structure which exists in any level above the first level, thereby flattening the structure to a standard linked list (with approximately half the original items).
- Re-insert all items that were removed back into the structure. Due to the random insertion, half will be inserted at level one only. At this point, approximately 0.75 of the items will be in the first level.
- Repeat steps 1 and 2 until nearly all items are re-inserted into level 1. This will yield a standard linked list structure with a search time of $O(n)$.

2.3 Hidden Channel Attack

In this scenario, the structure of the Skiplist is shared between two co-attackers. The first attacker (Transmitter) has read/write capabilities on the data structure

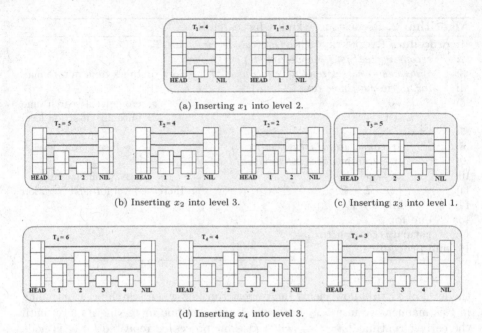

(a) Inserting x_1 into level 2.

(b) Inserting x_2 into level 3.　　　　(c) Inserting x_3 into level 1.

(d) Inserting x_4 into level 3.

Fig. 3. Example partial reconstruction of Skiplist - first 4 nodes.

while the second attacker (Receiver) only has read capabilities. By removing and re-inserting an item the Transmitter changes the max level of the node, thereby marking the item for the Receiver (note that due to the randomness of node insertion, the level may remain the same and the Transmitter may need to re-map and repeat this operation until the level has changed from the original). The Receiver can now re-scan the data structure and identify the item that has been re-inserted as its level has changed. This allows extra information to be shared between them without a casual observer noticing any anomaly. This method is closely related to the steganography methods of hiding data in digital images or videos by making minute changes that are not perceptible except to the two communicating parties. For example, assume each item in the structure represents anonymous information regarding a group of people who participated in a study, while hiding their sex (which is known only to the Transmitter). The Transmitter may change the levels of all items referring to male participants, allowing the Receiver to obtain additional information that is not available to other users reading the data (this entails publishing 2 versions of the same data set with the structure variation between them). We note that with this form of attack, since the Transmitter can only use the difference between 2 states of the Skiplist to convey information, they are limited to passing 1 bit of information per node to the Receiver. Using this method, one could either use that information to pass 1 bit of information regarding each node, or encode an $n-$bit message over the structure.

3   Defense from Timing Attacks: Splay Skiplist

In order to defend the Skiplist data structure from the attacks presented above, we must conceal the runtime of the search query so as to not enable the mapping algorithm. We propose a variation on the Skiplist structure, the Splay Skiplist. The Splay Skiplist is based on the Splay Tree concept of re-ordering nodes when a search function is performed (Sleator and Tarjan [12]). In a Splay Tree, this is done to optimize performance, we use this behavior in order to modify the Skiplist structure thus concealing the link between run-time and structure, rendering the mapping algorithm presented above useless.

3.1   Splay Skiplist Algorithm

The Splay Skiplist differs from a standard Skiplist in its search function. We use the SplayNode method defined in Algorithm 3 to change the level of a given node v inside a Skiplist. This method performs as follows:

- For each level l of the node v, denote u_l to be the node preceding v in that level, and w_l to be the node succeeding v in that level (i.e. the nodes linking to v and from v in level l respectively).
- Denote L_v and L'_v to be the original and new target levels of v respectively.
- If $L'_v > L_v$, we add node levels to v and connect each new node in level l to its predecessor node u_l and successor node w_l. This is similar to the process of adding a new node to the Skiplist, without needing to add all node levels.
- If $L'_v < L_v$, we remove node levels from v and at each level l connect the predecessor node u_l to the successor node w_l. This is similar to the process of removing a node from the Skiplist, without needing to remove all node levels.

Note that the run-time for the Splay algorithm is at most the same as the addition or removal of a node, $O(\log n)$.

The addition and deletion of nodes in the Splay Skiplist remains as in the original randomized Skiplist algorithm. Our change to the search function is as follows:

- When a search for a value x is performed, denote the corresponding node u_x (if x is not found, no further action is required).
- Select a random value, r, uniformly selected in the range X of the Splay Skiplist and denote the corresponding node u_r (if r does not exists in the Splay Skiplist, select the last node found in the search path).
- Denote the levels of u_x and u_r to be L_{u_x} and L_{u_r} respectively.
- Swap between the levels of u_x and u_r such that their new levels, L'_{u_x} and L'_{u_r}, will be $L'_{u_x} = L_{u_r}$ and $L'_{u_r} = L_{u_x}$ using the Splay algorithm described above.

Figure 4 gives an example of a search performed on the Splay Skiplist. It is easy to see that the mapping proposed in Sect. 2.1 cannot be performed on the Splay Skiplist, since any node previously found using the "search" function

---

Algorithm 3. Splay Skiplist node: change node level

```
 1: procedure SPLAYNODE(v, newLevel, prevNodesArray)
 2:     max ← maxLevel(v)                          ▷ Find current max level of v
 3:     if max > newLevel then
 4:         for l in max + 1 ... newLvel do
 5:             v_l ← newnodeLevel(v)              ▷ Create new level in node v
 6:             v_l.next ← nextNodesArray[l].next   ▷ Connect v to next node in level
 7:             prevNodesArray[l].next ← v_l        ▷ Connect v to previous node in level
 8:         end for
 9:     end if
10:     if max < newLevel then
11:         for l in newLevel + 1 ... max do
12:             prevNodesArray[l].next ← v_l.next
13:             v.deleteLevel(l)                    ▷ Remove level l from v
14:         end for
15:     end if
16: end procedure
```

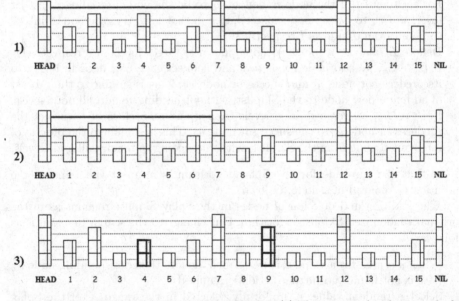

Fig. 4. Searching for the node 9 in the Splay Skiplist. (1) The requested node is searched for and found. (2) A random node to swap is selected (in our case 4) and found. (3) The top levels of each node, 9 and 4, are changed to the others'.

may have been relocated once the next "search" is performed. For each search function we are adding an additional search and running the Splay algorithm twice, the runtime for a search function is slightly increased, but remains within the bounds of $O(\log n)$.

4   Conclusions

We have shown the probabilistic Skiplist structure to be vulnerable to a timing attack exploit which allows mapping of the structure in direct opposition to what the probabilistic nature of Skiplist intended to avoid. This allows an attacker to manipulate the Skiplist into a structure that performs much worse than the desired $O(\log n)$, down to that of a standard List structure ($O(n)$). Another proposed attack allows two conspiring attackers to create a hidden channel between them using the structure of the Skiplist to hide the data itself. To fend off these and other "structure based" attacks, we offer a new variant of Skiplist called a Splay Skiplist, which randomly modifies its structure as it is being queried. We show that while performing slightly worse than the original Skiplist, this structure retains the $O(\log n)$ performance.

While our current research has focused on static releases of a Skiplist structure, there are interesting implications for consecutive releases of a Skiplist with changing data. Future work could be focused on modifying our method to follow such consecutive releases and attempt to maintain knowledge of the structure throughout. Multiple releases over time could allow for an attacker to create an ongoing hidden data channel similar to the method we detail in Sect. 2.3.

References

1. Aspnes, J., Shah, G.: Skip graphs. ACM Trans. Algorithms **3**(4) (2007). https://doi.org/10.1145/1290672.1290674
2. Bagchi, A., Buchsbaum, A.L., Goodrich, M.T.: Biased skip lists. Algorithmica **42**(1), 31–48 (2005). https://doi.org/10.1007/s00453-004-1138-6
3. Bethea, D., Reiter, M.K.: Data structures with unpredictable timing. In: Backes, M., Ning, P. (eds.) ESORICS 2009. LNCS, vol. 5789, pp. 456–471. Springer, Heidelberg (2009). https://doi.org/10.1007/978-3-642-04444-1_28
4. Chen, J., Jindel, S., Walzer, R., Sen, R., Jimsheleishvilli, N., Andrews, M.: The memSQL query optimizer: a modern optimizer for real-time analytics in a distributed database. Proc. VLDB Endow. **9**(13), 1401–1412 (2016). https://doi.org/10.14778/3007263.3007277
5. Crosby, S.A., Wallach, D.S.: Denial of service via algorithmic complexity attacks. In: Proceedings of the 12th Conference on USENIX Security Symposium - Volume 12, SSYM 2003, p. 3. USENIX Association, Berkeley (2003). http://dl.acm.org/citation.cfm?id=1251353.1251356
6. Ergun, F., Cenk Şahinalp, S., Sharp, J., Sinha, R.K.: Biased skip lists for highly skewed access patterns. In: Buchsbaum, A.L., Snoeyink, J. (eds.) ALENEX 2001. LNCS, vol. 2153, pp. 216–229. Springer, Heidelberg (2001). https://doi.org/10.1007/3-540-44808-X_18
7. Futoransky, A., Saura, D., Waissbein, A.: Timing attacks for recovering private entries from database engines, January 2008
8. Goodrich, M.T., Tamassia, R., Schwerin, A.: Implementation of an authenticated dictionary with skip lists and commutative hashing. In: Proceedings DARPA Information Survivability Conference and Exposition II, DISCEX 2001, vol. 2, pp. 68–82, June 2001. https://doi.org/10.1109/DISCEX.2001.932160

9. Goodrich, M.T., Kornaropoulos, E.M., Mitzenmacher, M., Tamassia, R.: More practical and secure history-independent hash tables. In: Askoxylakis, I., Ioannidis, S., Katsikas, S., Meadows, C. (eds.) ESORICS 2016. LNCS, vol. 9879, pp. 20–38. Springer, Cham (2016). https://doi.org/10.1007/978-3-319-45741-3_2

10. Messeguer, X.: Skip trees, an alternative data structure to skip lists in a concurrent approach. ITA **31**, 251–269 (1997)

11. Pugh, W.: Skip lists: a probabilistic alternative to balanced trees. Commun. ACM **33**(6), 668–676 (1990)

12. Sleator, D.D., Tarjan, R.E.: Self-adjusting binary search trees. J. ACM **32**(3), 652–686 (1985). https://doi.org/10.1145/3828.3835

13. Solis, J., Tsudik, G.: Simple and flexible revocation checking with privacy. In: Danezis, G., Golle, P. (eds.) PET 2006. LNCS, vol. 4258, pp. 351–367. Springer, Heidelberg (2006). https://doi.org/10.1007/11957454_20

DPM Workshop: Field/Lab Studies

A Study on Subject Data Access in Online Advertising After the GDPR

Tobias Urban[1,2](✉) ⓘ, Dennis Tatang[2], Martin Degeling[2] ⓘ, Thorsten Holz[2],
and Norbert Pohlmann[1]

[1] Institute for Internet Security, Westphalian University of Applied Sciences,
Gelsenkirchen, Germany
{urban,pohlmann}@internet-sicherheit.de
[2] Horst Görtz Institute, Ruhr University Bochum, Bochum, Germany
{dennis.tatang,martin.degeling,thorsten.holz}@rub.de

Abstract. Online tracking has mostly been studied by passively measuring the presence of tracking services on websites (i) without knowing what data these services collect, (ii) the reasons for which specific purposes it is collected, (iii) or if the used practices are disclosed in privacy policies. The European *General Data Protection Regulation* (GDPR) came into effect on May 25, 2018 and introduced new rights for users to access data collected about them.

In this paper, we evaluate how companies respond to *subject access requests* and *portability* to learn more about the data collected by tracking services. More specifically, we exercised our *right to access* with 38 companies that had tracked us online. We observe stark differences between the way requests are handled and what data is disclosed: Only 21 out of 38 companies we inquired (55%) disclosed information within the required time and only 13 (34%) companies were able to send us a copy of the data in time. Our work has implications regarding the implementation of privacy law as well as what online tracking companies should do to be more compliant with the new regulation.

Keywords: GDPR · Subject access request · Privacy · Online advertisement

1 Introduction

The business models of modern websites often rely—directly or indirectly—on the collection of personal data. The majority of websites tracks visitors and collects data on their behavior for the purpose of targeted advertising [12]. While in some cases, users knowingly and willingly share personal data, in many other cases, their data is collected without explicit consent or even goes without being noticed [26]. As a result, the imbalance of power over information between data processors (service providers) and data subjects (users) increased in the last years. Furthermore, attackers can also perform a malicious leakage of such data [27].

© Springer Nature Switzerland AG 2019
C. Pérez-Solà et al. (Eds.): DPM 2019/CBT 2019, LNCS 11737, pp. 61–79, 2019.
https://doi.org/10.1007/978-3-030-31500-9_5

The *European General Data Protection Regulation* (GDPR) aims to harmonize data protection laws through the EU and to regulate the collection and usage of personal data. Compliance with GDPR is required for any company that offers services in the European Union, regardless of where their headquarter is located. One of the law's goals is to allow users to (re)gain control of the immaterial wealth of their personal data by introducing additional possibilities like the right to request a copy of their data, the right to erasure, and the need for services to explicitly ask for consent before collecting or sharing personal information [14].

Previous work already passively measured the effects of the GDPR. For example, studies analyzed the adoption of privacy policies and cookie consent notices [10,23], while others focused on third parties embedded into websites [6,25].

In this paper, we make use of the new legislation and evaluate the subject access processes of several companies. We identify prominent third parties on popular websites that collect tracking data and exercise our *right to access* with these companies. Besides these two rights, the GDPR also grants the right to erasure, rectification and others that are not part of this work. We provide an in-depth analysis of the processes and show how different companies adopted the new legislation in practice. We analyze timings and success of our inquiries and report on obstacles, returned type of data, and further information provided by companies that help users to understand how personal data is collected. Asides from the detailed overview on different approaches how *subject access requests* (SARs) are implemented in practice; our work provides helpful pointers for companies, privacy advocates, and lawmakers how the GDPR and similar regulation could be improved.

To summarize, our study makes the following contributions:

- We requested access to our personal data from 38 companies and analyze the success of these *subject access requests*. We found that 58% of the companies did not provide the necessary information within the deadline defined in the GDPR and only a few actually granted access to the collected data.
- We analyzed the privacy policies of these companies regarding usage and sharing of collected personal data. We found that most policies fulfill the minimal requirements of the GDPR, but rarely contain additional information users might be interested in (e.g., partners with whom data is shared).
- Finally, we examined the *subject access request* process of each company and report on data (e.g., clickstream data) and obstacles users face when accessing their data. We found that the provided data is extremely heterogeneous and users sometimes have to provide sensitive information (e.g., copies of identity cards) to access their own data.

2   Background

Our study analyzes the effects of the GDPR as a relatively new legal regulation. We therefore first provide an overview of the GDPR's relevant rules before giving describing the technical background on tracking and data sharing in the online advertisement ecosystem.

2.1 Data Protection Law

The General Data Protection Regulation (GDPR or Regulation 2016/679) [14] is an initiative by the European Union (EU) to harmonize data protection law between its member states. After a transition period of two years, it went into effect on May 25, 2018. The GDPR specifies under which circumstances personal data may be processed, lists rights of data subjects, and obligations for those processing data of EU-citizens. Online advertising companies need to disclose, for example, in their privacy policy, for what purpose they collect and share data.

Besides other rights, the GDPR lists the *right to access* (Art. 15) and the *right to data portability* (Art. 20). The difference between those two is that Art. 15 grants users the right to request access to the personal data a company collected about them, while Art. 20 grants users the right to retrieve a copy of the data they provided. According to recital 68 of the GDPR (recitals describe the reasoning behind regulations), the *right to data portability* is meant to support an individual in gaining control over one's personal data by allowing access to the data stored about him or her *"in a structured, commonly used, machine-readable and interoperable format"*. For any information request, including those to data access/portability, the GDPR specifies that they must be answered within one month (Art. 12, No. 2), but can be extended by two months.

Some tracking companies claim that the data they use is not personal information because it is anonymized, while in fact is only pseudonymized (see Sect. 5.1). If the data was anonymous, it would free them from any data protection related obligations, while pseudonymous data that can be attributed to a person using additional information, still falls under the GDPR's rules (Recital 26). In addition, the Article 29 Working Group, a committee of European data protection officials, already made clear in 2010 that storing and accessing a cookie on a user's device is indeed processing of personal data since it *"enable[s] data subjects to be 'singled out', even if their real names are not known,"* and therefore requires consent [8]. Relevant for our study is the clarification that ad networks, and not those that embed the third-party scripts on their websites, are responsible for the data processing. Since advertisers rent the space on publisher websites and set cookies linked to their hosts, they are responsible for the data processing, and therefore have to respond to subject access requests.

2.2 Advertising Economy

Displaying ads is the most popular way to fund online services. In 2017, the online advertising industry generated $88.0 billion US dollars [19] in revenue in the US and €41.8 billion Euros in the EU [18]. The ecosystem behind this is complex and is, in a nutshell, composed out of three basic entities which are described in the following [13, 28]. On the one end, there are publishers and website owners that use *supply-side platforms* (SSP) to sell ad space (e.g., on websites or prior to videos). On the other end, the *demand-side platform* (DSP) is used by marketing companies to organize advertising campaigns, across a range of publisher. To do so, they not necessarily have to select a specific publisher they want to work with, but can define target users based on different criteria (e.g., geolocation, categories of websites visited, or personal preferences). A *data management platform* (DMP) captures and evaluates user data to organize digital ad campaigns. They can be used to merge data sets and user information from different sources to automate campaigns on DSPs.

To improve their reach, ad companies utilize *cookie syncing* (sometimes called ID syncing) [22] which allows them to exchange unique user identifiers. Using this method, companies can share information on specific users (e.g., sites on which they tracked them) and learn more about the user. While this is considered an undesirable, privacy-intrusive behavior by some, it is in practice a fundamental part of the online ad economy to perform *Real-time Bidding* (RTB). RTB involves that impressions and online ad space are sold in real-time on automated online marketplaces whenever a website is loaded in a browser.

3   Related Work

Most previous work analyzes online privacy through measurements (e.g., [12, 20]), but these studies have all been conducted prior to the GDPR. With the introduction of the GDPR, several research groups started measuring the effects of the legislation. Degeling etal. analyzed the adoption and effect of the GDPR regarding privacy policies and cookie notice banners [10]. Dabrowski etal. measure the effects of cookies set based on the location of a user and find that around 50% more cookies are being set if the users come from outside the EU [7]. In contrast, Sørensen etal. found that the number of third parties did slightly decline since the GDPR went into effect, but they conclude that the GDPR is not necessarily responsible for that effect [25]. Boniface etal. analyze the tension between authentication and security when users perform a SAR [4] and discuss measures used to identify users and discuss threats (e.g., denial of access) of too harsh measures. In line with our findings, they also report on disproportional identity checks. Most recently, two studies analyzed how adversaries could abuse subject access requests to get access to personal data of other individuals [5, 11]. Both studies spoof an identity and request access to personal data of the spoofed identity and by this they show that SARs are often not adequately verified and therefore, companies unintentionally leak personal data. De Hert etal. [17] discuss the right to data portability from a computer law point of view. They give a systematic interpretation of the new right and propose two approaches on how to interpret the legal term "data provided" in the GDPR. The authors argue that a minimal approach, where only data are directly given to the controller can be seen as "provided". They also describe a broad approach which also labels data observed by the controller (e.g., browser fingerprints) as "provided".

4   Study Design

To gain insights into the way how companies grant access to collected personal data, we first identified prominent companies often embedded into websites, and afterward, we exercised our right to access/portability with these companies.

4.1   Approach

Our study consist of two steps: First, we *passively* measure the most prominent companies used as third parties on websites and the companies most active in sharing personal identifiers. Afterward, we *actively* measure and analyze the information provided by companies to users that file a subject access request.

In order to identify the most prominent companies, we perform a three-step measurement (see Sect. 4.2): (1) We visit a number of websites, (2) extract the third parties embedded in these websites, and (3) extract all ID syncing activities from the observed requests. Based on the gathered information, we determine top companies that engage in ID syncing and top companies that are often embedded into websites. We did choose to focus on top embedded companies as these potentially affect most users and more users might issue a *subject access request* (SAR) to these companies. Furthermore, we choose the top syncing parties as these might share personal data of users without properly informing them—which would make it quite hard for users to actually regain control of their personal data if they do not know who holds their data. In order to learn more about the privacy practices from the companies themselves, we analyze privacy policies to see if the data sharing and other necessary information are made transparent to users (see Sect. 4.3). Then we use our right to access/portability to learn how companies respond to SARs and which data they provide to users.

In our experiments, we use the *openWPM* [12] platform and deployed it on two computers located at a European university. Thus, our traffic originates in the EU (and ultimately from an EU resident who started the crawl) and therefore the GDPR applies. *OpenWPM* was configured to log all HTTP request and response with the corresponding HTTP headers, HTTP redirects, and POST request bodies as well as various types of cookies (e.g., Flash cookies). We did not set the "Do Not Track" HTTP header and did allow third-party cookies.

To analyze the sharing of *digital identifiers* (IDs), we first have to define them. For every visited domain, we analyzed the HTTP GET and POST requests and split the requests at characters that are typically used as delimiters (e.g., '&' or ';'). As a result, we obtained a set of ID candidates that we stored as key-value pairs for later analysis. We identified IDs according to the rules previously defined by Acar etal. [1] (e.g., IDs have to be of a certain length or must be unique). To measure the syncing relations of third parties, it is necessary to identify URLs—that contain user IDs—inside a request (e.g., **foo.com/sync?partner=https://bar.com?/id=abcd-1234**). According to the named rules, we parsed all URLs and checked if an HTTP parameter contains an ID, Furthermore, we used the *WhoTracks.me* database [6] to cluster all observed third-party websites based on the company owning the domain.

4.2   Analysis Corpus

The complexity of the online advertising ecosystem was already highlighted in previous work [3,16]. To the best of our knowledge, there is no reliable public information on market shares in the online advertising ecosystem. Thus, we performed an empirical measurement and identified the top companies in that measurement. To identify the most popular companies, we visited the Alexa top 500 list [2] and randomly visited three to five subsites of each domain. We visited the selected websites using the *openWPM* setup described above.

We selected the 25 most embedded third parties as well as the top 25 third parties that engaged most in cookie syncing for in-depth analysis of what information they share with users. In total, we identified 36 different companies which we refer to as *analysis corpus*. In three cases, we were told during the SAR process to address our inquiry to another company so that our final corpus consists of 39 companies. In the remainder, if not stated otherwise, our analyses of privacy policies and information disclosure refer to this corpus.

The first company that we did *not* include in the corpus (i.e., the 26th most embedded company) was embedded by just 0.12% of the visited websites and the first ID syncing company *not* included in the corpus accounts for 0.58% of the syncing connections in the graph. The 39 companies in the corpus account for 66% of all ID syncing activities, while the reaming 33% are made up of 352 companies. The companies in the corpus represent 61% of the embedded third parties. Contacting ten more companies (an increase of 19%) would increase the amount of covered ID syncing by at most 5.8% or embedded websites by at most 1.2%. The corpus consists of six SSPs, nine DSPs, seven companies that specialized in targeted ads, four DMPs, and 13 companies whose primary business field is not directly tied to the advertising but instead utilizes ads to finance their services (e.g., *RTL Group*—a Luxembourg-based digital media group).

While most of the companies in our corpus operate globally and run multiple offices, 82% have their headquarters located in the United States. The remaining 18% are located in Europe. This distribution is likely based towards US/EU-based companies since we run our measurements from Europe. We discuss the limitations of our analysis corpus in Sect. 6.

4.3   Transparency Requirements

The privacy policies of all 39 companies described above were analyzed by a certified data protection expert with a computer science background to see whether they contain the information required by the GDPR (see Sect. 2). We specifically looked for information on data sharing practices and evaluated how data subjects can exercise their rights. As described above, data controllers are required to inform, besides other things, about the legal basis for their data collection, categories of companies they share the data with, and how long the data is stored. We do not report on observations that all policies had in common but focus on the differences. On the one hand, for example, the right to withdraw consent has been implemented through various opt-out mechanisms [10] that all services support and are therefore not listed. On the other hand, few services actually follow the "Do Not Track" signal, although it was designed as a common consent mechanism. Therefore, we listed statements about the latter. We were also interested in how companies deal with the requirements regarding profiling: If they use profiling, they are expected to describe *the logic involved* in this process, although the debate about what that should include is still ongoing [24]. Privacy policies should list the rights of the data subjects, e.g., to object to the processing and the possibility to access the data and they should describe how these rights can be exercised. While the policies should also specify whether data is shared with third parties, companies are not required to list them individually but can describe them in categories.

4.4   Assessing the SAR Process

In order to test to which extent users can actually exercise data access rights, we reached out to companies in the corpus after extracting contact information from their privacy policies. According to Articles 13 and 14 GDPR, contact details of a responsible person (e.g., the Data Privacy Officer) need to be provided for privacy-related questions. Most companies (27) named a general email address to handle such requests or referenced a web form to access the data.

In our requests, we referenced a profile that was generated specifically for this process. We used *openWPM* to randomly visits websites that include third parties, owned by the companies in our analysis corpus. From these websites, all internal links (subsites) were extracted and visited in random order. For this analysis, we kept the session active and continued visiting websites while we requested information about the profiles. This *openWPM* instance was left running until the end of our analysis in order to keep the cookies active.

When sending out inquiries, we included all cookie IDs and domains for which we observed ID syncing (with the corresponding IDs). If we could add custom text to our request (via email and in some web forms), we asked four questions regarding the usage of our data: (1) *What information about me/associated with that cookie do you store and process?*, (2) *Where did you get that information from? Did you get it from third parties?*, (3) *Do you use the data to perform profiling?*, and (4) *With whom do you share what information and how?*

We used informal language (e.g., we did not quote any articles from the GDPR nor did we use any legal terminology) because we wanted to assess the process when users with some technical understanding of online advertising (e.g., users who can read cookies from the cookie store), but no legal background, want to exercise their right to access/portability. Actual users might have trouble to access the information we added in our emails (e.g., the correct cookie values). However, some companies offer simplified ways to access the information to be included in requests (e.g., a web form that reads the user ID from the browser's cookie store). We assume that a user who has privacy concerns can obtain this information and usability improvements might follow in the near future.

We conducted two rounds of inquiries. The first round in June 2018, one month after the GDPR took effect, and the second round was starting three months later, in September 2018. We did so to make sure answers were not biased by being the first ones the companies received. We used two *GMail* accounts we created for this purpose (one for each round of contacting) to get in touch with the companies and did *not* disclose that we were conduction this survey to avoid biased responses. The response timings were evaluated in relation to two deadlines: The first deadline is the legal period defined in the GDPR, 30 days after the request, and a more relaxed deadline 30 *business* days after our requests.

5   Results and Evaluation

Companies are required to share certain information, e.g., who has access to the data and where it is transferred to, publicly in their privacy policies. Other information, e.g., what is stored about a user, has to be disclosed upon request.

5.1   Evaluation of Privacy Policies

We analyzed the privacy policies of the companies in our corpus to check whether they fulfill the requirements described in Sect. 4.3. The most relevant details are reported in Table 1 (see Appendix B). All but three policies fulfill the minimum requirements for privacy policies set by the GDPR, all companies offer the possibility to opt-out of their services, and all except one disclose that they share some information with others. At the same time, only three are transparent about who these third parties are and what type of information is actually shared. Only two of the policies disclosed and explain cookie syncing. Similarly, only eight mention whether or not they perform profiling. One company did not update its privacy policy since 2011 and it contained false claims, for example, that IP addresses are non-personal information. *Amazon*'s privacy policy was least transparent concerning the information we were looking for.

All policies except for four policies mention a legal basis for processing, which is now required. 31 claim that they rely on individual consent when processing data but at the same time only three mention that they adhere to the "Do Not Track" (DNT) standard, where information about whether or not users want to be tracked is conveyed in an HTTP header [21]. Instead, companies refer to implicit consent, which implies consent as long as a data subject has not manually objected to a data collection by opting-out.

Differences can be found on topics specific to GDPR, for example, regarding the question of whether a company processes data that contains sensitive information (e.g., about race or health). While 13 explicitly forbid to collect this information through their services, four acknowledge that some interest segments they provide might be health-related e.g., about beauty products. Three companies acknowledge that they process health-related information, but do not discuss how this data is better protected than the rest. The majority (17) does not make any statements about their practices in this area.

5.2   Subject Access Requests

In order to analyze the process how users can access personal data collected about them and to fill the blanks left by the privacy policies, we examined how third parties adopt the new requirements of the GDPR (see Sect. 2.1) and how they respond to *subject access requests* (SAR).

We contacted the companies in our analysis corpus and tried to exercise our right to access and right to portability of the data associated with a cookie ID to evaluate the SAR process of each company as described in Sect. 4.4. In the first round (June 20th, 2018), we sent out 32 emails and used six web forms to get in touch with each company. In the second round (September 21th, 2018), we sent 27 emails and used eleven web forms as the contact mechanisms had slightly changed. As part of this process, we extracted the cookie ID values and up to five domains associated with each company for which we observed ID syncing (with the ID key-value pairs) from the long-running profile in the email. The GDPR requires companies to grant users access to their data within 30 days after their initial request. Since it does not specify whether these referrers to business or calendar days, we marked two deadlines (dotted, gray lines in Figs. 1 and 2).

Response Types and Timing. We grouped responses in three types: (1) *automatic* responses, (2) *mixed* responses, and (3) *human* responses. A message was categorized as "automatic" if it was identifiable as sent automatically by a computer system (e.g., a message from a ticket system). We labeled a message *mixed* if the message did not directly refer to any of our questions but only included very generic information that responds to any privacy-related request. Messages that directly responded to our questions were labeled "human". To increase the accuracy of the classification, we compared the content from both inquiry rounds and if there was any doubt, we ruled in favor of the companies. Figure 1 shows amounts and type of responses we got during our analysis. We did not count status messages from ticket systems (e.g., a message stating that our email was received) but only looked at those messages that contained an actual reply.

In our second round of inquiries, we received fewer responses (approx half of the amount). This is partly because we did not have to report any broken data access forms, that we encountered in round one, to companies which explain the fewer human responses in weeks one and two. However, we observed that in our second round, companies did not follow up further questions as they did in round one (e.g., if we asked for further clarification about data sharing).

In round one, we received the most responses (51/100) during the first two weeks, where we labeled the majority as send by a *human* (57%) and 26% as *automatic*. While the share of response types stayed balanced, the number of responses significantly decreased (by 43%) in the following weeks, although we asked follow-up questions. In round two, these types of answers changed as we considered 17% of the responses as sent by a *human* and 61% *automatic*.

While we still received responses from human correspondents one week before the deadline in the first round, responses were lower in the second round (a third).

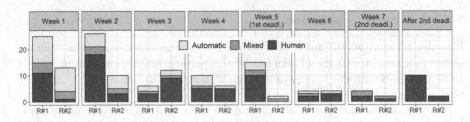

Fig. 1. Types and timings of the received responses.

Only one company told us (in both rounds) that due to the complexity of our inquiry that they would need more time.

Response Success. The effort necessary to obtain access to personal data differed depending on the inquired company. To asses the workload of the process of a company, we use a simple scoring mechanism that essentially takes four factors with different impact into account: (1) amount of emails sent to the company before getting access to data associated to the digital ID, (2) amount of emails sent after getting access to the data, (3) actions that the user has to perform online, and (4) actions that a user has to perform offline. These simple metrics do not account for the actual effort each obstacle might pose to an individual asking for access, but it is helpful to approximate the complexity of the process.

We differentiate between emails because we interpret access to collected data as the primary goal of the request. However, there might still be some open questions (e.g., if profiling is performed) that were not answered by the time the data was shared. An example of an action that a user must perform online is that the user has to enter additional data in an online form (e.g., legal name). On the contrary, scanning the user's official identification document (e.g., passport) is a typical example of a task a user has to perform offline. We created our "workload score" to measure (1) if companies set up obstacles, (2) if companies ask for additional information, and (3) the amount of interaction necessary.

In Sect. 5.2, we describe the procedure of how users can access personal data (of the companies in our analysis corpus) in more detail. The result of the workload determination and comparison between inquired companies is given in Fig. 2. The figure shows a clustered version of the SAR results. We computed the distance between all points of the same "response status" (e.g., "got access") and clustered the points that are close to each other. The larger and higher each point, the more companies asked for more effort to answer our requests.

Figure 3b shows the results of our inquiries by the time of the first deadlines (July 20th/October 31st). Note that is unlikely that we provided a wrong cookie ID, but it is possible that a company does not have any data on the record because of short retention times or when some events are not logged, because there was no further interaction. It is notable that some companies stated that if one does not have a user account on their website, they will not store any data related to a cookie ID. These companies did not respond to our SAR request within the legal deadline, in round two. One of these companies replied with our second deadline stating that they do not store any data related to the cookie ID.

Eight companies interpreted the start date of the process as the day on which they got all the administrative data they need to process the inquiry. In all cases, it was virtually impossible for users to know upfront that this data was needed since the companies only shared the needed documents via email and did not mention them in their privacy policies (e.g., one company replied after seven days and asked for a signed affidavit. After we provided the affidavit, they told us, five days later, that they would *"start the process"* and reply within 30 days.).

In total (after the second deadline of round one), only 21 of 36 companies (54%) shared data, or told that they do not store any data, 15 of 36 (42%) were still in the process (or did not respond), and one company said that it would not share the data with us because they cannot properly identify us. In round two, 64% granted access or told us that they do not store any data, 33% did not finish the process, and again one company declined to grant access since they could not identify us. In these numbers, we *excluded* companies that told us to address a subsidiary/parent company with our inquiry.

Figure 2 shows that if companies granted access, we see that the workload is often quite low (in both rounds). In one case with high workload, in round one, a long email exchange (in total 13 emails—six sent by us) was needed to get access, the other cases required a copy of the ID and in one case a signed affidavit. It is notable that the overall workload in round 2 lowered and companies usually wrapped up the process faster. The reduction of workload is because, on the one hand, we did not have to report broken SAR forms and on the other hand companies set up less "offline" obstacles.

Especially during round one, we observed that companies who claimed not to store any data still require multiple interactions prior to providing that information. Two companies required a signed affidavit and a photocopy of an ID. The third company, after a long email conversation, asked to call the customer support to explain our case in more detail, still coming to the result that they do not store any data. All three companies did not respond in round two.

Disclosed Information. Figure 3c gives an overview of the data we received as a result of the SARs. We categorized the received data in terms of readability and content. If data was presented in a way a human can easily read it (e.g., on a website), we labeled it *"human readable"* and otherwise *"raw"* (e.g., .csv files). If the data contained visited websites, we labeled it *"Tracking"*, if it contained segment information, associated with the profile, we labeled it *"Segment"*, and if it contained the location of the user, based on the used IP address, we labeled it *"Location"*. Otherwise, we labeled it *"Other"*.

The shared data was heterogeneous in format (e.g., .pdf, .csv, .htm, etc.), data contained (e.g., interest segments, clickstream data, IP addresses, etc.), and explanation of the data (examples of shared data are provided in Appendix A. One company shared an .csv file with headers named c_1 to c_{36} (sic.), while another company provided detailed explanations in an appended document and yet another told us that we should contact them if we had trouble

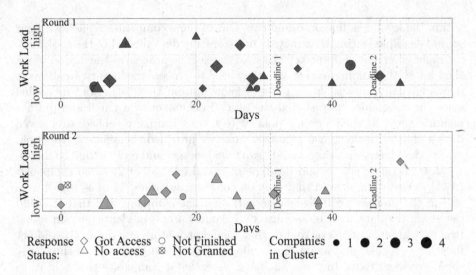

Fig. 2. Comparison of the workload to get access to personal data companies stored about a user.

understanding the data. If a company shared clickstream data (three in total), we manually checked if the data set contained additional or missing websites that we had observed. In all three cases, the data was accurate. Overall, the received data can be grouped into three categories: (1) technical data, (2) tracking data, and (3) segment data. *Technical data* is raw data, often presented in text files, the companies directly extracted from HTTP traffic (see Fig. 5). *Tracking data* is information on which websites the company has tracked the user, also typically presented in a text file (see Fig. 6). *Segment data* is data companies inferred from a user's online behavior (see Fig. 4), which was typically presented on a website (e.g., user interests). In terms of clarity of the provided data, we also found different approaches. Some companies shared segments they inferred from our (artificially) browsing behavior (e.g., Segment: *Parenting - Millennial Mom* (sic.)), others shared cryptic strings without explanation (e.g., *Company-Usersync-Global*), or data that was incorrectly formatted somewhere in the process to the point where it was almost unintelligible (e.g., *Your_hashed_IP_address: Ubuntu* (sic.)). However, we did not find any instance where data was provided that was not mentioned to be collected in the privacy policy and many instances (all but one) where not all data that might be collected was provided.

Subject Access Request Process. Companies handle inquiries very differently ranging from not responding at all, over simply sending the personal data via email, to sending (physical) letters which had to include a copy of a government-issued identification card and a signed affidavit, stating that the cookie and device belong to the recipient and only the recipient.

(a) Obstacles		
Status	R1	R2
Affidavit	4	3
ID card	6	5
Other	4	7
None	26	25

(b) Response success		
Status	R1	R2
Access	14 39 %	8 22 %
No Data	7 19 %	13 36 %
Denied	1 3 %	1 3 %
Not Finished	11 31 %	9 25 %
No Response	4 11 %	5 14 %

(c) Response data		
Type	R1	R2
Raw data	9	3
Human read.	5	5
Segments	4	4
Tracking	3	3
Location	4	4
Others	5	2

(d) Answers		
Question	R1	R2
Q1 (data)	21	23
Q2 (sources)	6	6
Q3 (profiling)	9	6
Q4 (sharing)	7	4

Fig. 3. Overview of the SAR process and responses for both rounds of inquiries.

Figure 3a gives an overview of the obstacles users face when filling a SAR. Most companies require the user to provide the digital identifier (or directly read it from the browser's cookie storage) in order to grant access to the data associated with it. Since most online forms do not provide all data, a company collected about the user (e.g., they provide the ad segments associated with the user but not the used IP addresses or visited websites) it is reasonable to grant access to this data if the cookie ID is provided. However, online forms come with the risk that an adversary might fake the cookie ID to get access to personal data that is associated with another individual. An affidavit is a way to counter this sort of misuse, and one company stated this as the reason for this step.

The GDPR states companies *"should use all reasonable measures to verify the identity of a data subject who requests access"*, to make sure they do not disclose data to the wrong person. Asking for identifying information is supposed to add a layer of security when data subjects request a copy of their data. The ad industry association emphasizes the possibility of this additional safeguard [15], but official interpretations state that data processors should have *"reasonable doubts"* before asking for additional data [9]. Those that request an ID card did not explain their doubt and did not describe how the ID helps them to verify that the person requesting the data actually owns the cookie ID.

Answers to Our Questions. Finally, we want to discuss the answers to the four questions we asked in the inquiries (see Sect. 5.2). Only a few companies did answer the additional questions we asked. Most of them referred to their privacy policy or did not provide further details. Figure 3d gives an overview of the responses we got to our questions. Note that companies were not obliged to answer the question and that we could not check if they answered truthfully—if there is no public information in e.g., the privacy statements that say otherwise (see Sect. 6). With respect to Q1 and Q2, most answers contained references to or parts of the privacy policy. As Table 1 (Appendix B) shows, only a few companies (nine/seven) disclose whether or not they perform profiling. Only one of the answers, where the privacy policy was unspecific, clearly stated that the data is not used for profiling. Six answers described in more detail how the data is processed and would suffice the GDPR rule that *"meaningful information about the logic involved"* should be provided. One company stated in their email that they do not perform profiling, although their privacy policy mentions it. Unfortunately, only seven/five companies listed their actual sharing partners. When companies stated with whom specifically they shared our data (i.e., not a general list of partners), we could confirm this through our measurement, but in three cases companies stated that they did share data with specific companies that were not listed in their privacy policy. The low amount of companies that named partners with whom they share data poses a problem for users that want to understand who has received a copy of their personal information.

6   Limitations and Ethical Considerations

We contacted 39 companies, which represents only a small subset of all online advertising companies. However, we showed that the contacted companies come from different market areas and that they represent the most prominent companies (in our measurement). Future work should focus on the usability of SARs in a user study and include more companies. Similarly, our scale to visualize the complexity of the subject access requests (Fig. 2), should be validated with user experiments. Right now it serves only as an approximation.

Since our research includes human subjects (the persons exercising their rights and the persons responding to our requests), ethical considerations need to be taken into account. In this work, we analyze the SAR process of different companies and not the *persons* replying in detail. Hence, we do not see any particular reason why we have to disclose that we conduct this survey. Note that after our second deadline (in our first measurement), we contacted the companies that did not respond at all or had a poorly designed process, without any responses. When contacting the companies, we did not disclose we conduct a scientific survey, but we did disclose the real names of two authors in each mail and on the photocopied IDs. We also answered all of the companies questions truthfully (e.g., if we had been in contact with a company in any other way aside from this survey) and reported all problems (e.g., broken data access forms) that we noticed during the process.

7   Conclusion

Our work shows that while most companies offer easy ways to access the collected personal data, few disclose all the information they have and some companies create significant obstacles for users to access it. The obstacles range from signed affidavits over providing additional information (e.g., phone numbers) to copies of official ID documents. Some larger companies do not disclose data to users that are not registered with their services. The different approaches of how access to personal data is granted show the different interpretations of the new law. Looking into the response behavior, we see that over 58% of the companies did not respond within the legal period of 30 days, but only one company extended the deadline by two more months.

Acknowledgment. This work was partially supported by the Ministry of Culture and Science of the German State of North Rhine-Westphalia (MKW grants 005-1703-0021 "MEwM" and Research Training Group NERD.nrw). We would like to thank the anonymous reviewers for their valuable feedback.

A   Provided Data

In this section, we provide examples of different types of data we received when performing the subject access requests. The data can be grouped into three categories: (1) "interest segments"—inferred from the user's online activities (Fig. 4), (2) "technical data"—extracted from HTTP traffic (Fig. 5), and (3) "tracking data"—websites on which users were tracked (Fig. 6).

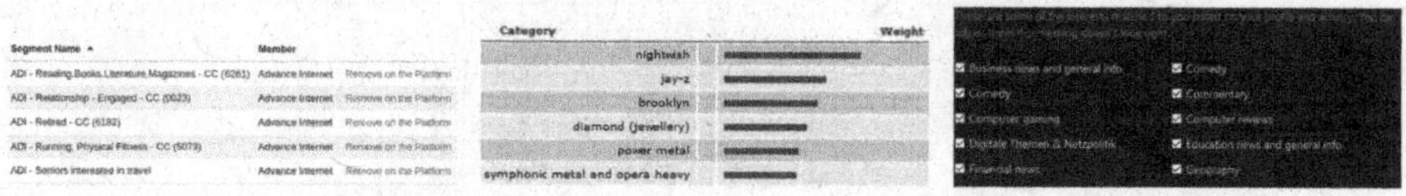

Fig. 4. Inferred *interest segments* provided by different companies (anonymized).

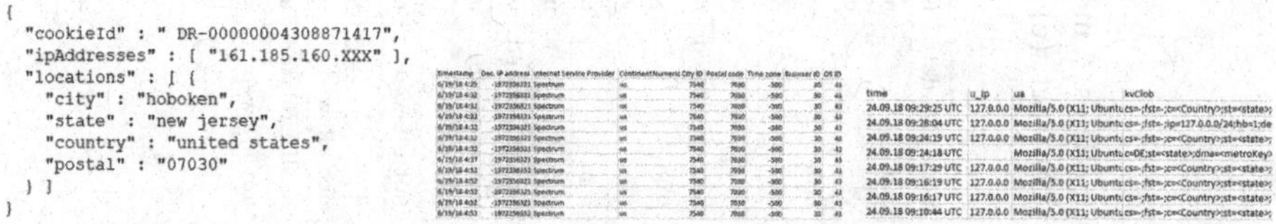

Fig. 5. *Technical data* provided by different companies (anonymized).

Fig. 6. *Tracking data* provided by different companies (anonymized).

B   Privacy Policy Overview

Table 1 provides a summary of the privacy policies of the companies in our data set. It lists the most important tracking and GDPR-related attributes and what information is disclosed.

Table 1. Overview of information available in privacy policies. * marks information that is required by the GDPR. *Legal Basis* refers to the sections in Article 6 of the GDPR: (a) consent, (b) contract, (c) legal obligation, (d) vital interests, (e) public, (f) legitimate interest; n.m. = not mentioned

Company	Legal basis*	Shared data	3rd CO*	Sensitive data	Profiling	Retention*	Partners*	Data access*	DNT	Version
Google	a, b, c, f	unspecified	y	n.m	n.m	unspecified	7	account	n.m	05/2018
Facebook	a, b, c, d, e, f	unspecified	y	y	n.m	differs	categories	account	n.m	04/2018
Amazon	n.m	unspecified	n.m	n.m	n.m	n.m	categories	n.m	n.m	08/2017
Verizon	a, b, c, f	unspecified	y	n.m	n.m	unspecified	329	website, email	n.m	05/2018
AppNexus	a, f	unspecified	y	n.m	n.m	3–60d, up to 18m	2309	website	n.m	05/2018
Oracle	a,c,f	unspecified	y	health related	n.m	12–18m	categories	website	y	05/2018
Adobe	a, b, c, f	unspecified	y	n.m	y	until opt-out	categories	email, form	n	05/2018
Smart AdServer	a, f	unspecified	y	n.m	y	1d–13m	categories	email	n.m	05/2018
RTL Group	a, c, f	unspecified	y	n.m	n.m	as long as necessary	categories	email	n.m	unclear
Improve Digital	a	listed	y	n.m	n.m	90d	categories	email	y	05/2018
MediaMath	f	unspecified	y	health related	y	up to 13m	categories	email	n.m	05/2018
Triplelift	a, f	unspecified	y	ask to avoid	n.m	as long as necessary	categories	website	n	05/2018
RubiconProject	a, b, c, f	unspecified	y	n.m	n.m	90–366d	categories	form	n.m	05/2018
The Trade Desk	a, f	unspecified	US	not allowed	n.m	18m, 3y aggregated	categories	website	n.m	10/2018
ShareThrough	a, b, c, f	unspecified	y	n.m	y	13m	categories	email	n.m	05/2018
Neustar	n.m	IDs, segments	US	not allowed	n	13m + 18m aggregated	categories	email	n.m	08/2018
Drawbridge	n.m	IDs, segments	US	health related	n.m	n.m	categories	email	n	08/2018
Adform	a, f	unspecified	y	not allowed	n.m	13m	33	form/email	n.m	unclear

(continued)

Table 1. (*continued*)

Company	Legal basis*	Shared data	3rd CO*	Sensitive data	Profiling	Retention*	Partners*	Data access*	DNT	Version
Bidswitch	a, b, c, f	unspecified	y	n.m	n.m	"as long as necessary"	categories	n.m	n.m	05/2018
Harris I & A	a, c	listed	y	y	n.m	purpose fulfilled	categories	email	n.m	07/2018
Acxiom	a, f	categories	y	no	n.m	unspecified	categories	register	n.m	05/2018
IndexExchange	n.m	aggregated only	US	no	no	13m	categories	website	n	09/2018
Criteo	a	aggregated	y	no	13m	13m	61	email/mail	n	05/2018
OpenX	a, f	unspecified	US	n.m	n.m	unspecified	categories	email	y	05/2018
DataXU	a, b, c, f	behavioural	y	not in EU	n.m	13m	categories	email	n	06/2018
Lotame	n.m	unspecified	US	health related	n.m	13m	categories	website	y	09/2018
FreeWheel	a, b, f	unspecified	Y	n.m	n.m	18m	categories	email	n	05/2018
Amobee	a, f	unspecified	US	n.m	y	13m	categories	website	n.m	06/2018
comScore	a, b, c, f	unspecified	y	n.m	n.m	n.m	categories	website	n.m	12/2017
spotX	a, f	listed	n.m	n.m	n.m	18m	65	website	y	unclear
Sovrn	a, c, f	n.m	y	n.m	y	n.m	unspecifc	webform	n.m	05/2018
Sizmek	a, b, c, f	segments	y	not knowingly	n.m	13m	unspecified	website	mixed	05/2018
Twitter	a, b, c, f	listed	y	not allowed	n.m	18m	16	account	n	05/2018
Microsoft	a, b, c, f	unspecified	y	y	n.m,	13m	>9	account	n	10/2018
Media Innovation	a	unspecified	US	n	n.m	14m	partners	n.m	n.m	09/2011
Quantcast	a, f	listed	y	not in EU	n.m	13m	33	website	n.m	05/2018

References

1. Acar, G., Eubank, C., Englehardt, S., Juarez, M., Narayanan, A., Diaz, C.: The web never forgets. In: Proceedings of the 21st ACM Conference on Computer and Communications Security, CCS 2014, pp. 674–689. ACM Press (2014)
2. Alexa: Top sites for countries (2018). https://www.alexa.com/topsites/countries. Accessed 05 Feb 2019
3. Barford, P., Canadi, I., Krushevskaja, D., Ma, Q., Muthukrishnan, S.: Adscape: harvesting and analyzing online display ads. In: Proceedings of the 23rd World Wide Web Conference, WWW 2014, pp. 597–608. ACM Press (2014)
4. Boniface, C., Fouad, I., Bielova, N., Lauradoux, C., Santos, C.: Security analysis of subject access request procedures how to authenticate data subjects safely when they request for their data. In: Naldi, M., Italiano, G.F., Rannenberg, K., Medina, M., Bourka, A. (eds.) APF 2019. LNCS, vol. 11498, pp. 182–209. Springer, Cham (2019). https://doi.org/10.1007/978-3-030-21752-5_12
5. Cagnazzo, M., Holz, T., Pohlmann, N.: Gdpirated-stealing personal information on- and offline. In: Proceedings of the 2019 European Symposium on Research in Computer Security, ESORICS 2019. Springer (2019)
6. Cliqz: Whotracks.me data - tracker database (2018). https://whotracks.me/blog/gdpr-what-happened.html. Accessed 24 Apr 2019
7. Dabrowski, A., Merzdovnik, G., Ullrich, J., Sendera, G., Weippl, E.: Measuring cookies and web privacy in a post-GDPR world. In: Choffnes, D., Barcellos, M. (eds.) PAM 2019. LNCS, vol. 11419, pp. 258–270. Springer, Cham (2019). https://doi.org/10.1007/978-3-030-15986-3_17
8. Data Protection Working Party: Opinion 2/2010 on online behavioural advertising (2010)
9. Data Protection Working Party: Article 29–guidelines on the right to data portability. Technical report 16 /EN WP 242, European Commission, December 2016
10. Degeling, M., Utz, C., Lentzsch, C., Hosseini, H., Schaub, F., Holz, T.: We value your privacy ... now take some cookies: measuring the GDPR's impact on web privacy. In: Proceedings of the 2019 Symposium on Network and Distributed System Security, NDSS 2019. Internet Society (2019)
11. Di Martino, M., Robyns, P., Weyts, W., Quax, P., Lamotte, W.L., Andries, K.: Personal information leakage by abusing the GDPR "right of access". In: Proceedings of the 15th Symposium on Usable Privacy and Security, SOUPS 2019. USENIX Association (2019)
12. Englehardt, S., Narayanan, A.: Online tracking: A 1-million-site measurement and analysis. In: Proceedings of the 2016 ACM Conference on Computer and Communications Security, CCS 2016, pp. 1388–1401. ACM Press (2016)
13. Estrada-Jiménez, J., Parra-Arnau, J., Rodríguez-Hoyos, A., Forné, J.: Online advertising: analysis of privacy threats and protection approaches. Comput. Commun. **100**, 32–51 (2017)
14. European Union: Regulation (EU) 2016/679 of the European Parliament and of the Council (2016). http://eur-lex.europa.eu/legal-content/EN/TXT/?uri=OJ:L:2016:119:TOC
15. GDPR Implementation Working Group: Data subject requests. Technical report Working Paper 04/2018 v1.0, IAB Europe, April 2018. https://www.iabeurope.eu/wp-content/uploads/2018/04/20180406-IABEU-GIG-Working-Paper04_Data-Subject-Requests.pdf

16. Guha, S., Cheng, B., Francis, P.: Challenges in measuring online advertising systems. In: Proceedings of the 10th Internet Measurement Conference, IMC 2010, pp. 81–87. ACM Press (2010)
17. Hert, P.D., Papakonstantinou, V., Malgieri, G., Beslay, L., Sanchez, I.: The right to data portability in the GDPR: towards user-centric interoperability of digital services. Comput. Law Secur. Rev. **34**(2), 193–203 (2018)
18. IAB Europe: European digital advertising market has doubled in size in 5 years (2017). https://www.iabeurope.eu/research-thought-leadership/resources/iab-europe-report-adex-benchmark-2017-report/. Accessed 05 Feb 2019
19. Interactive Advertising Bureau: Internet advertising revenue report (2017). https://www.iab.com/wp-content/uploads/2018/05/IAB-2017-Full-Year-Internet-Advertising-Revenue-Report.REV2_.pdf. Accessed 24 Apr 2019
20. Karaj, A., Macbeth, S., Berson, R., Pujol, J.M.: Whotracks.me: Monitoring the online tracking landscape at scale. CoRR arXiv:abs/1804.08959 (2018)
21. McDonald, A., Peha, J.M.: Track gap: policy implications of user expectations for the 'Do Not Track' internet privacy feature. SSRN Scholarly Paper, Social Science Research Network, Rochester, NY (2011)
22. Papadopoulos, P., Kourtellis, N., Markatos, E.P.: The cost of digital advertisement. In: Proceedings of the 2018 World Wide Web Conference, WWW 2018, pp. 1479–1489. International World Wide Web Conference Committee (2018)
23. Sanchez-Rola, I., et al.: Can I opt out yet?: GDPR and the global illusion of cookie control. In: Proceedings of the 2019 ACM Symposium on Information, Computer and Communications Security, pp. 340–351. ACM Press (2019)
24. Selbst, A.D., Powles, J.: Meaningful information and the right to explanation. Int. Data Priv. Law **7**(4), 233–242 (2017)
25. Sørensen, J.K., Kosta, S.: Before and after GDPR: the changes in third party presence at public and private European websites. In: Proceedings of the 2019 World Wide Web Conference, WWW 2019. International World Wide Web Conferences Steering Committee (2019)
26. TRUSTe and Harris Interactive: Consumer research results - privacy and online behavioral advertising (2011). https://www.eff.org/files/truste-2011-consumer-behavioral-advertising-survey-results.pdf. Accessed 24 Apr 2019
27. Urban, T., Tatang, D., Holz, T., Pohlmann, N.: Towards understanding privacy implications of adware and potentially unwanted programs. In: Lopez, J., Zhou, J., Soriano, M. (eds.) ESORICS 2018. LNCS, vol. 11098, pp. 449–469. Springer, Cham (2018). https://doi.org/10.1007/978-3-319-99073-6_22
28. Yuan, Y., Wang, F., Li, J., Qin, R.: A survey on real time bidding advertising. In: Proceedings of the 2014 Conference on Service Operations and Logistics, and Informatics, SOLI 2014, pp. 418–423. IEEE (2014)

On Privacy Risks of Public WiFi
Captive Portals

Suzan Ali, Tousif Osman, Mohammad Mannan$^{(\boxtimes)}$, and Amr Youssef

Concordia University, Montreal, Canada
{a_suzan,t_osma,mmannan,youssef}@ciise.concordia.ca

Abstract. Open access WiFi hotspots are widely deployed in many public places, including restaurants, parks, coffee shops, shopping malls, trains, airports, hotels, and libraries. While these hotspots provide an attractive option to stay connected, they may also track user activities and share user/device information with third-parties, through the use of trackers in their captive portal and landing websites. In this paper, we present a comprehensive privacy analysis of 67 unique public WiFi hotspots located in Montreal, Canada, and shed light on the web tracking and data collection behaviors of these hotspots. Our study reveals the collection of a significant amount of privacy-sensitive personal data through the use of social login (e.g., Facebook and Google) and registration forms, and many instances of tracking activities, sometimes even before the user accepts the hotspot's privacy and terms of service policies. Most hotspots use persistent third-party tracking cookies within their captive portal site; these cookies can be used to follow the user's browsing behavior long after the user leaves the hotspots, e.g., up to 20 years. Additionally, several hotspots explicitly share (sometimes via HTTP) the collected personal and unique device information with many third-party tracking domains.

1 Introduction

Public WiFi hotspots are growing in popularity across the globe. Most users frequently connect to hotspots due to their free-of-cost service, (as opposed to mobile data connections) and ubiquity. According to a Symantec study [24] conducted among 15,532 users across 15 global markets, 46% of participants do not wait more than a few minutes before connecting to a WiFi network after arriving at an airport, restaurant, shopping mall, hotel or similar locations. Furthermore, 60% of the participants are unaware of any risks associated with using an untrusted network, and feel their personal information is safe.

A hotspot may have a captive portal, which is usually used to communicate the hotspot's privacy and terms-of-service (TOS) policies, and collect personal identification information such as name and email for future communications, and authentication if needed (e.g., by asking the user to login to their social media sites). Upon acceptance of the hotspot's policy, the user is connected

© Springer Nature Switzerland AG 2019
C. Pérez-Solà et al. (Eds.): DPM 2019/CBT 2019, LNCS 11737, pp. 80–98, 2019.
https://doi.org/10.1007/978-3-030-31500-9_6

to the internet and her web browser is often automatically directed to load a landing page (usually the hotspot brand's webpage).

Several past studies (e.g., [6,23]) focus on privacy leakage from browsing the internet or using mobile apps in an open hotspot, due to the lack of encryption, e.g., no WPA/WPA2 support at the hotspot, and the use of HTTP, as opposed to HTTPS for connections between the user device and the web service. However, in recent years, HTTPS adoption across web servers has increased dramatically, mitigating privacy exposure through plain network traffic. For example, according to the Google Transparency Report [11], as of Apr. 6, 2019, 82% of web pages are served via HTTPS for Chrome users on Windows. On the other hand, in the recent years, there have also been several comprehensive studies on web tracking on regular web services and mobile apps with an emphasis on most popular domains/services (see e.g., [3,4,9]).

In contrast to past hotspot and web privacy measurement studies, we analyze tracking behaviors and privacy leakage in WiFi captive portals and landing pages. We design a data collection framework (CPInspector)[1] for both Windows and Android, and capture raw traffic traces from several public WiFi (in Montreal, Canada) that require users to go through a captive portal before allowing internet access. Challenges here include: manual collection of captive portal data by physically visiting each hotspot; making our test environment separate from the regular user environment so that we do not affect the user's browsing profiles; ensuring that our tests remain unaffected by the user's past browsing behaviors (e.g., saved tracking cookies); and creating and monitoring several test accounts in popular social media or email services as some hotspots mandate such authentication. CPInspector does not include any real user information in the collected dataset, or leak such information to the hotspots (e.g., by using fake MAC addresses).

From each hotspot, we collect traffic using both Chrome and Firefox on Windows. In addition to the default browsing mode, we also use private browsing, and deploy two ad-blockers to check if such privacy-friendly environments help against captive portal trackers—leading to a total of eight datasets for each hotspot. We also use social logins if required by the captive portal, or provided as an option; we again use both browsers for social login tests (two to six additional datasets as we have observed at most three social login options per hotspot). Some hotspots also require the user to complete a registration form that collects the user's PII—in such cases, we collect two more datasets (from both browsers).

On Android, we collect traffic only from the custom captive portal app (as opposed to Chrome/Firefox on Windows) as the cookie store of this app is separate from browsers. Consequently, tracking cookies from the Android captive portal app cannot be used by websites loaded in a browser. Recent Android OSes also use dynamic MAC addresses, limiting MAC address based tracking. However, we found that cookies in the captive portal app may remain valid for up to 20 years, allowing effective tracking by hotspot providers.

---

[1] https://github.com/MadibaLab/CPInspector.

We also design our framework to detect ad/content injection by hotspots; however, we observed no content modification attempts by the hotspots. Furthermore, we manually evaluate various privacy aspects of some hotspots, as documented in their privacy/terms-of-service policies, and then compare the stated policies against what happens in practice.

Note: *By default all our statistics refer to the measurements on Windows; we explicitly mention when results are for Android (mostly in Sect. 5).*

Contributions and Summary of Findings

1. We collected a total of 679 datasets from the captive portal and landing page of 80 hotspot locations between Sept. 2018 to Apr. 2019. 103 datasets were discarded due to some errors (e.g., network failure). We analyzed over 18.5 GB of collected traffic for privacy exposure and tracking, and report the results from 67 unique hotspots (576 datasets), making this the largest such study to characterize hotspots in terms of their privacy risks.

2. Our hotspots include cafes and restaurants, shopping malls, retail businesses, banks, and transportation companies (bus, train and airport), some of which are local to Montreal, but many are national and international brands. 40 hotspots (59.7%) use third-party captive portals that appear to have many other business customers across Canada and elsewhere. Thus our results might be applicable to a larger geographical scope.

3. 27 hotspots (40.3%) use social login or a registration page to collect personal information (19 hotspots make this process mandatory for internet access). Social login providers may share several privacy-sensitive PII items—e.g., we found that LinkedIn shares the user's full name, email address, profile picture, full employment history, and the current location.

4. Except three, all hotspots employ varying levels of user tracking technologies on their captive portals and landing pages. On average, we found 7.4 third-party tracking domains per captive portal (max: 34 domains).
   40 hotspots (59.7%) create persistent third-party tracking HTTP cookies (validity up to 20 years); 4.2 cookies on average on each captive portal (max: 34 cookies). Surprisingly, 26 hotspots (38.8%) create persistent cookies even *before* getting user consent on their privacy/TOS document.

5. Several hotspots explicitly share (sometimes even without HTTPS) personal and unique device information with many third-party domains. 40 hotspots (59.7%) expose the user's device MAC address; five hotspots leak PII via HTTP, including the user's full name, email address, phone number, address, postal code, date of birth, and age (despite some of them claiming to use TLS for communicating such information). Two hotspots appear to perform cross-device tracking via Device Co-op [2].

6. Two hotspots (3.0%) state in their privacy policies that they explicitly link the user's MAC address to the collected PII, allowing long-term user tracking, especially for desktop OSes with fixed MAC.

7. From our Android experiments, we reveal that 9 out of 22 hotspots can effectively track Android devices even though Android uses a separate captive portal app and randomizes MAC address as visible to the hotspot.

2   Related Work

In this section, we briefly review related previous studies on hotspots, web tracking, and ad injection.

Several prior studies have demonstrated the possibility of eavesdropping WiFi traffic to identify personal sensitive information in public hotspots. Cheng et al. [6] collected WiFi traffic from 20 airports in four countries, and found that two thirds of the travelers leak private information while using airport hotspots for web browsing and smartphone app usage. Sombatruang et al. [23] conducted a similar study in Japan by setting up 11 experimental open public WiFi networks. The 150 h experiment confirmed the exposure of private information, including photos, and users' credentials— transmitted via HTTP. In contrast, we analyze web tracking and privacy leakage within WiFi captive portals and landing pages.

Web tracking, a widespread phenomenon on the internet, is used for varying purposes, including: targeted advertisements, identity checking, website analytics, and personalization. Eckersley [7] showed that 83.6% of the panopticlick.eff.org visitors could be uniquely identified from a fingerprint composed of only 8 attributes. Laperdrix et al. [14] showed that AmIUnique.org can uniquely identify 89.4% of fingerprints composed of 17 attributes, including the HTML5 canvas element and the WebGL API. In a more recent large-scale study, Gómez-Boix et al. [10] collected over 2 million real-world device fingerprints (composed of 17 attributes) from a top French website; they found that only 33.6% device fingerprints are unique, raising questions on the effectiveness of fingerprinting in the wild. Note that developing advanced fingerprinting techniques to detect the so-called *golden image* (the same software and hardware as often deployed in large enterprises), is an active research area—see e.g., [13,22]. Several automated frameworks have also been designed for large-scale measurement of web tracking in the wild; see e.g., FPDetective [1] and OpenWPM [9]. In this work, we measure tracking techniques in captive portals and landing pages, and use OpenWPM to verify the prevalence of the found trackers on popular websites.

Previous work has also looked into ad injection in web content, see e.g., [21, 25]. We use similar methods for detecting potential similar content injection in hotspots since such incidents have been reported in the past (e.g., [16,20]).

3   CPInspector on Windows: Design and Data Collection

In this section, we describe CPInspector, the platform we develop for measuring captive portal web-tracking and privacy leakages; see Fig. 1 for the Windows variant. As Android uses a special app for captive portal, we modify CPInspector accordingly; see Sect. 5. The main components of CPInspector include: a browser automation framework, a data migration tool and an analysis module. Selenium is used to visit the hotspot captive portal and perform a wide range of measurements. It collects web traffic, HTTP cookies, WebStorage, fingerprints, browsing profiles, page source code, privacy policy, and screen shots of rendered pages (used to verify the data collection process). CPInspector utilizes Wireshark to capture traffic between the instrumented browser and the hotspot access point. CPInspector uses WebExtensions APIs[2] to collect relevant data (e.g., HTTP cookies, JavaScript calls) from the instrumented browser. Selenium is also used to isolate the test environment from the regular user environment, ensuring that our tests remain unaffected by the user's past browsing behaviors. We also save a copy of the privacy policy, if available. The datasets collected from the hotspots are parsed and committed to a central SQLite database. CPInspector's analysis module then examines the recorded data for tracking behaviors or privacy leaks.

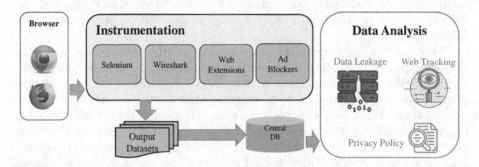

Fig. 1. CPInspector components.

Identifying Third-Parties. We identified the corporate websites for each hotspot. Then, we use the WHOIS registration records to identify third-party domains by comparing the domain owner name to the hotspot corporate website owner. In cases where the domain information is protected by the WHOIS privacy policy, we visit the domain to detect any redirect to a parent site; we then lookup the parent site's registration information. If this fails, we manually review the domain's `Organization` in its TLS certificate, if available. Otherwise, we try to identify the domain owner based on its WHOIS registration email. We also use Crunchbase.com and Hoovers.com to determine if the organizations are subsidiaries or acquisitions of larger companies.

Identifying Third-Party Trackers. We use EasyList, EasyPrivacy, and Fanboy to identify known third-party trackers. These lists rely on blacklisted script names, URLs, or domains, which may fail to detect new trackers or variations of known trackers. For this reason, we classify third-party trackers as follows: (a) A *known tracker* is a third-party that has already been identified in the above blacklists. (b) A *possible tracker* is any third-party that can potentially track the user's browsing activities but not included in a blacklist. We observed variations of well-known trackers such as Google Analytics, were missed by the blacklists.

---

[2] https://wiki.mozilla.org/WebExtensions.

Table 1. List of evaluated hotspots

Category	Count	Hotspot name
Cafe and Restaurant	19	A&W, Bombay Mahal Thali, Burger King, Cafe Osmo, Copper Branch, Domino's Pizza, Harvey's, Hvmans Cafe, Juliette Et Chocolat, Mcdonalds, Moose BAWR, Nespresso, Pizza Hut, Pizz izza, Starbucks, Sushi STE-Catherine, The Second Cup, Tim Hortons, Vua Sandwiches
Retail business	17	Canadian Tire, Dynamite, ECCO, Fossil, GAP, Garage, H& M, Home Depot, IGA, Ikea, Laura, Maison Simmons, Michael Kors, Roots, SAQ, Sephora, Walmart
Shopping Mall	12	Atrium 1000, Carrefour Angrignon, Carrefour Laval, Carrefour iA, Centre Eaton, Centre Rockland, Complexe Desjardins, Fairview Pointe-Claire, Mail Champlain, Place Montreal Trust, Place Vertu, Place Ville Marie
Bank	5	CIBC Bank, Desjardins 360, RBC Bank, ScotiaBank, TD Bank
Art and Entertainment	4	Grevin Montreal, YMCA, Montreal Science Centre, Place Des Arts
Transportation	3	Gare d'Autocars de Montreal, Via Rail Station, YUL Airport
Telecom kiosk	2	Fido, Telus
Car rental	1	Discount Car Rental
Gymnasium	1	Nautilus Plus
Hospital	1	CHU Sainte-Justine
Hotel	1	Fairmont Hotel
Library	1	Westmount Public Library

Ad Injection Detection. Our framework also includes a module to detect modifications to user traffic, e.g., for ad injections. We visit two decoy websites (i.e., honeysites in our control and hosted on AmazonAWS) and BBC.com, via a home network and a public hotspot, and then compare the differences in the retrieved content (i.e., DOM trees [8]). The use of honeysites allows us to avoid any false positive issues due to the website's dynamic content (e.g., dynamic ads). The first honysite is a static web page while the second is comprised of dynamic content that has four fake ads. The fake ads were created based on source code snippets from Google Adsense, Google TagManager, Taboola.com, and BuySellAds.com.

Data Collection. We collected a total of 679 datasets from the captive portal and landing page of 80 hotspots (12 hotspots are measured at multiple physical locations) between Sept. 2018 to Apr. 2019. We discarded 103 datasets due to some errors (e.g., network failures). We analyzed over 18.5 GB of collected traffic for privacy exposure and tracking measurements, and report the results from 67 unique hotspots (576 datasets). We discuss the results in Sect. 4. For the ad injection experiments, we collected a total of 368 datasets from crawling the two honey websites and the BBC.com website at 98 hotspot locations. We analyzed over 8.7 GB of collected traffic for ad injection, and report the results from 87 unique hotspots (368 datasets). We did not observe any content modification attempts.

4   Analysis and Results for Windows

In this section, we present the results of our analysis on collected personal information, privacy leaks, web trackers, HTTP cookies, and fingerprinting, and the effectiveness of two anti-tracking extensions and private browsing mode.

4.1   Personal Information Collection, Sharing, and Leaking

Personal Identifiable Information (PII) Collection. Most hotspots (40; 59.7%) allow internet access without seeking any explicit personal data. The remaining 27 (40.3%) hotspots use social login, or a registration page to collect significant amount of personal information; 19 (28.4%) of these hotspots mandate social login or user registration, see Table 2.

Sharing with Third-Parties. Most hotspots share personal information and browser/device information with third-parties via the referrer header, the request-URL, HTTP cookie or WebStorage. We identified 40 hotspots (59.7%) that use third-party captive portals where they share personal information, including 18 (26.9%) share email address; 15 (22.4%) share user's full name; 12 (17.9%) share profile picture; 5 (7.5%) share birthday, current city, current employment and LinkedIn headline; see Table 2. We also found some captive portals leak device/browser information to third-parties, including 40 (59.7%) leak MAC address and last visited site; 18 (26.9%) leak screen resolution; 26 (38.8%) leak user agent; 24 (35.8%) leak browser Information and language; and 15 (22.4%) leak plugins. Moreover, some hotspots leak the MAC address to third-parties, e.g., Pizza Hut to 11 domains, and H&M, Place Montreal Trust and Discount Car Rental to six domains each. Top organizations that receive the MAC addresses include: Network-auth.com from 21 hotspots, Alphabet 18, Openh264.org 12, Facebook 10, Datavalet 8, and Amazon 6.

PII Leaks via HTTP. We searched for personal information of our used accounts in the collected HTTP traffic, and record the leaked information, including the HTTP request URL, and source (captive portal vs. landing page). Three hotspots transmit the user's full name via HTTP (Place Montreal Trust,

Nautilus Plus and Roots). In Place Montreal Trust, the user's full name is saved in a cookie (valid for five years), and each time the user connects to the captive portal, the cookie is automatically transmitted via HTTP. Moreover, three hotspots leak the user's email address via HTTP (Dynamite, Roots, and Garage). In Nautilus Plus, a user must enter her membership number in the captive portal. For partially entered membership numbers, the captive portal verifies the identity by displaying personal information of five people in a scrambled way (first and last names, postal codes, ages, dates of birth, and phone numbers),

Table 2. Personal information collected via social login, registration, or optional surveys. The "Powered By" column refers to third-parties that provide hotspot services (when used/identified). F refers to Facebook, L: LinkedIn, I: Instagram, G: Google, T: Twitter, R: registration form, and S: survey; *: personal information is mandatory to access the service.

Hotspot	Powered By	Name	Email	Gender	Birthday	Phone Number	Current City	Profile Picture	Home Town	Country	Facebook Likes	Facebook Friends	LinkedIn Headline	Current Employment	Postal Code	# of Children	Basic Profile	Instagram Media	Tweets	People You Follow
Bombay Mahal Thali*	Sy5	FR	FR	F			F													
Carrefour Laval*	Aislelabs	FR	FR	FR		F		F	F	F				F						
Fairview Pointe-Claire*	Aislelabs	FR	FRT	FR		F		F	F	F				F					T	T
Carrefour Angrignon	Eye-In	FGL	FGL						FGL						L	L	L			
Centre Eaton	Eye-In	F	F						F											
Centre Rockland	Eye-In	FL	FL						FL						L	L	L			
Desjardins 360*	JoGoGo	F	F	F	F		R		F				F							
Domino's Pizza			R																	
Dynamite*			R																	
GAP			R																	
Garage*			R																	
Grevin Montreal	Eye-In	FL	FL					F	FL						L	L	S	S		L
Harvey's*	Colony Networks	F	FR					F												
Hvmans Cafe*	Purple	FR	FR				F	F	F					F				I	I	
Mail Champlain	Eye-In	FL	FL						FL						L	L	L			
Maison Simmon*			R																	
Michael Kors*	Purple	R	R	R	R	R											R			
Montreal Science Centre*	Telus		R																	
Moose BAWR*	Sticky WiFi		R																	
Nautilus Plus*			R																	
Nespresso*	Orange						R													
Place Montreal Trust			R							R							R			
Roots*	Yelp WiFi	R	R																	
Telus*			R																	
Sushi STE-Catherine*	MyWiFi		R																	
Vua Sandwiches*	Coolblue	FR	FR			R		F												
YUL Airport*	Datavalet	FL	FRL						FL						L	L	L			

over HTTP. The user then chooses the right combination corresponding to her personal information. We also confirmed that some of this data belongs to real people by authenticating to this hotspot using ten randomly generated partial membership numbers. Then, we used the reverse lookup in canada411.ca to confirm the correlation between the returned phone numbers, names, and addresses.

4.2 Presence of Third-Party Tracking Domains and HTTP Cookies

Tracking Domains. We detect third-party tracking domains using: EasyList, EasyPrivacy, and Fanboy's List. On average, each captive portal hosts 7.4 third-party tracking domains (max: 34 domains, including 10 known trackers); see Fig. 2(a). We noticed that the hotspots that use the same third-party captive portal still have a different number of third-parties. For example, for the Datavalet hotspots (YUL Airport, McDonald's, Starbucks, Via Rail Station, Tim

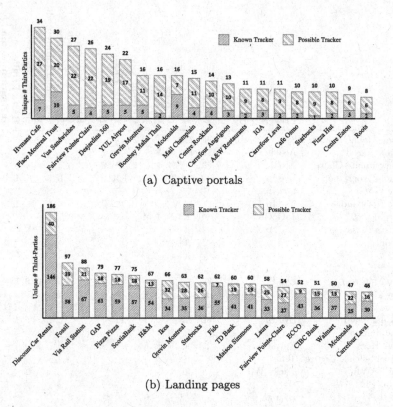

(a) Captive portals

(b) Landing pages

Fig. 2. Number of third-party domains on captive portals and landing pages (top 20). For example, Hvmans Cafe captive portal hosts a total of 34 tracking domains, including 7 known trackers. Note that for all reported tracking/domain statistics, we accumulate the distinct tracking domains as observed in all the datasets collected for a given hotspot (e.g., from both browsers and for different social logins, if required). For list of evaluated hotspots see Table 1.

Hortons, CIBC Bank, Place Vertu), the number of third-parties are 22, 16, 10, 8, 5, 5, and 2 respectively. The hotspots (46; 68.7%) that redirect users to their corporate websites, host more known third-party tracking domains—on average, 30.6 domains per landing page; see Fig. 2(b). We also analyzed the organizations with the highest known-tracker representations. We group domains by the larger parent company that owns these domains. Alphabet, Facebook, and Datavalet are present on over 10% of the captive portals. Alphabet and Facebook are also present on over 50% of the landing pages.

HTTP Tracking Cookies on Captive Portals. We found 40 (59.7%) hotspots create third-party cookies valid for various duration—e.g., over 5 years from 10 (14.9%) hotspots, six months to five years from 23 (34.3%) hotspots, and under six months from 38 (56.7%) hotspots; see Fig. 3(a). Via Rail Station, Fairview Pointe-Claire, Carrefour Laval, Roots, McDonald's, Tim Hortons, and Harvey's have a third-party cookie from network-auth.com, valid for 20 years.

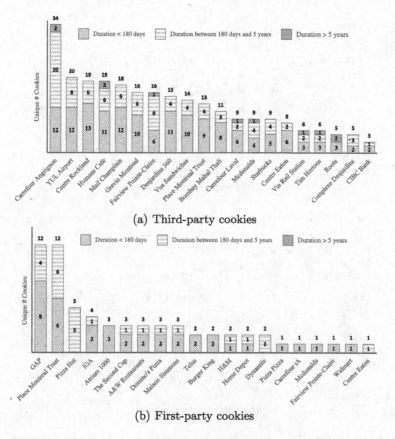

(a) Third-party cookies

(b) First-party cookies

Fig. 3. Number of third-party and first-party cookies on captive portals (top 20). Note that for all reported cookies/domain statistics, we accumulate the distinct cookies as observed in all the datasets collected for a given hotspot.

Moreover, YUL Airport, Via Rail Station, Complexe Desjardins, McDonald's, Starbucks, Tim Hortons, CIBC Bank have a common 1-year valid cookie from Datavalet, except for CIBC (17 days). This cookie uniquely identifies a device based on the MAC address (set to the same value unless the MAC address is spoofed). Some hotspots save the MAC address in HTTP cookies, including CHU Sainte-Justine, Moose BAWR, and Centre Rockland.

We also analyze first-party cookies on captive portals; see Fig. 3(b). 22 (32.8%) hotspots create first-party cookies valid for various durations; 14 (20.9%) hotspots include cookies valid for periods ranging from six months to five years, and 17 (25.4%) hotspots for less than 6 months. Place Montreal Trust saves the user's full name in a first-party cookie valid for five years; this cookie is transmitted via HTTP. Finally, we analyzed hotspots that create persistent cookies

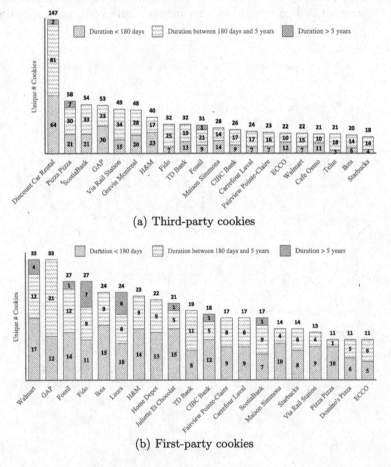

(a) Third-party cookies

(b) First-party cookies

Fig. 4. Number of third-party and first-party cookies on landing pages (top 20). Note that for all reported cookies/domain statistics, we accumulate the distinct cookies as observed in all the datasets collected for a given hotspot.

before explicit consent from the user, we found 26 (38.8%) hotspots create cookies that are valid for periods varying from 30 min to a year, including Domino's Pizza, Fido, GAP, H&M, McDonald's, Roots, Starbucks, and Tim Hortons.

HTTP Tracking Cookies on Landing Pages. We found 48 (71.6%) hotspots create third-party cookies valid for various durations—e.g., over 5 years from 4 (6.0%) hotspots, six months to five years from 47 (70.1%) hotspots, and under six months from 42 (62.7%) hotspots, see Fig. 4(a). Prominent examples include the following. Fossil has a 25-year valid cookie from pbbl.com; CIBC Bank has two 5-year valid cookies from stackadapt.com, a known tracker. We also analyzed the first-party cookies on landing pages; see Fig. 4(b). 41 (62.7%) hotspots create first-party cookies valid for various durations—e.g., over 5 years from 10

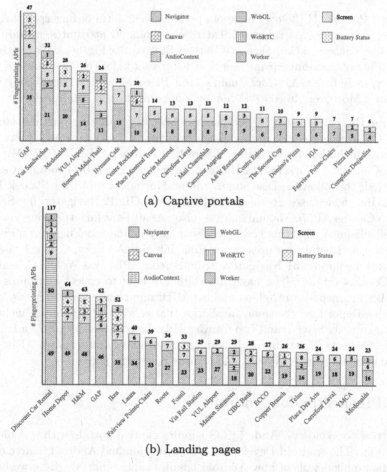

(a) Captive portals

(b) Landing pages

Fig. 5. Number of fingerprinting APIs on captive portals and landing pages (top 20). Note that for all fingerprinting statistics, we accumulate the distinct APIs as observed in all the datasets collected for a given hotspot.

(14.9%) hotspots, six months to five years from 42 (62.7%) hotspots, and under six months from 41 (61.2%) hotspots. Notable examples: Fossil has a 99-year valid cookie, Fido has three cookies valid for 68–81 years, CHU Sainte-Justine has a 20-year valid cookie, CIBC Bank has a 19-year cookie, and Walmart has four cookies valid for 9–20 years.

4.3   Device and Browser Fingerprinting

We analyzed fingerprinting attempts in captive portals and landing pages. We use Don't FingerPrint Me (DFPM [12]) for detecting known fingerprinting techniques, including the screen object, navigator object, WebRTC, Font, WebGL, Canvas, AudioContext, and Battery Status [9,17–19]. We use attribute and API interchangeably, when referring to fingerprinting JavaScript APIs.

Captive Portal. 24 (35.8%) hotspots perform some form of fingerprinting. On average, each captive portal uses 5.9 attributes (max: 47 attributes, including 35 Navigator, 6 Screen, 3 Canvas, and 3 Battery Status); see Fig. 5(a). We also found 10 (14.9%) hotspots fingerprint user device/browser before explicit consent from the user, including GAP, McDonald's, and Place Montreal Trust, using 6–46 attributes. Moreover, 46 (68.7%) hotspots fingerprint MAC addresses.

Landing Pages. 51 (76.1%) hotspots perform fingerprinting on their landing pages. On average, each landing page fingerprints 19.4 attributes (max: 117 attributes, including 49 Navigator, 9 Screen, 2 Canvas, 3 WebRTC, 50 WebGL, 1 AudioContext, 1 Worker and 2 Battery Status); see Fig. 5(b). Prominent examples include the following. Discount Car Rental includes script from Sizmek Technologies Inc., which uses a total of 67 APIs (48 WebGL, 12 Navigator, five Screen, and two Canvas APIs). Manual analysis also reveals Font fingerprinting via sidechannel inference [18]; this script is also highly similar to FingerprintJS [26]. Discount Car Rental also uses script from Integral Ad Science, which uses 41 attributes, including: 31 Navigator, seven Screen APIs, two WebRTC, and one AudioContext (cf. [9]). The navigator APIs are used to collect attributes such as the USB gamepad controllers, and list MIDI input and output devices. H&M and Home Depot host the same JavaScript that collects 42 attributes, including 34 Navigator, six Screen, and two Canvas APIs. Laura has a script from PerimeterX that collects 27 attributes, including 21 Navigator and 6 Screen APIs; code manual analysis reveals WebGL and Canvas fingerprinting.

5   CPInspector for Android

In contrast to Windows, Android OS handles captive portals with a dedicated application. The Android Developers documentation and Android Source documentation omit details of how Android handles captive portals. Here we briefly document the inner working of Android captive portals, and discuss our preliminary findings, specifically on tracking cookies on Android devices.

Android Captive Portal Login App. Using Android `ps` (Process Status), we observe that a new process named `com.android.captiveportallogin` appears whenever the captive portal is launched. The Manifest file for CaptivePortalLogin explicitly defines that its activity class will receive all captive portal broadcasts by any application installed on the OS and handle the captive portal. We observe that files in the data folder of this application are populated and altered during a captive portal session; we collect these files from our tests.

Capturing Network Traffic. To capture traffic from Android apps, several readily-available VPN apps from Google Play can be used (e.g., Packet Capture, NetCapture, NetKeeper). However, Android does not use VPN for captive portals. On the other hand, using an MITM Proxy server such as mitmproxy (https://mitmproxy.org/) requires the server to run on a desktop environment, which would make the internet traffic come out of the desktop OS, i.e., the mobile device would not be visible to the hotspot. To overcome this, we set up a virtual Linux environment within the Android OS by using Linux Deploy (https://github.com/meefik/linuxdeploy), enabling us to run Linux desktop applications within Android with access to the core component of Android OS, e.g., Android OS processes, network interfaces, etc. We use Debian and mitmproxy on the virtual environment, and configure Android's network settings to proxy all the traffic going through the WiFi adapter to the mitmproxy server. The proxy provides us the shared session keys established with a destination server, enabling us to decrypt HTTPS traffic. We use tcpdump to capture the network traffic.

Data Collection and Analysis. We visited 22 hotspots and collected network traffic from their captive portals. First, we clear the data and cache of the CaptivePortalLogin app and collect data from a given hotspot. Next, we change the MAC address of our test devices (Google Pixel 3 with Android 9 and Nexus 4 with Android 5.1.1) and collect data again without clearing the data and cache. From the proxy's request packets, we confirm that the browser agent correctly reflects our test devices, and the traffic is being originated from the CaptivePortalLogin app. Next, we analyze the data extracted from the app. The structure of the data directory is similar to Google Chrome on Android. We locate the `.\app_webview \ Cookies` SQLite file in the data directory, storing the CaptivePortalLogin app's cookies.

We observe that 9 out of 22 hotspots store persistent cookies in the captive portal app; see Fig. 6. These cookies are not erased when the portal app is closed, or when the user leaves the hotspot. Instead, the cookies remain active as set in their validity periods, although they are unavailable to the regular browser apps. Prominent examples include: Tim Hortons inserts a 20-year valid cookie from network-auth.com, and Hvmans Cafe stores a 10-year valid cookie from Instagram. In the captive portal traffic, we confirm that these cookies are indeed present and shared in subsequent visits, and follow the Same-Origin Policy. Hotspots can use these cookies to uniquely identify and authenticate user devices even when the device MAC address is dynamically changed; Tim Hortons hotspot uses its cookies for authentication. However, McDonald's did not authenticate the device even though the cookies were present but the MAC was new.

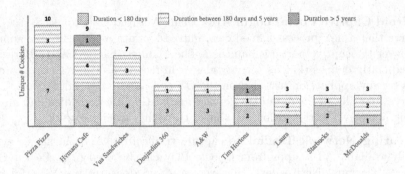

Fig. 6. Number of cookies stored on the Android captive portal app

6   Privacy Policy and Anti-Tracking

We performed a preliminary manual analysis of privacy policy and TOS documents from hotspots that appear to be most risky. Roots states clearly in their privacy policy that they use SSL to protect PII, but their captive portal transmits a user's full name and email address via HTTP. Place Montreal Trust transmits the user's full name via HTTP, and they explicitly state that transmission of information over the public networks cannot be guaranteed to be 100% secure. Nautilus Plus has a very basic TOS that omits important information such as the laws they comply with and privacy implications of using their hotspot. They state clearly that the assurance of confidentiality of the user's information is of great concern to Nautilus Plus, but they use HTTP for all communications, leaking personal information while they attempt to verify the customer's identity; see Sect. 4.1. Their privacy policy is also inaccessible from the captive portal and omits any reference to WiFi. Dynamite and Garage transmit the user's email address via HTTP despite claiming to use SSL. Their privacy policy is inaccessible from the captive portal and omits any reference to the WiFi. GAP explicitly mentions their collection of browser/device information, and they indeed collect 46 such attributes, *before* the user accepts the hotspot's policies.

Although McDonald's tracks users in their captive portal (9 known trackers, 28 fingerprinting attributes), the captive portal itself lacks a privacy policy stating their use of web tracking. Carrefour Laval and Fairview Pointe-Claire perform cross device tracking by participating in the Device Co-op [2], where they may collect and share information about devices linked to the user. Two hotspots link the user's MAC address to the collected personal information, including Roots, and Bombay Mahal Thali. Sharing the harvested personal data with subsidiaries and third-party affiliates is also the norm. Eight hotspots (including Hvmans Cafe, Fairview Pointe-Claire, and Carrefour Laval) state that PII may be stored outside Canada. Ten hotspots omit any information about the PII storage location, including Dominos's Pizza and Roots. However, five hotspots have their captive portal domain in the US, including Bombay Mahal Thali, Carrefour Angrignon, Domino's Pizza, Grevin Montreal, and Roots. We found

34 (50.7%) hotspots have a TOS document but lack a privacy policy on their captive portal, including TD Bank, and Burger King. Three hotspots lack both the privacy policy and TOS document on their captive portals, including Laura, ECCO, and Maison Simmons.

The Same Hotspot Captive Portal in Different Locations. 12 hotspots are measured at multiple physical locations. We stopped collecting datasets from different locations of the same chain-business as the collected datasets were largely the same. We provide an example where some minor differences occur: Starbucks' captive portal domain varies in the two evaluated locations (am.datavalet.io vs. sbux-j2.datavalet.io). However, the number of known trackers remained the same, while the number of third-parties increased by one domain. Moreover, the --sf-device cookie validity increased from 17 days to 1 year, and the --sf-landing cookie was not created in the second location.

Effectiveness of Privacy Extensions and Private Browsing. To evaluate the effectiveness anti-tracking solutions against hotspot trackers, we collected traffic from both Chrome and Firefox in private browsing modes, and by enabling Adblock Plus, and Privacy Badger extensions—leading to a total of six datasets for each hotspot. Then, we use the EasyList, EasyPrivacy, and Fanboy's lists to determine whether known trackers remain in the collected datasets; see Table 3.

Table 3. The number of unique known trackers not blocked by our anti-tracking solutions.

	W/O Ad blockers	AdBlock plus	Privacy badger	Private browsing
Firefox	382	33	180	315
Chrome	488	117	212	356

Hotspot Trackers in the Wild. We measured the prevalence of trackers found in captive portals and landing pages, in popular websites—to understand the reach and consequences of hotspot trackers. We use OpenWPM [9] between Feb. 28–Mar. 15, 2019 to automatically browse the home pages of the top 143k Tranco domains [15] as of Feb. 27, 2019. We extract the tracking persistent cookie domains from captive portals or landing pages; we define such cookies to have validity ≥ 1 day and the sum of the value lengths from all the cookies from the same third-party website longer than 35 characters—cf. [5]. Then, we counted those tracking domains in the OpenWPM database; see Table 4. For example, the doubleclick.net cookie as found in 4 captive portals and 30 landing pages, appears 160,508 times in the top 143k Tranco domains (multiple times in some domains). Overall, hotspot users can be tracked across websites, even long time after the user has left a hotspot .

Table 4. Count of tracking domains from captive portals and landing pages in Alexa 143k home pages (top 10).

Captive Portal		Landing Page	
Tracker	Count	Tracker	Count
doubleclick.net	160508	pubmatic.com	326991
linkedin.com	48726	rubiconproject.com	257643
facebook.com	37107	doubleclick.net	160508
twitter.com	14874	casalemedia.com	131626
google.com	13676	adsrvr.org	116438
atdmt.com	5198	addthis.com	83221
instagram.com	3466	demdex.net	83160
gap.com	295	contextweb.com	82965
maxmind.com	294	rlcdn.com	75295
gapcanada.ca	64	livechatinc.com	69919

7   Conclusion and Future Work

Many people across the world use public WiFi offered by an increasing number of businesses and public/government services. The use of VPNs, and the adoption of HTTPS in most websites and mobile apps largely secure users' personal/financial data from a malicious hotspot provider and other users of the same hotspot. However, device/user tracking as enabled by hotspots due to their access to MAC address and PII, remains as a significant privacy threat, which has not been explored thus far. Our analysis shows clear evidence of privacy risks and calls for more thorough scrutiny of these public hotspots by e.g., privacy advocates and government regulators.

Our study covers hotspots in Montreal, Canada, and we are currently working on collecting data from other parts of the world. Our recommendations for hotspots users include the following: avoid sharing any personal information with the hotspot (social media or registration forms); use private browsing and possibly some other anti-tracking browser addons, and software programs that may allow to use a fake MAC address on Windows; and clear the browser history after visiting a hotspot if private browsing mode is not used. Additional suggestions are available at: https://madiba.encs.concordia.ca/reports/OPC-2018/.

Acknowledgement. This work was partly supported by a grant from the Office of the Privacy Commissioner of Canada (OPC) Contributions Program. We thank the anonymous DPM 2019 reviewers for their insightful suggestions and comments, and all the volunteers for their hotspot data collection. We also thank the members of Concordia's Madiba Security Research Group, especially Nayanamana Samarasinghe, for his help in running OpenWPM to automatically browse the home pages of the top 143k Tranco domains.

References

1. Acar, G., et al.: FPDetective: dusting the web for fingerprinters. In: ACM CCS 2013. Berlin, Germany, November 2013
2. Adobe.com: Adobe experiance cloud: Device Co-op privacy control. https://cross-device-privacy.adobe.com
3. Binns, R., Zhao, J., Kleek, M.V., Shadbolt, N.: Measuring third-party tracker power across web and mobile. ACM Trans. Internet Technol. **18**(4), 52:1–52:22 (2018)
4. Brookman, J., Rouge, P., Alva, A., Yeung, C.: Cross-device tracking: measurement and disclosures. In: Proceedings on Privacy Enhancing Technologies (PETS). Minneapolis, MN, USA, July 2017
5. Bujlow, T., Carela-Español, V., Sole-Pareta, J., Barlet-Ros, P.: A survey on web tracking: mechanisms, implications, and defenses. Proc. IEEE **105**(8), 1476–1510 (2017)
6. Cheng, N., Wang, X.O., Cheng, W., Mohapatra, P., Seneviratne, A.: Characterizing privacy leakage of public WiFi networks for users on travel. In: 2013 Proceedings IEEE INFOCOM. Turin, Italy, April 2013
7. Eckersley, P.: How unique is your web browser? In: International Symposium on Privacy Enhancing Technologies Symposium (2010)
8. Elifantiev, O.: NodeJS module to compare two DOM-trees. https://github.com/Olegas/dom-compare
9. Englehardt, S., Narayanan, A.: Online tracking: A 1-million-site measurement and analysis. In: Proceedings of the 2016 ACM SIGSAC Conference on Computer and Communications Security. Vienna, Austria, October 2016
10. Gómez-Boix, A., Laperdrix, P., Baudry, B.: Hiding in the crowd: an analysis of the effectiveness of browser fingerprinting at large scale. In: TheWebConf (WWW 2018). Lyon, France, April 2018
11. Google: HTTPS encryption on the web. https://transparencyreport.google.com/https/overview?hl=en
12. Klafter, R.: Don't FingerPrint Me. https://github.com/freethenation/DFPM
13. Klein, A., Pinkas, B.: DNS cache-based user tracking. In: Network and Distributed System Security Symposium (NDSS 2019). San Diego, CA, USA, February 2019
14. Laperdrix, P., Rudametkin, W., Baudry, B.: Beauty and the beast: diverting modern web browsers to build unique browser fingerprints. In: IEEE Symposium on Security and Privacy (SP). San Jose, CA, USA (2016)
15. Le Pochat, V., Van Goethem, T., Tajalizadehkhoob, S., Korczyński, M., Joosen, W.: Tranco: a research-oriented top sites ranking hardened against manipulation. In: NDSS 2019. San Diego, CA, USA, February 2019
16. Medium.com: My hotel WiFi injects ads. does yours?, news article (25 March 2016). https://medium.com/@nicklum/my-hotel-WiFi-injects-ads-does-yours-6356710fa180
17. Mowery, K., Shacham, H.: Pixel perfect: fingerprinting canvas in HTML5. In: Proceedings of W2SP, pp. 1–12 (2012)
18. Nikiforakis, N., Kapravelos, A., Joosen, W., Kruegel, C., Piessens, F., Vigna, G.: Cookieless monster: Exploring the ecosystem of web-based device fingerprinting. In: 2013 IEEE Symposium on Security and Privacy. Berkeley, CA, USA, May 2013
19. Olejnik, L., Acar, G., Castelluccia, C., Diaz, C.: The leaking battery. In: Garcia-Alfaro, J., Navarro-Arribas, G., Aldini, A., Martinelli, F., Suri, N. (eds.) DPM/QASA -2015. LNCS, vol. 9481, pp. 254–263. Springer, Cham (2016). https://doi.org/10.1007/978-3-319-29883-2_18

20. PCWorld.com: Comcast's open WiFi hotspots inject ads into your browser, news article, 09 September 2014. https://www.pcworld.com/article/2604422/comcasts-open-wi-fi-hotspots-inject-ads-into-your-browser.html
21. Reis, C., Gribble, S.D., Kohno, T., Weaver, N.C.: Detecting in-flight page changes with web tripwires. In: NSDI 2008, San Francisco, CA, USA (2008)
22. Sanchez-Rola, I., Santos, I., Balzarotti, D.: Clock around the clock: time-based device fingerprinting. In: ACM CCS 2018, Toronto, Canada, October 2018
23. Sombatruang, N., Kadobayashi, Y., Sasse, M.A., Baddeley, M., Miyamoto, D.: The continued risks of unsecured public WiFi and why users keep using it: evidence from Japan. In: Privacy, Security and Trust (PST 2018), Belfast, UK, August 2018
24. Symantec: Norton WiFi risk report: Summary of global results, technical report, 5 May 2017. https://www.symantec.com/content/dam/symantec/docs/reports/2017-norton-wifi-risk-report-global-results-summary-en.pdf
25. Tsirantonakis, G., Ilia, P., Ioannidis, S., Athanasopoulos, E., Polychronakis, M.: A large-scale analysis of content modification by open HTTP proxies. In: Network and Distributed System Security Symposium (NDSS 2018) (2018)
26. Valve: Fingerprintjs by Valve. https://valve.github.io/fingerprintjs/

User Perceptions of Security and Usability of Mobile-Based Single Password Authentication and Two-Factor Authentication

Devriş İşler[1]([✉]), Alptekin Küpçü[2], and Aykut Coskun[2]

[1] imec-COSIC, KU Leuven, Leuven, Belgium
devris.isler@kuleuven.be
[2] Koç University, İstanbul, Turkey
{akupcu,aykutcoskun}@ku.edu.tr

Abstract. Two-factor authentication provides a significant improvement over the security of traditional password-based authentication by requiring users to provide an additional authentication factor, e.g., a code generated by a security token. In this decade, single password authentication (SPA) schemes are introduced to overcome the challenges of traditional password authentication, which is vulnerable to the offline dictionary, phishing, honeypot, and man-in-the-middle attacks. Unlike classical password-based authentication systems, in SPA schemes the user is required to remember only a single password (and a username) for all her accounts, while the password is protected against the aforementioned attacks in a provably secure manner.

In this paper, for the first time, we implement the state-of-the-art mobile-based SPA system of Acar et al. (2013) as a prototype and assess its usability in a lab environment where we compare it against two-factor authentication (where, in both cases, in addition to the password, the user needs access to her mobile device). Our study shows that mobile-based SPA is as easy as, but less intimidating and more secure than two-factor authentication, making it a better alternative for online banking type deployments. Based on our study, we conclude with deployment recommendations and further usability study suggestions.

Keywords: Password-based authentication · Usability ·
Two-factor authentication · Single password authentication

1 Introduction

Password-based authentication that is widely deployed today is vulnerable to many attacks including offline dictionary, phishing, honeypot, and man-in-the-middle attacks. Unfortunately, the server password databases get hacked, and

D. İşler—Work done at Koç University.

millions of users are affected because their passwords are not complicated enough to resist offline dictionary attacks or because of server misconfiguration that stores plaintext passwords [10,11]. The damage of these attacks on the password becomes dramatically dangerous when the user reuses the same password for multiple sites, which is common in practice [16].

Two-factor authentication (2FA) has emerged as a way to improve security by requiring the user to provide one more authentication factor in addition to the password. To be successful, the attacker has to exploit both authentication factors. 2FA has been employed mostly in finance, government and enterprise areas due to the sensitivity of the users' information. However, the servers employing 2FA still keep password databases as in the traditional password-based authentication systems. Therefore, the users' passwords are still vulnerable to offline dictionary, honeypot, phishing, man-in-the-middle attacks even when they employ 2FA. Even though such an attacker cannot gain access to the servers employing 2FA by attacking the password, password re-use is still problematic.

In this decade, another approach called single password authentication (SPA) based on cryptographic building blocks is presented. SPA systems (first shown by [4] (with their patent application dating 2010 [7]), [8,20,28], and [18]) ensure provable security even when the user re-uses the same password on multiple sites. SPA methods achieve this by introducing an additional party to store a secret (e.g., mobile device). A secret independent of the password (e.g., a cryptographic key) is generated and stored on this storage device protected by the user's single password. The associated verification information is shared with the login server during the registration. Whenever the user wants to log in to the server, the user communicates with both the storage device and the login server. She securely retrieves the secret information from the storage device in a way that only the legitimate user can reconstruct the secret using her single password. Then, the user signs in to the server with the reconstructed secret. For full cryptographic details, we refer the reader to the cited papers.

In this setting, similar to 2FA, the attacker also needs to guess the user's password, and additionally access the secret storage device (e.g., the mobile device of the user). But, differently from 2FA, in SPA systems, when any one of the parties (i.e., storage provider and login server) is compromised, the user's single password is still kept secure from attackers. Compared to 2FA, SPA solutions provide provable security against all the aforementioned attacks.

In this paper, we study the usability of the first mobile-based SPA mechanism (where the storage device is the mobile device of the user) by Acar et al. [4] and compare it against 2FA commonly used for online banking because both approaches similarly employ a random one-use challenge via the mobile device, in addition to the password. Acar et al. [4] solution can be implemented as only a mobile device application (unlike [28] that requires both a mobile phone application and a browser extension) and is the *only* existing SPA proposal that protects the user's single password against malware-infected computers (e.g., at internet cafes or public computers at laboratories, libraries, etc.).

Conducting user studies on a new system is important for determining whether the system is suitable for its end users and its purpose. These studies ascertain any difficulties that the users may have while using the system in real life. In this work, we measure the usability considering various standardized aspects [36]: **effort expectation**(perceived ease of use), **anxiety, behavioral intention to use the system, attitude towards using technology, performance expectancy**, and **perceived security**. Our expectation is to observe significant benefits of mobile-based SPA systems regarding effort expectation, attitude towards using technology, and perceived security compared to the 2FA counterpart. On the other hand, we do not expect to see a significant difference in behavioral intention to use the system and anxiety. While it is not the main goal of our usability study, we also provide some average success and failure metrics but leave precise timing-related measurements as future work. **Our contributions** can be summarized as follows:

1. We implement a unique and state-of-the-art mobile device based single password authentication system (mobile SPA method of [4], see also Appendix A).
2. We conduct a comparative usability study of this mobile-based SPA solution for the first time[1] in the literature against its commonly-employed counterpart authentication system: two-factor authentication.
3. We provide our findings based on both quantitative and qualitative data. We discuss the advantages that the mobile-based SPA system provides relative to an existing commonly-employed 2FA solution. While we did not observe any disadvantages, we include important recommendations for possible deployment of a mobile SPA system in practice.

Scope of the Work: SPA systems could be mobile-based and/or cloud-based depending on the employed storage provider(s). Their security guarantees were already analyzed in their respective papers. In our study, we focus on various usability aspects and the perceived security of a mobile-based SPA system. Additionally, our analysis ends up with some suggestions that can be applied to all SPA systems in general (e.g., password reset in SPA system, see Sect. 4.1).

Our study is conducted at a laboratory environment using fake websites because Acar et. al [4] method requires changes at the server-side, which makes it impossible to conduct the study on real online banking sites (see Sect. 3 for the details of our methodology and a discussion of the limitations).

This study is the first study comparatively analyzing a unique state-of-the-art mobile-based SPA solution (namely the Acar et al. [4] work) against a commonly-employed 2FA solution. Both constructions that we analyzed are similar in the sense that the users experiences are alike (e.g., a token generation via the mobile device in addition to the user's password).

---

[1] The only previous work on mobile SPA usability compared SPHINX mobile-based SPA system against password managers [28], and hence their work is complementary and incomparable.

2   Related Work

We explain studies exploring usability of various authentication systems.

Traditional Password Authentication: In these schemes, the username and the output of a deterministic function (e.g., hash) of the password is stored at the server. For authentication, the user types her username and password, and the server compares this information against its database. The user has to remember the corresponding password for each server registered with. This approach is vulnerable to offline dictionary attacks and the effect of these attacks increases dramatically if the user uses the same password for multiple servers, which is common in practice [16]. SPA systems, on the other hand, ensure security even under server database compromise. [37] discussed the traditional password authentication usability. [37] provided a quantitative point of reference for the difficulty of remembering random passwords, which is necessary to employ traditional solutions securely.

Two-Factor Authentication (2FA): These schemes generally employ any combination of two of what you know (e.g., password), what you have (e.g., token), who you are (e.g., biometric), and who you know (see [12,34]). 2FA aims to strengthen the security of traditional password authentication by deploying a secondary authentication token (e.g., SMS sent to mobile device). To pass the authentication, the user needs to provide a valid password and token. Despite the widespread use in banking, these systems still suffer from users' negative influence such as reusing the same password. [14] conducted a comparative study of the usability of two-factor authentication technologies, where they found that 2FA is perceived as usable, regardless of motivation or use. [17] showed that 2FA provides more security but lower level of usability. [33] proposed a 2FA solution, where they found their system is reliable and usable. [27] analyzed different communication channels in 2FA (e.g., QR code, bluetooth). They concluded that their full bandwidth WiFi to WiFi system provides the highest security and usability when a browser extension and radio interface exist. [21] proposed a different 2FA called Sound-Proof to reduce the communication between the user and device. It authenticates the user based on proximity to a mobile device. Their user study concluded that their new system was more usable than the Google Authenticator application. Another 2FA is the FIDO Alliance and the protocol proposed by them Universal 2nd Factor protocol (U2F) [30]. The U2F is currently implemented by security keys (a piece of hardware authenticating the user after pressing a button on the key [23]). [26] conducted a user study on a U2F security key called YubiKey comparing it with 2FA for non-experts. They discovered that the setup phase is unusable and suggested an improvement on the design.

Single Password Authentication (SPA): SPA systems (first shown by [4] (with their patent application dating 2010 [7]), [8,20,28], and [18]) ensure provable security even when the user re-uses the same password on multiple sites. SPA methods achieve this by introducing an additional party to store a secret (e.g., mobile device). Similar to 2FA, the attacker also needs to guess the user's password, and additionally access the secret storage device (essentially the mobile device of the user). But, differently from 2FA, in SPA systems, when any one of the parties (i.e., storage provider and login server) is compromised, the user's single password is still kept secure from attackers. Compared to 2FA, SPA solutions provide provable security against all the aforementioned attacks.

SPHINX [28] is a mobile-phone-based SPA solution that uses cryptographic tools to ensure password security against the aforementioned attacks, whose usability was analyzed in the same paper. It is efficient, relatively simple to use, and provides better security capabilities compared to password managers, such as security in the case of mobile device compromise. Similarly, Acar et al. [4] mobile-based SPA solution is also secure in such a case, but has a different design goal: SPHINX ensures that the password is input to the client computer and not the mobile device, whereas Acar et al. intentionally use the mobile device for inputting the password, rather than the computer (considering a potentially malware-infected public terminal scenario). Since the usability of SPHINX is already examined in [28], we studied the Acar et al. [4] mobile-based SPA solution in this paper, which does not require client-side browser extension installation that SPHINX requires (useful for public terminal scenarios). We compare it against 2FA commonly used for online banking because both approaches employ a random one-use challenge via the mobile device, in addition to the password.

[19] proposes an SPA framework and suggests various secure SPA systems based on different cryptographic building blocks. It would be useful to study if a low entropy password can be replaced with other authentication mechanisms such as biometrics. We leave such an analysis as future work.

3   Methodology

Our tests were conducted in the Koç University's Media and Visual Arts Lab, and the methodology was reviewed and approved by the university ethics committee (IRB). Written consent of the participants were taken, and the questionnaire data was kept anonymous. We took precautions according to the European Union General Data Protection Regulation [1] and local data protection laws [2,3] to protect personally-identifiable information of the participants. We did not collect such information unnecessarily, and used the names only for the consent forms. We gifted each participant with a mug with the logo of our research group on it. Each participant was allocated a 30-min time slot.

Demographics: Before conducting the study, participants were first asked to complete an online demographics and technical background questionnaire, whose data is kept anonymous, where they were given a general idea about single password authentication. In addition to sex, age interval, and education level, the users were also asked about their experience with mobile and online banking, password managers, and whether or not they have prior knowledge of password security (see Table 1). Based on the information provided, there were

25 participants[2] (11 male, 14 female) with an age distribution: 18–25 years (6 users), 25–35 years (15 users), 35–45 years (1 user), 45–55 years (1 user) and 55+ years (2 users). The participants had diverse educational backgrounds such as post-graduate (10 users), graduate (7 users), undergraduate (6 users), and high-school (2 users) degrees. They were university students, faculty, and staff from various departments (both technical and non-technical).

Table 1. Responses of the participants regarding technical information

How often do you use your mobile device?		Do you have prior knowledge of password security?	
So often (Daily)	24	I heard from news, social media etc.	16
Few times in a day	1	I had a course	6
Weekly	0	Not me but someone I know had experience	3
How often do you use mobile banking?		**How often do you use online banking?**	
Daily	4	Daily	4
Weekly	11	Weekly	9
Monthly	5	Monthly	7
Rarely	0	Rarely	3
Never	5	Never	2
Have you ever used a browser extension?		**Have you ever used a password manager?**	
Yes	16	Yes	4
No	4	No	17
Never Heard	5	Never Heard	4
How often do you change your password?			
Weekly	1	Monthly	4
Every 3 months	4	Every 6 months	2
Once a year	0	If I have to	14

---

[2] Despite the fact that deciding how many participants are needed for the user study remains vague, [15] justifies that even 20 users can be enough to have certainty on finding the usability problems in the testing.

3.1   Study Design

At the beginning of the study, the participants were provided with a ready setup: a pre-installed desktop computer[3] and an Android mobile phone[4]. For mobile-based SPA, we used our own SIM card and configured our servers to send SMS messages to our number using NEXMO online service; hence, we did not need to collect participants' phone numbers. For the 2FA implementation, we used Google authenticator[5] to provide the smart codes the server asks for, as it is a commonly-employed and well-known app. We did not enforce the participants to install the mobile-based SPA application and Google Authenticator from scratch, since their setup is the same as a regular mobile application installation.

In our study, the pre-determined tasks were carefully constructed to preserve the reality as much as possible, though we accept that this is a lab study and therefore our findings should be interpreted as an important first step, rather than the final verdict. For our user study, the participants did not need any training to use the system as they will not in real life. Since the mobile-based SPA solution requires server-side changes, we created our own websites just for the purposes of the study. Three websites created were framed as *online banking sites*. This choice was intentional: 2FA is widely employed for online banking (among the participants 80% employed mobile banking and 92% employed online banking). No website had any data; we just created registration and login pages, and displayed success or failure messages. The only information these websites collected were usernames and (hashed) passwords (which were deleted after data evaluation was completed), and success/failure logs, for this study.

The participants were presented with the aforementioned three online-banking type websites (e.g., Bank A), and were asked to register with and login to these websites using the mobile-based SPA technique and separately using the 2FA. The order of which password authentication system a participant started with was random, where either they began with 2FA and then continued with mobile-based SPA, or vice versa. Per technique, after registering with the three websites in random order, they logged in to these websites in random order. If a participant failed to login to a website three times, we counted it as a login failure and asked user continue to login to the next website. This represented a realistic scenario where if a user enters an incorrect password three times, the user is asked to go through a CAPTCHA process or the user's account is blocked temporarily. The tasks followed by the participants in each authentication technique are described as follows:[6]

---

[3] A desktop computer running 64-bit Windows 8 on Intel Core i7-3770 3.4 GHz CPU and 16 GB RAM.

[4] A Samsung Galaxy J1 with Android version 4.4.4.

[5] Google Authenticator Android app. https://goo.gl/Q4LU7k.

[6] Note that the list of tasks were not given to the participants; instead, such instructions were clarified on the web pages and mobile applications that we created (see, for example, Fig. 2(d)). The users simply followed those instructions.

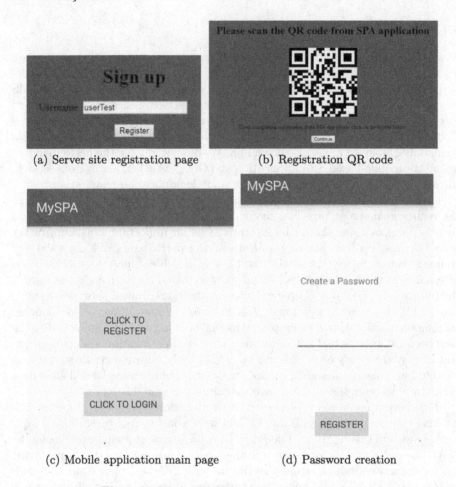

(a) Server site registration page

(b) Registration QR code

(c) Mobile application main page

(d) Password creation

Fig. 1. Mobile-based SPA registration screenshots

Two-Factor Password Authentication Registration: The user

(1) selects a strong[7] password, where they are asked to choose a different password for each website, [**Remark:** Ideally users are expected not to use a password for more than one website for security[8], and previous studies show that an average user has approximately 7 unique passwords [16]. The username may be chosen the same or differently for each website.]

(2) types her username and password, (3) clicks the signup button,

---

[7] One with at least eight characters containing at least one of each category: lower case and upper case letters, numerical character, and special character.

[8] 2FA does not protect the user password against dictionary attacks when the password database is compromised. Therefore, such an attacker may impersonate the user on other websites that do *not* employ 2FA. Such offline dictionary and impersonation attacks are prevented by SPA systems.

(4) opens Google Authenticator app, (5) scans the QR code, (6) confirms 6 digit numerical code with the website,

(7) is informed whether the registration is successful or not.

Two-Factor Password Authentication Login: The user

(1) types her username and password on the server site,
(2) is shown a message by the server site to type the code generated by Google Authenticator if the user types the correct username and password,
(3) opens the Google authenticator application on the phone,
(4) types the application-generated six-digit numerical code to the site,
(5) is informed whether the login attempt is successful.

Mobile-Based SPA Registration: The user

(1) selects a strong[7] password, where the participant is told to use the *same* password during all three account registrations,
(2) types her username (Fig. 1(a)),
(3) presses the signup button,
(4) opens mobile-based SPA application on the phone as it is told on the site,
(5) clicks the register button on mobile-based SPA application (Fig. 1(c)),
(6) scans the QR code shown on the website (Fig. 1(b)),
(7) types her password on the mobile application (Fig. 1(d)),
(8) clicks the register button on the mobile application,
(9) is informed whether the registration is successful.

Mobile-Based SPA Login: The user

(1) types the username on the website (Fig. 2(a)),
(2) is shown on the website that an SMS code is sent to the mobile phone and should open SPA mobile application,
(3) opens the mobile application and clicks the login button (Fig. 1(c)),
(4) types the single password on the mobile application (Fig. 2(b)),
(5) types the 8-digit alphanumeric code displayed by the mobile application to the website (Fig. 2(d)), [**Remark:** The application automatically retrieves the SMS code and generates the code for the user; the user did not need to type SMS into the application (Fig. 2(c)).]
(6) is informed whether the login attempt is successful.

3.2   Measures

We measure usability considering various standardized aspects from [36] as in various studies [14,26,28,31] and added some SPA-specific questions. To collect the data for observation, we had two different methods:

Post-Questionnaire: Measures from the post-questionnaire were 4-point Likert-scale (strongly disagree, disagree, agree, strongly agree)[9]. Participants answered 23 questions per phase (e.g., 23 questions once they completed the two-factor authentication phase and 23 questions after completing the mobile-based SPA phase). We followed the standard questions in [36] because it is a commonly used standardized questionnaire measuring system usability, and added single-password specific questions ourselves to measure the perceived security, where we were inspired by previous work on password usability [9,13,28]. The questions in the post-questionnaire formed six sets that considered different aspects of the systems: **effort expectation, anxiety, behavioral intention to use the system, attitude towards using technology, performance expectancy,** and **perceived security** (see Table 2). For quantitative evaluation, we first converted the participants' responses to their numerical values from 1 to 4. For each aspect, we then calculated means, standard deviations, and t-test values based on the numerical values of users' responses. Dependent t-test (paired t-test)[10], which is common in usability studies on password authentication systems [13,22,24], is applied to compare the systems, since each participant tested both systems (mobile-based SPA and two-factor authentication).

Comments to the Observer: There was an observer in the room who observed the user actions and received feedback from each participant. Since it is important not to give any additional information influencing the participants' actions, the observer provided the same standard information to all participants. At the end of the study, the observer had a discussion with each participant, where the users freely commented about their feelings and concerns about the studied systems, as well as password and system security in general.

Limitations: Our results were limited by the self-reported nature of surveys and natural selection bias. Since our experiments were held in a laboratory setting using prototype implementation (mobile-based SPA Android application and the simple websites), the participants may not have behaved the same as they would in the real world. One would hope to obtain more realistic results if users could be examined in real-life, while connecting to real websites, and for a longer period of time rather than 30 min. Yet, this makes it very hard to conduct such a user study, especially because currently there is no deployed mobile-based SPA solution that is widely-used for online banking. Thus, we expect the reader to take our results as an important first step rather than the final verdict, and we hope to help future deployment of mobile-based SPA solutions with our discussion and suggestions based on user feedback.

---

[9] We intentionally used 4-point Likert scale as it allows accounting for exact responses [5,6].

[10] [25] argues that parametric statistics can be used with Likert data without reaching to the wrong conclusion.

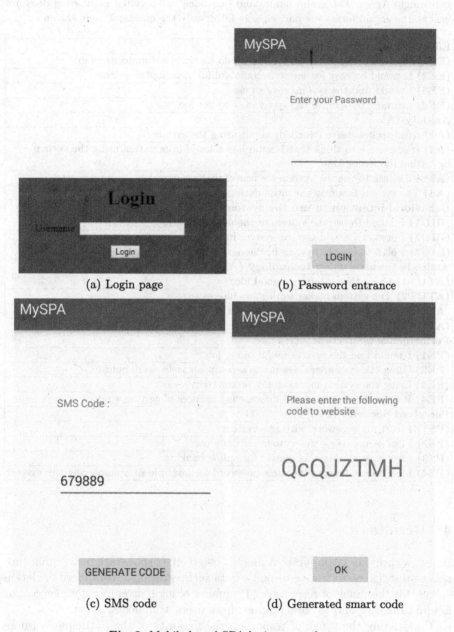

(a) Login page

(b) Password entrance

(c) SMS code

(d) Generated smart code

Fig. 2. Mobile-based SPA login screenshots.

Table 2. Post-questionnaire form questions asked to the participants. The form employed a 4-point scale, where 1=Strongly Disagree, 2=Disagree, 3=Agree, and 4=Strongly Agree. The group names and questions' abbreviated numbering does not exist in the actual forms the participants filled; only the questions were shown.

Effort Expectation (EE)
(EE1) My interaction with the system would be clear and understandable
(EE2) It would be easy for me to become skillful at using the system
(EE3) I would find the system easy to use
(EE4) Learning to operate the system is easy for me
Anxiety (A)
(A1) I feel apprehensive (worried) about using the system
(A2) It scares me to think that I could lose a lot of information using the system by hitting the wrong key
(A3) I hesitate to use the system for fear of making mistakes I cannot correct
(A4) The system is somewhat intimidating to me
Behavioral intention to use the system (BIU)
(BIU1) I intend to use the system in the next 6 months
(BIU2) I predict I would use the system in the next 6 months
(BIU3) I plan to use the system in the next 6 months
Attitude towards using technology (ATUT)
(ATUT1) Using the system is a good idea
(ATUT2) The system makes work more interesting
(ATUT3) Working With the system is fun
(ATUT4) I like working with the system
Performance Expectancy (PE)
(PE1) I would find the system useful in my job
(PE2) Using the system enables me to accomplish tasks more quickly
(PE3) Using the system increases my productivity
(PE4) If I use the system, I will increase my chances of getting a raise
Perceived Security (PS)
(PS1) I trust my password with this system
(PS2) I feel secure using this system for daily use
(PS3) I feel secure using this system for online banking
(PS4) I feel secure reusing the same password for multiple sites employing this system

4   Results

Below, we provide a comparative analysis based on: (1) the statistical significance using t-test, (2) quantitative response data such as mean and standard deviation values, (3) the range of responses, (4) number of login attempts until success or failure (Table 3), and (5) observations from users' comments.

Considering the range of responses, the majority of the participants (more than 50% per question) agreed (or strongly agreed) that mobile-based SPA is easy to use, useful, trustworthy, and not intimidating to use, as well as they have a positive attitude towards and intention to using this system. This holds for all 20 questions out of 23 asked. The three questions that the majority did not agree were *"I plan to use the system in the next 6 months"* (**BIU3**), *"Using the system increases my productivity"*(**PE3**), and *"If I use the system, I will increase my chances of getting raise"* (**PE4**). This holds for both the mobile-based SPA and two-factor authentication responses, since the participants did not feel like an authentication system is tied to their salary or productivity.

As for the usability of mobile-based SPA compared to two-factor authentication, we found significant differences in terms of three dimensions: **anxiety**, **perceived security**, and **attitude towards using technology**. There was no significant difference between mobile-based SPA and 2FA regarding **effort expectancy** ($t(24) = 1.10$ and $p = 0.28$), **behavioral intention to use the system** ($t(24) = 0.00$ and $p = 1.00$), and **performance expectancy** ($t(24) = 1.04$ and $p = 0.30$).

Anxiety: Mobile-based SPA was less threatening than two-factor authentication ($t(24) = 2.77$ and $p = 0.01$). 70% of the comments (14 out of 20 participants who commented) stated that the participants were not worried while using mobile-based SPA because they typed the password on their mobile phone (conceived as a personal device) rather than the website. 96% of the participants (24 out of 25) were not scared to lose a lot of information by hitting the wrong key in mobile-based SPA. A participant explained that there was nothing to worry, since he did not give any important information to the websites.

Perceived Security: 80% of the participants (20 out of 25) felt secure while using mobile-based SPA based on the range of responses. The users trusted mobile-based SPA more than they trust 2FA ($t(24) = 3.25$ and $p = 0.003$), including all sub-statements. 80% of the comments (16 out of 20 participants who commented) stated that typing the password on the mobile device (conceived as a personal item) made the user feel more secure, whereas they needed to type their passwords on the websites in standard 2FA. One participant commented that seeing all works (computations) carried out on the mobile device made her feel more secure, and she felt as though she had the control of her password security, since she could see the steps (e.g., SMS challenge, smart code generated). Another participant pointed out that he was aware of the danger if he used the same password for multiple websites, just as 56% of participants (14 out of 25) agreed that they would feel insecure to use the same password for multiple websites in password-based authentication.

Attitude Towards Using Technology: Mobile-based SPA performed statistically significantly better compared to 2FA ($t(24) = 2.71$ and $p = 0.01$), including all sub-statements. The users are required to remember only a single password and used it all the time, while they need to remember each one of the passwords in the two-factor approach. One of the participants stated that she found two things she wanted at the same time, which are usability (easing her job by remembering one password) and more security (via employing a personal device and challenge).

Even though mobile-based SPA and 2FA did not have a significant difference regarding **effort expectation**, 80% of the participants (20 out of 25) agreed

Table 3. Mobile-based SPA (SPA Mobile) and 2FA (Two Factor): The percentage distribution of password attempts to login. μ: mean, σ: standard deviation.

	Login trial		Success percent at trial number			
	μ	σ	1	2	3	Failure(%)
SPA mobile	1.00	0	100	0	0	0
Two factor	1.17	0.5	82	5	4	9

that mobile-based SPA was easy to use. The users reported a high satisfaction with mobile-based SPA, even though the tasks of the mobile-based SPA study were a little bit more complex (such as typing an 8-character alphanumerical code versus a 6-digit numerical code in the 2FA). 84% of the participants (21 out of 25) found that the mobile-based SPA is easy to learn, and they were fine with the steps they need to follow, since it was for online banking.

Success/Failure Rates: We measured that 100% of the time the participants successfully remembered their passwords without any trials using mobile-based SPA. Therefore, the average number of password attempts by a user is 1 (see Table 3). However, we measured a 20% overall login failure rate, due to the participants' inability to type the correct authentication code within 3 attempts. This indicates that simpler smart codes should be employed in the future.

For 2FA, we measured that 82% of the time the participants successfully remembered their passwords at the first attempt, out of which 91% of the time the participants could enter the authentication code (generated by the Google Authenticator) at their first attempt and 9% of the time at their second attempt. 5% of the time the participants remembered their passwords at their second attempt, out of which 80% of the time the participants could enter the authentication code at their first attempt and 20% of the time at their second attempt. 4% of the time the participants remembered their passwords at their third attempt, out of which 67% of the time the participants could enter the authentication code at their first attempt and 33% of the time at their second attempt. 9% of the time the participants did not remember their passwords within the first three attempts, resulting in a login failure. The average number of password attempts by a user is 1.17 (see Table 3).

We conclude that for both 2FA and mobile SPA, the participants had high login success rates. Using mobile-based SPA, the participants did not have problems with the password, but they had issues with the smart codes. On the other hand, using 2FA, the users did not have problems with the authentication codes, but they had issues remembering the password. We deduce that simpler smart codes should be employed in such systems, as they may make things as bad as remembering passwords.

4.1 Further Discussion

The participants mentioned valuable statements and discussed their habits while creating, securing, and recalling the passwords. [16,32,35] observe how users manage, create, and secure their passwords and points out some challenges users face such as password creation (with the intent of reuse) and recall in traditional password authentication schemes. We observed how an SPA method overcomes some of the challenges users face.

Password Creation and Recall: 88% of the study participants (22 out of 25) were aware of password security. 85% of the comments (17 out of 20 participants who commented) stated that the participant always struggled while coming up with a password satisfying the requirements (e.g., at least one lowercase and one uppercase letter, a number, and a special character). The participants usually came up with a password after a number of trials. Once they created it, remembering the password was another struggle they bear. Thus, they created their own way to recall the passwords. More than 50% of comments (10 out of 20) noted that the participants wrote down their passwords to remember. One of the users commented that he stored password reminders (as hints helping him to recall the passwords) in a file, while he emphasized that anyone who obtained the file could not learn the passwords. When we questioned why he needed this storage, he responded that it is hard for him to remember the password for some sites he rarely used and he came up with this solution. However, even this solution did not stop him from re-using the same password for multiple sites.

Password Reset: While there is a functionality to reset a password in traditional approaches, a participant found it cumbersome, since the password reset procedure requires steps such as logging in to a backup e-mail, which requires remembering another password, or memorizing and entering all necessary information (such as security questions) to reset. Another participant shared his experience when he lost the paper where he noted a password for a site and wanted to reset the password. Unfortunately, he needed to follow a long official password reset procedure because of system requirements (e.g., personal application was required and he waited for a week). He stated that everything would be easier if he could use a secure SPA system that minimizes password remembering problems. As in password creation and recall discussion, similar comments support that SPA systems are easing the burden on users by requiring them to remember only one password (in addition to the cryptographic benefits they provide such as provable security against offline dictionary attacks). In the light of these comments, we recommend that the SPA systems should investigate how a secure single password reset can be efficiently carried out.

Widespread Use: While this idea might require further and detailed research all by itself, users may feel more secure when a new system is collectively used. 52% of the participants (13 out of 25) shared that they would use the SPA system and trust it if it is commonly used and advertised by a "trusted" authority (rather than university researchers) such as Facebook, Google, etc. One of the participants said that *"I feel secure while I am using WhatsApp, since WhatsApp* *is employed for secure messaging. They use something like encryption."* The participant was not aware of the cryptographic scheme employed in WhatsApp and had no idea what it was, but stated that it "feels" secure since WhatsApp was widely advertised and employed.

Complexity of the Solution: We found some insights about online banking which is commonly used for financing [29]. 90% of the comments (18 out of 20 participants who commented) stated that mobile-based SPA provided a better security for online banking, and users felt secure in the online banking scenario because it was "complex" enough. Interestingly, the participants stated that a "complex" solution using the mobile device (i.e., mobile-based SPA) feels secure for banking since the password is typed on the phone. On the other hand, the mobile-based SPA system was found unproductive for email type daily purposes due to its complexity, while it was considered more secure by the participants. Considering such feedback on security and usability, there might be an inverse relationship between the perceived security and ease of use, since mobile-based SPA was found more secure for online banking. This interpretation is worth exploring for future research.

Our user study concluded that SPA systems provide usability benefits. The main reasoning is that it is not convenient to expect users to create different complex passwords for each website and remember them. While this approach would be secure, it is not usable. On the other hand, SPA systems enable single password re-use securely.

5   Conclusion, Recommendations, and Future Work

We implemented mobile-based single password authentication method of [4] and conducted its usability analysis for the first time. It has two unique properties apart from being the first such proposal: it can be implemented as only a mobile application, and it protects the user's password from malware-infected computers at public locations. We compared it against 2FA in a fake online banking scenario. Quantitative and qualitative results support that the mobile-based SPA solution has usability and security advantages compared to its counterpart.

Our findings suggest that the smart code mechanism should be simpler and the SPA branding should provide more trust to the users. Based on the feedback reported by the participants, we suggest that mobile-based SPA solution(s) should be deployed for online banking type of settings, where more complicated solutions are expected (at least seemingly more complicated, regardless of the underlying cryptography). Observations also indicate that there is potentially a trade-off between usability and perceived security, which is worth exploring as future work.

We believe our study constitutes an important step in understanding the usability of SPA systems regarding their future deployment. Yet, to obtain more generalizable results, we recommend to conduct future studies taking into account timing information, taking place in a natural settings instead of a lab environment and examining other dimensions of user experience of SPA systems beyond usability such as emotional satisfaction, increasing the number of participants, and considering privacy of SPA systems.

Acknowledgments. We thank İlker Kadir Öztürk and Arjen Kılıç for their efforts on implementation. This work has been supported in part by TÜBİTAK (the Scientific and Technological Research Council of Turkey) under the project number 115E766, by the Royal Society of UK Newton Advanced Fellowship NA140464, by ERC Advanced Grant ERC-2015-AdG-IMPaCT, and by the FWO under an Odysseus project GOH9718N.

A   Mobile-Based Single Password Authentication Scheme of Acar et al. [4]

We briefly present Acar et al. [4] mobile-based SPA solution here for completeness. In their mobile-based SPA, there are three parties; a user holding a password pwd, a trusted mobile device of the user, and a server, with which the user wishes to register. The protocol is roughly as follows:

Registration:

1. The **user**:
   - generates a Message Authentication Code (MAC) key K.
   - sends the key K and her username UID to the server.
   - encrypts the MAC key K where the encryption key is derived using the hash of her password $H(pwd)$ as $ctext \leftarrow Encrypt(H(pwd), K)$. [**Remark:** The user also sends an identifier with ciphertext.]
2. The **trusted mobile device** stores the ciphertext $ctext$.
3. The **server** stores the username UID and the MAC key K.

Authentication:

1. The **user** sends her username UID to the server.
2. The **server** generates a random challenge $chal$ and sends it to the mobile device. [**Remark:** The server can send the challenge in various ways such as via SMS, or via a QR code where the user scans the code with her mobile device.]
3. The **user** types her single password on the mobile device.
4. The **trusted mobile device:**
   - decrypts the ciphertext and retrieves the MAC key K as $K \leftarrow Decrypt(H(pwd), ctext)$.
   - generates a MAC $resp$ as a response to the challenge $chal$ using the retrieved key K as $resp \leftarrow MAC(K, chal)$. [**Remark:** To resist man-in-the-middle attacks, as [19] notes, preferable usage is $resp \leftarrow MAC(K, chal\|domain)$.]
   - applies trimming function $Trim$ on the generated response $resp$ to get a short one-time code/password $resp'$ as $resp' \leftarrow Trim(resp)$.

5. The **user** types the short one-time code $resp'$ on the user machine and sends it to the server.
6. The **server** checks if the $resp'$ is generated based on a valid MAC of the challenge $chal$ with the corresponding user MAC key K in his database as $Trim(MAC(K, chal)) \stackrel{?}{=} resp'$.
7. The **server** informs the user whether the login attempt is successful or not.

References

1. European Union General Data Protection Regulation 2016/679 (GDPR) (2016)
2. Turkish Personal Data Protection Law no. 6698 (KVKK) (2016)
3. Turkish Personal Data Deletion and Anonymization Regulation no. 30224 (2017)
4. Acar, T., Belenkiy, M., Küpçü, A.: Single password authentication. Comput. Netw. **57**(13), 2597–2614 (2013)
5. Allen, I.E., Seaman, C.A.: Likert scales and data analyses. Qual. Prog. **40**(7), 64–65 (2007)
6. Behnke, K.C., Andrew, O.: Creating programs to help latino youth thrive at school: the influence of latino parent involvement programs. J. Extension **49**(1), 1–11 (2011)
7. Belenkiy, M., Acar, T., Morales, H., Küpçü, A.: Securing passwords against dictionary attacks. US Patent 9,015,489 (2015)
8. Bicakci, K., Atalay, N.B., Yuceel, M., van Oorschot, P.C.: Exploration and field study of a browser-based password manager using icon-based passwords. In: RLCPS (2011)
9. Bicakci, K., Yuceel, M., Erdeniz, B., Gurbaslar, H., Atalay, N.B.: Graphical passwords as browser extension: implementation and usability study. In: Ferrari, E., Li, N., Bertino, E., Karabulut, Y. (eds.) IFIPTM 2009. IAICT, vol. 300, pp. 15–29. Springer, Heidelberg (2009). https://doi.org/10.1007/978-3-642-02056-8_2
10. Bicchierai, L.F.: Another day, another hack: 117 million linkedin emails and passwords (2016). https://bit.ly/2Nq1b9M
11. Bicchierai, L.F.: Hacker tries to sell 427 milllion stolen myspace passwords for $2,800 (2016). https://bit.ly/2GBnu9S
12. Brainard, J., Juels, A., Rivest, R.L., Szydlo, M., Yung, M.: Fourth-factor authentication: somebody you know. In: ACM CCS (2006)
13. Chiasson, S., van Oorschot, P.C., Biddle, R.: A usability study and critique of two password managers. In: USENIX Security Symposium (2006)
14. De Cristofaro, E., Du, H., Freudiger, J., Norcie, G.: A comparative usability study of two-factor authentication. In: NDSS USEC (2014)
15. Faulkner, L.: Beyond the five-user assumption: benefits of increased sample sizes in usability testing. Instrum. Comput. Behav. Res. Meth. **35**(3), 379–383 (2003)
16. Florencio, D., Herley, C.: A large-scale study of web password habits. In: ACM WWW (2007)
17. Gunson, N., Marshall, D., Morton, H., Jack, M.: User perceptions of security and usability of single-factor and two-factor authentication in automated telephone banking. Comput. Secur. **30**(4), 208–220 (2011)
18. İşler, D., Küpçü, A.: Threshold single password authentication. In: ESORICS DPM (2017)
19. İşler, D., Küpçü, A.: Distributed single password protocol framework. Cryptology ePrint Archive, Report 2018/976 (2018). https://eprint.iacr.org/2018/976

20. Jarecki, S., Krawczyk, H., Shirvanian, M., Saxena, N.: Device-enhanced password protocols with optimal online-offline protection. In: ACM ASIACCS (2016)
21. Karapanos, N., Marforio, C., Soriente, C., Capkun, S.: Sound-proof: usable two-factor authentication based on ambient sound. In: USENIX (2015)
22. Karole, A., Saxena, N., Christin, N.: A comparative usability evaluation of traditional password managers. In: ICISC (2010)
23. Lang, J., Czeskis, A., Balfanz, D., Schilder, M., Srinivas, S.: Security keys: practical cryptographic second factors for the modern web. In: FC (2016)
24. McCarney, D., Barrera, D., Clark, J., Chiasson, S., van Oorschot, P.C.: Tapas: design, implementation, and usability evaluation of a password manager. In: ACSAC, ACM (2012)
25. Norman, G.: Likert scales, levels of measurement and the "laws" of statistics. Adv. Health Sci. Educ. : Theor. Pract. **15**(5), 625–632 (2010)
26. Reynolds, J., Smith, T., Reese, K., Dickinson, L., Ruoti, S., Seamons, K.: A tale of two studies: the best and worst of yubikey usability. In: IEEE SP (2018)
27. Shirvanian, M., Jarecki, S., Saxena, N., Nathan, N.: Two-factor authentication resilient to server compromise using mix-bandwidth devices. In: NDSS (2014)
28. Shirvanian, M., Jareckiy, S., Krawczykz, H., Saxena, N.: Sphinx: a password store that perfectly hides passwords from itself. In: IEEE ICDCS (2017)
29. Smith, S.: Digital banking users to reach 2 billion this year, representing nearly 40% of global adult population (2018). https://bit.ly/2GPRhdE
30. Srinivas, S., Balfanz, D., Tiffany, E., Czeskis, A.: Universal 2nd factor (u2f) overview. FIDO Alliance Proposed Standard (2015)
31. Stobert, E., Biddle, R.: The password life cycle: user behaviour in managing passwords. In: ACM SOUPS (2014)
32. Stobert, E., Biddle, R.: The password life cycle. In: ACM TOPS (2018)
33. Sun, H.-M., Chen, Y.-H., Lin, Y.-H.: oPass: a user authentication protocol resistant to password stealing and password reuse attacks. In: IEEE TIFS (2012)
34. Taheri-Boshrooyeh, S., Küpçü, A.: Inonymous: anonymous invitation-based system. In: ESORICS DPM (2017)
35. Ur, B., et al.: I added '!' at the end to make it secure: Observing password creation in the lab. In: USENIX SOUPS (2015)
36. Venkatesh, V., Morris, M.G., Davis, G.B., Davis, F.D.: User acceptance of information technology: toward a unified view. MIS Q. **27**, 425–478 (2003)
37. Zviran, M., Haga, W.J.: A comparison of password techniques for multilevel authentication mechanisms. Comput. J. **36**, 227–237 (1993)

DPM Workshop: Privacy by Design and Data Anonymization

Graph Perturbation as Noise Graph Addition: A New Perspective for Graph Anonymization

Vicenç Torra[1,2] and Julián Salas[3(✉)]

[1] Hamilton Institute, Maynooth University, Maynooth, Ireland
vtorra@ieee.org
[2] University of Skövde, Skövde, Sweden
[3] CYBERCAT-Center for Cybersecurity Research of Catalonia, Internet
Interdisciplinary Institute (IN3), Universitat Oberta de Catalunya, Barcelona, Spain
jsalaspi@uoc.edu

Abstract. Different types of data privacy techniques have been applied to graphs and social networks. They have been used under different assumptions on intruders' knowledge. i.e., different assumptions on what can lead to disclosure. The analysis of different methods is also led by how data protection techniques influence the analysis of the data. i.e., information loss or data utility.

One of the techniques proposed for graph is graph perturbation. Several algorithms have been proposed for this purpose. They proceed adding or removing edges, although some also consider adding and removing nodes.

In this paper we propose the study of these graph perturbation techniques from a different perspective. Following the model of standard database perturbation as noise addition, we propose to study graph perturbation as *noise graph addition*. We think that changing the perspective of graph sanitization in this direction will permit to study the properties of perturbed graphs in a more systematic way.

Keywords: Data privacy · Graphs · Social networks · Noise addition · Edge removal

1 Introduction

Data privacy has emerged as an important research area in the last years due to the digitalization of the society. Methods for data protection were initially developed for data from statistical offices, and methods focused on standard databases.

Currently, there is an increasing interest to deal with big data. This includes (see e.g. [10,38,40]) dealing with databases of large volumes, streaming data, and dynamic data.

C. Pérez-Solà et al. (Eds.): DPM 2019/CBT 2019, LNCS 11737, pp. 121–137, 2019.
https://doi.org/10.1007/978-3-030-31500-9_8

Data from social networks is a typical example of big data, and we often need to take into account the three aspects just mentioned. Data from social networks is usually huge (as the number of users in most important networks are on the millions), posts can be modeled in terms of streaming, and is dynamic (we can consider multiple releases of the same data set).

As data from social networks is highly sensitive, a lot of research efforts have been devoted to develop data privacy mechanisms. Data privacy tools for social networks include disclosure risk models and measures [2,35], information loss and data utility measures, and perturbation methods (masking methods) [21].

Tools have been developed for the different existing privacy models. We review the most relevant ones below.

1. Differential privacy. In this case, given a query, we need to avoid disclosure from the outcome of the query. Privacy mechanisms are tailored to queries.
2. k-Anonymity. Different competing definitions for k-anonymity for graphs exist, depending on the type of information available to the intruder. Then, for each of these definitions, several privacy mechanisms have been developed. One of the weakest condition is k-degree anonymity. Strongest conditions include nodes neighborhoods.
3. Reidentification. Methods introduce noise into a graph so that intruders are not able to find a node given some background information. Again, we can have different reidentification models according to the type of information available to the intruders (from the degree of a node to its neighborhood). Here noise is usually understood as adding and removing edges from a graph. Nevertheless, adding and removing nodes have also been considered. Methods for achieving k-anonymity are also available for avoiding reidentification.

Note that while differential privacy and k-anonymity are defined as Boolean constraints (i.e., a protection level is fixed and then we can check if this level is achieved or not), this is not the case for reidentification. The risk of reidentification is defined in terms of the proportion of records that can be reidentified.

In this work we focus on methods for graphs perturbation. More particularly, we consider algorithms for graph randomization. In our context, they are the methods that modify a graph to avoid re-identification (and also to achieve k-anonymity). The goal is to formalize data perturbation for graphs, as a model for perturbation of social networks.

Methods for graph randomization usually consist on adding and removing edges. Performance of these methods is evaluated with respect to how the modification modifies the graph (the information loss caused to the graph), and the disclosure risk after perturbation. Often, the execution time is also considered.

The structure of the paper is as follows. In Sect. 2 we review noise addition and random graphs, as the formalization introduced in this paper is based on the former and uses the latter. In Sect. 3 we formalize noise addition for graphs. In Sect. 4, we discuss existing methods for graph randomization from this perspective. In Sect. 5, we present some results related to our proposal. The paper finishes with some conclusions.

2   Preliminaries

In this section we review some concepts and definitions that are needed later on in this paper. We discuss noise addition for standard numerical databases and random graphs.

2.1   Noise Addition

Noise addition has been applied for statistical disclosure control (SDC) for a long time. An overview of different algorithms for noise additiorf for microdata can be found in [7]. Simple application of noise addition corresponds to replace value x for variable V with mean $\bar{V}$ and variance σ^2 by $x + \epsilon$ with $\epsilon \sim N(0, \sigma^2)$. Correlated noise has also been considered so that noise do not affect correlations computed from perturbated data. See [39] (Chapter 6) for details.

Noise addition is known because it can be used to describe other perturbative masking techniques. In particular, rank swapping, post-randomization and microaggregation are special cases of matrix masking (see [12]). i.e., multiplying the original microdata set X by a record-transforming mask matrix A and by an attribute transforming mask matrix B and adding a displacing mask C to obtain the masked dataset $Z = AXB + C$.

Such operations are well defined for matrices and numbers, here we will define an equivalent operation for graphs.

2.2   Random Graphs

Probability distributions over graphs are usually referred by random graphs. There are different models in the literature. We review some of them here.

The Gilbert Model. This model is denoted by $\mathcal{G}(n, p)$. There are n nodes and each edge is chosen with probability p. With mean degree $d_m = p(n - 1)$ for a graph of n nodes, the probability of a node to have degree k is approximated by $d_m^k e^{-d_m}/k!$. I.e., this probability follows a Poisson distribution.

The Erdös-Rényi Model. This model is denoted by $G(n, e)$, and represents a uniform probability of all graphs with n nodes and e edges. This model is similar to the previous one (basically asymptotically equivalent as [1] writes it).

Most usual networks we find in real-life situations (including all kind of social networks) do not fit well with these two models. For example, it is known that most networks have degree distributions that follow a power-law. Other properties usually present in real-life networks is transitivity (also known as clustering) which means that if nodes v_1 and v_2 are connected, and nodes v_2 and v_3 are also connected, the probability of having v_1 and v_3 connected is high.

Some models extend the previous ones considering degree sequences that follow other distributions than the Poisson. One of such models is based on having a given degree sequence.

Models Based on a Given Degree Sequence. We denote this model by $\mathcal{D}(n, d^n)$, where n is the number of nodes and d^n is a degree sequence. That is,

$d^n \in \mathbb{N}^n$. $\mathcal{D}(n, d^n)$ represents a uniform probability of all graphs with n nodes and degree sequence d^n. Not all degree sequences are allowed (e.g., there is no graph for $\mathcal{D}(5, d^5 = (1, 0, 0, 0, 0))$). A graphical degree sequence is one for which there is at least one graph that satisfies it.

The Havel-Hakimi algorithm [18, 19] is an example of algorithm that builds a graph for a graphical degree sequence. In order to have a graph *drawn* from a uniform distribution $\mathcal{D}(n, d^n)$ we can use [23]. The authors discuss a polynomial algorithm for generating a graph from a distribution arbitrarily close to a uniform distribution, once the degree sequence is known. The algorithm is efficient (see [22]) for p-stable graphs. This type of approach is known as the Markov chain approach, as algorithms are based on switching edges according to given probabilities. See e.g. [26, 27].

Another way to draw a graph according to a uniform distribution from the degree sequence, is to assign to each node in the graph d_i stubs (i.e., edges still to be determined or out-going edges not yet assigned to the in-going ones). Then, choose at random pairs of stubs to settle a proper edge. Nevertheless this does not always work as we may finish with multiple edges and loops (see e.g. [3]).

The Configuration Model. In this case it is considered that the degree sequence of a graph follows a given distribution. To build such a graph we can first randomly choose a graphical degree sequence and then apply the machinery for building a graph from the degree sequence as explained above.

In this configuration model, the degree sequence adds a constraint to the possible graphs. All previous models can be combined with the fact that the degree sequence is given. In fact, other types of constraints can be considered in a model. For example, temporal and spatial.

For example, Spatial graphs may be generated in which the probability of having an edge is proportional to the distance between the nodes [39]. Random dot product graph models, a model proposed by Nickel et al. [29], is another example. In a random dot product graph, each node has associated a vector in an $\mathbb{R}^d$ space, and then the probability of having an edge between two nodes is proportional to the random dot product of the vectors associated to the two nodes.

Given a probability distribution over graphs $\mathcal{G}$ we will denote that we draw a graph G from $\mathcal{G}$ using $G \sim \mathcal{G}$. E.g., $G \sim \mathcal{G}(100, 0.2)$ is a graph G drawn from a Gilbert model with 100 nodes and probability 0.2 of having an edge between two nodes.

3   Noise Addition for Graphs

The main contribution of this paper is to formalize graph perturbation in terms of noise addition. For this, we introduce two different definitions for this type of perturbation. Both of them are based on selecting a graph according to a probability distribution over graphs and *adding* the selected graph to the original one.

We first define what we mean with graph *addition*.

In order to add two graphs, we must align the set of nodes in which we are going to add them. For example if G_1 and G_2 do not have any node in common, then the addition $G_1 \oplus G_2$ is equivalent to the disjoint union $G_1 \cup G_2$.

The next simplest case is when one of the graphs is a subset of the other. Then, the resulting graph will have the largest set of nodes (i.e., the union of both sets of nodes), and the final set of edges is an *exclusive-or* of the edges of both graphs.

Using an *exclusive-or* we can model both addition and removal of edges. Whether we remove or add an edge will depend on whether this edge is in only one graph or in both. Note also that this process is somehow analogous to noise addition where $X' = X + \epsilon$ with ϵ following a normal distribution. In particular, adding noise may correspond to increasing or decreasing the values in X. We formalize below graph addition for subgraphs.

Definition 1. *Let $G_1(V, E_1)$ and $G_2(V', E_2)$ be two graphs with $V \subset V'$; then, we define the addition of G_1 and G_2 as the graph $G = (V', E)$ where E is defined*

$$E = \{e | e \in V \wedge e \notin V'\} \cup \{e | e \notin V \wedge e \in V'\}$$

We will denote that G is the addition of G_1 and G_2 by

$$G = G_1 \oplus G_2.$$

Note that for any graph G_1 and G_2, G is a valid graph (i.e., G has no cycles and no multiedges).

We discuss now the definitions of noise graph addition. As one of the graphs, say G, is drawn from a distribution of graphs, the set of vertices created in G are not necessarily connected with the ones in the original graph. Nevertheless, even in this case, we can align (or, better, we still need to align) the two set of nodes. The difference between alternative definitions is about how the two sets of nodes are aligned. So, the general definition will be based on an alignment between the two set of nodes. This alignment corresponds to a homomorphism $A : V' \rightarrow V \cup \{\emptyset\}$. Here, $\emptyset$ corresponds to a null node, or dummy node. This notation comes from graph matching terminology (see e.g., [4]).

Definition 2. *Let $G_1(V, E_1)$ and $G_2(V', E_2)$ be two graphs with $|V| \leq |V'|$. An alignment is a homomorphism $A : V' \rightarrow V \cup \{\emptyset\}$.*

A matching algorithm builds such homomorphism for a given pair of graphs.

Definition 3. *Let $G_1(V, E_1)$ and $G_2(V', E_2)$ be two graphs with $|V| \leq |V'|$. A matching algorithm M is a function that given G_1 and G_2 returns an alignment.*

We denote it by $M(G_1, G_2)$. Then, given $A = M(G_1, G_2)$, $A(v)$ for any V' is either a node in V or $\emptyset$.

Definition 4. *Let $G_1(V, E_1)$ be a graph, let $\mathcal{G}$ be a probability distribution on graphs, and let M be a matching algorithm; then, noise graph addition for G_1*

following the distribution $\mathcal{G}$ and based on M consists of (i) drawing a graph $G_2(V', E_2)$ from $\mathcal{G}$ such that $|V| \leq |V'|$ (i.e., $G_2 \sim \mathcal{G}$)), (ii) aligning it with M obtaining $A = M(G_1, G_2)$, (iii) defining G'_2 as G_2 after renaming the nodes V' into V'' as follows

$$V'' = V \cup \{v' | v' \in V' \text{ and } A(v') = \emptyset\}$$

where $B_V : V' \to V''$ is, as expected, $B_V(v') = v'$ if $A(v') = \emptyset$ and $B_V(v') = A(v')$ otherwise, (iv) defining $B_E(E_2) = \{(B_V(v_1), B_V(v_2)) | (v_1, v_2) \in E_2\}$ and (v) defining the protected graph $G'(V', E')$ as

$$G' = G_1(V, E_1) \oplus G_2(V'', B_E(E_2)).$$

This definition allows for adding noise graphs that are not necessarily subgraphs of each other.

Another equivalent way, is to think of graphs as labeled graphs, that is, a graph $G = (V, E, \mu)$, where $\mu : V \to L_V$ is a function assigning labels to the vertices of the graph G. For simplicity we will assume that L_V is a set of integers. Then, given two labeled graphs G_1 and G_2, their edges would be defined by pairs of integers corresponding to the labels of their corresponding nodes. In such way, an alignment will be given naturally by the nodes that have the same label in both graphs.

Otherwise, if the nodes from two graphs G_1 and G_2 are labeled respectively as $[n_1] = 1, \ldots, n_1$ and $[n_2] = 1, \ldots, n_2$. To define a different alignment, we may relabel their nodes in the following way.

First, choose two subsets $S_1 \subset [n_1]$ and $S_2 \subset [n_2]$ of the same size, assume that $n_1 \leq n_2$, define a bijection $f : S_2 \mapsto S_1$, then relabel the nodes in $V(G_2)$ as $f(j)$ if $j \in S_2$ and as $n_1 + 1, \ldots, n_2$ for the remaining nodes, whose new labels $f(j)$ are not in S_2. In this way we obtain an alignment. Now we can identify a node with its label and then use Definition 1 for graph addition.

3.1 Graph Matching and Edit Distance

On the other side of noise addition, we find the concept record linkage. In the context of graphs this concept is similar to graph matching. Graph matching is a key task in several pattern recognition applications [41]. However, it has a high computational cost. Indeed, finding isomorphic graphs is NP-complete. Thus, several methods have been devised for efficient approximate graph matching, some use tree structures and strings such as [28,30,42], other are able to process very large scale graphs such as [24,37] or [36].

A commonly used distance for graph matching and finding isomorphic subgraphs is graph edit distance (GED). It was formalized mathematically in [34].

For defining GED, a set of edit operations is introduced, for example, deletion, insertion and substitution of nodes and edges. Then, the similarity of two graphs is defined in terms of the shortest or least cost sequence of edit operations that transforms one graph into the other.

Of course when $GED(G_1, G_2) = 0$ this means that such graphs are isomorphic. Therefore for computing the similarity (or testing an isomorphism) between graphs an exhaustive solution is to define all possible labelings (alignments) and testing for similarity, this is quite inefficient.

However, we may use our definition of addition for defining a distance (that we call edge distance) between two labeled graphs as:

$$ed(G_1, G_2) = |E(G_1 \oplus G_2)|$$

We will prove that ed is a metric in the following theorem.

Theorem 1. *Let $\mathcal{G}$ be the space of labeled graphs without isolated nodes, $G_1, G_2 \in \mathcal{G}$. Let $ed(G_1, G_2) = |E(G_1 \oplus G_2)|$, then the following conditions hold:*

1. *$ed(G_1, G_2) = 0 \Leftrightarrow G_1 = G_2$*
2. *$ed(G_1, G_2) = ed(G_2, G_1)$*
3. *$ed(G_1, G_3) \leq ed(G_1, G_2) + ed(G_2, G_3)$*

Proof. First, $ed(G_1, G_2) = 0$ implies that every $uv \in G_1$ is also in G_2 and the other way around. Hence $E(G_1) = E(G_2)$. Since they do not have isolated nodes, also $V(G_1) = V(G_2)$ and $G_1 = G_2$. Second, $ed(G_1, G_2) = ed(G_2, G_1)$ comes from $G_1 \oplus G_2 = G_2 \oplus G_1$.

Third, for the triangle inequality we calculate two cases for an edge $uv \in E(G_1) \oplus E(G_3)$.

(i) $uv \in E(G_1)$ and $uv \notin E(G_3)$ In this case, if $uv \in E(G_2)$ then $uv \in E(G_2 \oplus G_3)$. If $uv \notin E(G_2)$ then $uv \in E(G_1 \oplus G_2)$.
(ii) $uv \in G_3$ and $uv \notin G_1$ This case is equivalent to case $i)$ only substituting G_1 by G_3.

Therefore we conclude that ed is a metric.

In fact, our metric $ed(G_1, G_2)$ actually counts the number of edges we have to modify (erase/add) to transform a graph G_1 to obtain G_2. Moreover, if we sum $G_1 \oplus (G_1 \oplus G_2)$ we obtain G_2 and $G_2 \oplus (G_1 \oplus G_2) = G_1$.

Therefore we may use our definition for graph addition not only for adding noise, but also for measuring it. This is useful for comparing any published graph with other graphs published under a different privacy protection algorithm (such as a k-anonymous graph).

Moreover, we can use the edge distance to find the median graph G_{median} similar to the median graph obtained in [13] for the graph edit distance.

For a family of graphs $\mathcal{F}$, we define:

$$G_{median}(\mathcal{F}) = arg \, min_{\tilde{G}} \sum_{G_i \in \mathcal{F}} ed(\tilde{G}, G_i)$$

Note that G_{median} is not necessarily unique for a given family of graphs $\mathcal{F}$, so G_{median} is also a set of graphs.

In the following result we will show that the original graph G is such that is at minimum distance from m protected graphs in $G \oplus \mathcal{G}$ if there is not any edge that belongs to more than $m/2$ of the graphs in $\mathcal{G}$.

Theorem 2. *Let* $\mathcal{F} = G \oplus \mathcal{G}$ *where* $\mathcal{G} = \{G'_1, \ldots, G'_m\}$. *If* $|\{G'_i \in \mathcal{G} : e \in G'_i\}| \leq \frac{|\mathcal{G}|}{2}$ *for all* $e \in E(\mathcal{G})$, *then* $G \in G_{median}(\mathcal{F})$.

Proof. For each $G'_i \in \mathcal{G}$ denote $G_i = G \oplus G'_i$. Let $\tilde{G}$ any graph different from G. Then, $E(\tilde{G} \oplus G) = A \cup B$, where A denotes the edges in $\tilde{G} \setminus G$ and B the edges in $G \setminus \tilde{G}$. Therefore $\tilde{G} \oplus G_i = \tilde{G} \oplus G \oplus G'_i = (A \cup B) \oplus G'_i$.

For $e \notin A \cup B$, then its either in both $E(G \oplus G_i)$ and $E(\tilde{G} \oplus G_i)$ or in none of both.

For $e \in A \cup B$ and $e \notin E(G'_i)$ then $e \in E((A \cup B) \oplus G'_i) = E(\tilde{G} \oplus G_i)$, in this case, $e \notin E(G'_i) = E(G \oplus G \oplus G'_i) = E(G \oplus G_i)$, no matter whether e was in G or not. So e is not counted in the sum of $|E(G \oplus G_i)|$ but it is counted in $|E(\tilde{G} \oplus G_i)|$

For $e \in A \cup B$ and $e \in E(G'_i)$ then $e \notin E((A \cup B) \oplus G'_i)$. So, $e \notin E(\tilde{G} \oplus G_i)$ but $e \in E(G'_i) = E(G \oplus G_i)$.

We conclude for each edge $e \in A \cup B$ that the cases when $e \in E(G'_i)$, the edge e is counted in $ed(G, G_i)$ but not in $ed(\tilde{G}, G_i)$. While, in the cases when $e \notin E(G'_i)$, the edge e is not counted in $ed(G, G_i)$ but it is counted in $ed(\tilde{G}, G_i)$.

Therefore, since all edges e belong to at most half of the graphs in $\mathcal{G}$, then
$$\sum_{G_i \in \mathcal{F}} ed(G, G_i) \leq \sum_{G_i \in \mathcal{F}} ed(\tilde{G}, G_i), \text{ or equivalently, } G \in G_{median}(\mathcal{F}).$$

This provides us with a similar result as in noise addition for SDC in which the expected value $E(x + \epsilon) = x$ for $\epsilon \in N(0, \sigma^2)$.

4   Edge Noise Addition, Local Randomization and Degree Preserving Randomization

In this section we explore the relation of graph noise addition with previous randomization algorithms.

4.1   Edge Noise Addition

The method for graph matching from [28], uses the structure of the neighbors at increasing distance from each node in the graph to grow a tree that is later used to codify the entire graph. Such queries are mentioned in [21] and can be iteratively refined to re-identify nodes in the graph. Hay et al. [20] propose as a solution a graph generalization algorithm similar to [8] and recently improved in [31]. To measure privacy, they sample a graph from all the possible graphs that are consistent with such generalization.

In [20], they have suggested a random perturbation method that consisted on adding m edges and removing m edges. It was later studied from an information-theoretic perspective in [6]. They also suggested a random sparsification method. Other random perturbation methods may be found in [9].

All these perturbation approaches are included in our definition. Observe that they correspond to constraints on the definition of the set of noise graphs ($\mathcal{G}$) that we are using.

For example, the obfuscation by random sparsification is defined in [6] as follows. The data owner selects a probability $p \in [0, 1]$ and for each edge $e \in E(G)$ the data owner performs an independent Bernoulli trial, $B_e \sim B(1, p)$. And leaves the edge in the graph in case of success (i.e., $B_e = 1$) and remove it otherwise ($B_e = 0$). Letting $E_p = \{e \in E | B_e = 1\}$ be the subset of edges that passed this selection process, the data owner releases the subgraph $G_p = (U = V, E_p)$. Then, they argue that such graph will offer some level of identity obfuscation for the individuals in the underlying population, while maintaining sufficient utility in the sense that many features of the original graph may be inferred from looking at G_p.

Note that random sparsification is obtained with our method by using $\mathcal{G} = \mathcal{G}(n; 1 - p) \cap G$, then adding $G \oplus G'$ for some $G' \in \mathcal{G}$.

Another relevant example is when we define $\mathcal{G} = \{G' : |E(G')| = m\}$, then $G \oplus G'$ where $G' \sim \mathcal{G}$ is equivalent to adding x and deleting $m - x$ edges, in total changing m edges from the original graph. If we restrict $\mathcal{G}$ to be the family of graphs G' such that $|E(G')| = 2m$ and $|E(G') \cap E(G)| = m$, then we are adding m edges and deleting m other edges, as in [20].

Note that, in this case all the m edges may be incident to the same node or they may be all independent, however such restrictions could be added with our method simply by specifying that the graphs G' in $\mathcal{G}$ do not have degree greater than $m - 1$, or that their maximum degree is at most 1.

We will show an example of such a local restriction in the following subsection.

4.2   Local Randomization

In [43] a comparison between edge randomization and k-anonymity is carried out, considering that the adversary knowledge is the degree of the nodes in the original graph. There, they calculate the prior and posterior risk measures of the existence of a link between node i and j (after adding k and removing k edges) as:

$$P(a_{ij} = 1) = \frac{m}{N}$$

$$P(a_{ij} = 1 | \tilde{a}_{ij} = 1) = \frac{m - k}{m} \tag{1}$$

$$P(a_{ij} = 1 | \tilde{a}_{ij} = 0) = \frac{k}{N - m}$$

We define $G \oplus G_u^t$ to be the local t-randomization which adds the graph G_u^t with vertex set $V(G_u^t) = u, u_1, \ldots, u_t$ and edge set $E(G_u^t) = uu_1, \ldots, uu_t$ in which u is a fixed vertex from G and $u_1, \ldots, u_t \subset V(G \setminus u)$.

Then $G \oplus G_u^t$ changes t random edges incident to u in G. So we can apply local t-randomization for all u in $V(G)$ to obtain the graph

$$G^t = G \bigoplus_{u \in V(G)} G_u^t$$

Following the notation from [43], we compare the privacy guarantees of local t-randomization to the equivalent that would be $k = tn/2$ in their case. This value comes from the fact that each node has t randomized edges and therefore are counted twice.

We consider an adversary that knows about a given node and its degree. Then tries to infer if the adjacent edges in the published graph come from the original.

For a given node i we denote its degree as d_i and as $\overline{d_i}$ the value $n-1-d_i$, in general for any value t, we denote as $\overline{t}$ the value $n-1-t$, this is the complement of the degree d_i and the complement of the edges in $E(G_u^t)$ respectively.

Theorem 3. *The adversary's prior and posterior probabilities to predict whether there is a sensitive link between two target individuals $i, j \in V(G)$ by exploiting the degree d_i and accessing to the randomized graph G^t are the following:*

(i) *the probability* $P(a_{ij} = 1) = \dfrac{d_i}{n-1}$

(ii) *the probability* $P(a_{ij} = 1 | a_{ij}^t = 1)$ *is equal to:*

$$\frac{d_i(\overline{t}^2 + t^2)}{d_i(\overline{t}^2 + t^2) + 2\overline{d_i}\overline{t}t}$$

(iii) *the probability* $P(a_{ij} = 1 | a_{ij}^t = 0)$ *is equal to:*

$$\frac{2d_i\overline{t}t}{2d_i\overline{t}t + \overline{d_i}(\overline{t}^2 + t^2)}$$

Proof. (i) First of all, $P(a_{ij} = 1) = \dfrac{d_i}{n-1}$ because the adversary knows d_i and there are only $n - 1$ possible neighbors for i

(ii) Now we calculate $P(a_{ij} = 1 | a_{ij}^t = 1)$ in our graph G^t.

There are only two possibilities for an edge to belong to $E(G^t)$ (i.e., $a_{ij}^t = 1$) and to G (i.e., $a_{ij} = 1$). The edge was already in $E(G)$ and it does not belong to $E(G_i^t)$ neither to $E(G_j^t)$ or it is in $E(G)$ and at the same time belongs to both $E(G_i^t)$ and $E(G_j^t)$.

For an edge to belong to $E(G^t)$ and not to G (i.e., $a_{ij} = 0$). There are also two cases, $a_{ij} = 0$ and the edge belongs to exactly one of $E(G_i^t)$ or $E(G_j^t)$.

Considering that the probability that $ij \in E(G^t) = \frac{t}{n-1}$, the probability that $ij \notin E(G^t) = \frac{\overline{t}}{n-1}$, $P(a_{ij} = 1) = \frac{d_i}{n-1}$ and $P(a_{ij} = 0) = \frac{\overline{d_i}}{n-1}$ we obtain the following:

$$\frac{d_i\overline{t}^2 + d_it^2}{d_i\overline{t}^2 + d_it^2 + \overline{d_i}\overline{t}t + \overline{d_i}t\overline{t}}$$

which yields (ii).

For (iii) we follow a similar argument as in (ii). For $P(a_{ij} = 1 | a_{ij}^t = 0)$ there are also two possibilities for an edge that was in $E(G)$ to be removed from $E(G^t)$. The edge ij is in $E(G)$ and either $ij \in E(G_i^t) \setminus E(G_j^t)$ or $ij \in E(G_j^t) \setminus E(G_i^t)$. For an edge to belong to G and not to $E(G^t)$. There are also two more cases, $a_{ij} = 1$ and the edge belongs to both $E(G_i^t)$ and $E(G_j^t)$, or to none of them. This implies that:

$$P(a_{ij} = 1 | a_{ij}^t = 0) = \frac{d_i \bar{t} t + d_i t \bar{t}}{d_i \bar{t} t + d_i t \bar{t} + \overline{d_i} t^2 + \overline{d_i} t^2}$$

Which simplifies to (iii) and finishes the proof.

4.3   Degree Preserving Randomization

As we have discussed in the introduction, graphs with a given degree sequence can be uniformly generated by starting from a graph $G \in \mathcal{D}(n, d^n)$ and swapping some of its edges. The method of swap randomization is also used for generating matrices with given margins for representing tables and assessing data mining results in [14].

Recall that a swap in a graph consists of choosing two edges $u_1 u_2, u_3 u_4 \in E(G)$, such that $u_2 u_3, u_4 u_1 \notin E(G)$ to obtain the graph $\tilde{G}$ such that $V(\tilde{G}) = V(G)$ and $E(\tilde{G}) = E(G) \setminus \{u_1 u_2, u_3 u_4\} \cup \{u_2 u_3, u_4 u_1\}$. It is also called alternating circuit in [39] were it is used for proving that all multigraphs may be transformed into graphs via alternating circuits, and used for generating synthetic spatial graphs.

In [32] and [33] this is used to prove that scale-free sequences with parameter $\gamma > 2$ are P-stable and graphic. Note that the P-stability property guarantees that a graph with a given degree sequence may be uniformly generated by choosing a graph $G \in \mathcal{D}(n, d^n)$ by applying sufficient random swaps.

Then for a given graph G with n nodes labeled as $[n] = 1, \ldots, n$ We define the family $\mathcal{S}_G = G'$ such that $G' = \{i, j, k, l\} \subset [n]$, where $ij, kl \in E(G)$ and $jk, li \notin E(G)$. In other words $\mathcal{S}_G$ is the set of alternating 4-circuits of G.

We may choose an arbitrary number m of such noise graphs $G_1', \ldots, G_m'$ and add them to G, that is $G = G \oplus G_1' \oplus \ldots \oplus G_m'$.

Following this procedure for m large enough is equivalent to randomizing G to obtain all the graphs $\mathcal{D}(n, d^n)$. Actually, for any two graphs with the same degree sequence $G_1, G_2 \in \mathcal{D}(n, d^n)$, $G' = G_1 \oplus G_2$ must have at each node i the same number of edges ij that belong to $E(G)$ as the edges il that do not belong to $E(G)$, since the node i has the same degree in G_1 and in G_2. Thus, any graph in $\mathcal{D}(n, d^n)$ may be generated by starting at some graph G and adding G' from the set $\mathcal{G}$ of all the graphs that are a union of alternating circuits of G.

5   The Most General Approach for Noise Addition for Graphs

In all the previous sections we were restrictive with respect to the noise graphs that we were adding, in this last section we apply the Gilbert model which was discussed in Sect. 2.2, without any restriction.

This has as a consequence that we may not be able to obtain strong conclusions on the structure of the graphs obtained after noise addition, in contrast to the results in previous sections. However, by reducing the restrictions, we are increasing the possible noise graphs that we add, hence we increase uncertainty and therefore protection. In Table 1 we present the noise addition methods discussed in this paper, together with some of their properties and requirements.

Table 1. Graph perturbation methods in terms of graph noise addition.

Noise addition method	Definition of $\mathcal{G}$	Additional requirements for $G' \in \mathcal{G}$	Properties of $G \oplus \mathcal{G}$
Random perturbation [20]	$|E(G')| = 2m$	$|E(G') \cap E(G)| = m$ $|E(G') \cap E(\overline{G})| = m$	G' adds m edges and removes m edges
Random sparsification [6]	$G' \in \mathcal{G}(n; 1-p) \cap G$	None	The edges of G remain with probability p, no added edges
Local t-randomization	$G' = G_u^t$	Applied to every node in G	Every node has t modified incident edges
Degree preserving randomization [5]	$G' \in \mathcal{S}_G$	$\mathcal{S}_G$ is the set of swaps of G	$G, G \oplus G' \in \mathcal{D}(n, d^n)$
Gilbert model	$G' \in \mathcal{G}(n; 1-p)$	None	Every edge is added or removed with probability p

Proposition 1. Let $G_1(V, E_1)$ be an arbitrary undirected graph, and let $G_2(V, E_2)$ be an undirected graph generated from a Gilbert model $\mathcal{G}(|V|, p)$. Then, the expected number of edges of G_2 is $p \cdot |V| \cdot (|V| - 1)/2$.

Proposition 2. Let $G_1(V, E_1)$ be an arbitrary undirected graph with $n_1 = |E_1|$ edges, and let $G_2(V, E_2)$ be an undirected graph generated from a Gilbert model with n_2 edges. Then, the noise graph addition of G_1 and G_2

$$G' = G_1(V, E_1) \oplus G_2(V, E_2)$$

will have on average $|E|' = n_2 \cdot (t - n_1)/t + (t - n_2) \cdot n_1/t$ edges, where $t = |V| * (|V| - 1)/2$.

Proof. Let the graph G_1 have $n_1 = |E_1|$ edges, and let the graph G_2 have n_2 edges. Being the graph undirected, the number of edges for a complete graph is $t = v * (v-1)/2$ where $v = |V|$ is the number of nodes. Then, we will have on average that

- $n_2 \cdot n_1/t$ edges of G_2 link two nodes also linked by G_1,
- $n_2 \cdot (t - n_1)/t$ edges of G_2 link two nodes not linked by G_1,
- $(t - n_2) \cdot n_1/t$ edges of G_1 link two nodes not linked by G_2, and
- $(t - n_2)(t - n_1)/t$ corresponds to pairs of nodes neither linked in G_1 nor in G_2.

So, addition of graphs G_1 and G_2 result into a graph with $n_2 \cdot (t - n_1)/t + (t - n_2) \cdot n_1/t$ edges on average.

From this, it follows that only when $n_2 = 0$, the number of edges of G' is the same as the ones in G_1; or when $n_1 = t/2$. This is expresses as a lemma below.

Lemma 1. *Let $G_1(V, E_1)$ and $G_2(V, E_2)$ be as above with n_1 and n_2 the corresponding number of edges. Then, the noise graph addition of G_1 and G_2, $G' = G_1(V, E_1) \oplus G_2(V, E_2)$ has n_1 nodes when $n_2 = 0$ or when $n_1 = t/2$.*

Proof. The solutions of $n_1 = n_2 \cdot (t - n_1)/t + (t - n_2) \cdot n_1/t$ are such that $tn_1 = n_2 t - n_2 n_1 + tn_1 - n_2 n_1$ or, equivalently, $0 = n_2 t - n_2 n_1 + -n_2 n_1$. So, one solution is that $n_2 = 0$, and if this is not the case, $n_1 = t/2$.

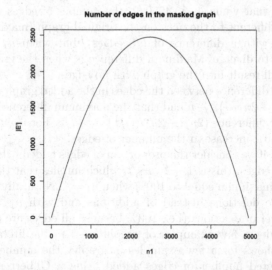

Fig. 1. Average number of edges in G' when the graph G_1 has 100 nodes and, thus, $t = 4950$, and n_1 is in the range $[0, 4950]$. G_2 generated using the Gilbert model $\mathcal{G}(|V|, p)$ with p as in the graph $G(V, E_1)$ (i.e., the expected number of edges satisfying $|E_2| = |E_1|$).

Let us consider the case that we use our original graph to build a Gilbert model $\mathcal{G}(|V|, p)$ from the graph $G_1(V, E_1)$. That is, we generate a graph with

Table 2. Number of nodes and edges in some of small graphs in the literature using the Gilbert model $\mathcal{G}(|V|, p)$ with p as in the graph $G(V, E_1)$ (i.e., the expected number of edges satisfying $|E_2| = |E_1|$).

Name	Reference	Nodes	Edges	Edges added	% of edges added
Jazz	[16]	198	5484	2400	43.7
Karate	[44]	34	156	69	44.3
Football	[15]	115	1226	767	62.5
Erdos971	[45]	472	2628	2503	95.2
Urvemail	[17]	1133	10902	10531	96.5
CElegans	[11]	453	4050	3729	92.0
Caida	[25]	26475	106762	106697	99.9

$p = |E_1|/(|V| * (|V| - 1)/2)$. Then, the expected value for n_2 is $n_2 = n_1 = |E_1|$. Therefore, noise graph addition of G_1 and G_2 will have $n_1 \cdot (t - n_1)/t + (t - n_1) \cdot n_1/t = (2n_1 t - 2n_1 n_1)/t$ edges.

Figure 1 illustrates this case for the case of 100 nodes and edges between 0 and 4950 (i.e., the case of a complete graph). As proven in the lemma, only when $n_1 = 0$ or when $n_1 = t/2$ we have that the number of edges in the resulting graph is exactly the same as in the original graph (i.e., n_2 is 0 or $t/2$, respectively). We can also see that while $n_1 \leq 4950/2$, the number of edges in the resulting graph is not so different to the ones on the original graph (maximum difference is at $n_1 = t/4$ with a difference of $t/8$ edges), but when $n_1 > 4950/2$ the difference starts to diverge. Maximum difference is when the graph is complete that addition will result into the graph with no edges.

Observe that difference between the edges in the added graph and the original one will be $(2n_1 t - 2n_1 n_1)/t - n$ and that the maximum difference is at $n_1 = t/4$. For $n_1 = t/4$ the difference $(2n_1 t - 2n_1 n_1)/t - n = t/8$. For $n_1 = t/4$ this means a maximum of 50% increase in the number of edges.

Nevertheless, if we consider the proportion of edges added with respect to the total number of edges, this is $1 - 2 * n_1/t$, which implies that the proportion is decreasing with maximum equal to 100% when $n_1 = 1$. Naturally, when $n_1 \geq t/2$ we start to have deletions instead of additions and with $n_1 = t$ we have a maximum number of deletions (i.e., 100% because all edges are deleted).

Let us consider a few examples of graphs used in the literature. For each graph, Table 2 shows for a few examples of graphs, the number of nodes and edges, the expected number of edges added using a Gilbert model, and the expected increase in the number of edges. It can be seen that for large number of nodes, as the actual number of edges is rather low with respect to the ones in a complete graph, the % of increment of edges is around 100%.

6   Conclusions

In this paper we have introduced a formalization for graph perturbation based on noise addition. We have shown that some of the existing approaches for graph randomization can be understood from this perspective. We have also proven some properties for this approach, noted that it encompasses many of the previous randomization approaches while also defining a metric for graph modification. Additionally, we defined the local t-randomization approach which guarantees that every node has t-neighboring edges modified. It remains as future work to study new families of noise graphs and test their privacy and utility guarantees.

Acknowledgments. This work was partially supported by the Swedish Research Council (Vetenskapsrådet) project DRIAT (VR 2016-03346), the Spanish Government under grants RTI2018-095094-B-C22 "CONSENT" and TIN2014-57364-C2-2-R "SMARTGLACIS", and the UOC postdoctoral fellowship program.

References

1. Aiello, W., Chung, F., Lu, L.: A random graph model for power law graphs. J. Exp. Math. **10**, 53–66 (2001)
2. Balsa, E., Troncoso, C., Díaz, C.: A metric to evaluate interaction obfuscation in online social networks. Int. J. Unc. Fuzz. Knowl.-Based Syst. **20**, 877–892 (2012)
3. Bannink, T., van der Hofstad, R., Stegehuis, C.: Switch chain mixing times through triangle counts. arXiv:1711.06137 (2017)
4. Bengoetxea, E.: Inexact graph matching using estimation of distribution algorithms, PhD Dissertation, Ecole Nationale Supérieure des Télécommunications, Paris (2002)
5. Berge, C.: Graphs and Hypergraphs. North-Holland, Netherlands (1973)
6. Bonchi, F., Gionis, A., Tassa, T.: Identity obfuscation in graphs through the information theoretic lens. Inf. Sci. **275**, 232–256 (2014)
7. Brand, R.: Microdata protection through noise addition. In: Domingo-Ferrer, J. (ed.) Inference Control in Statistical Databases. LNCS, vol. 2316, pp. 97–116. Springer, Heidelberg (2002). https://doi.org/10.1007/3-540-47804-3_8
8. Campan, A., Truta, T.M.: Privacy, security, and trust in KDD, pp. 33–54 (2009)
9. Casas-Roma, J., Herrera-Joancomartí, J., Torra, V.: A survey of graph-modification techniques for privacy-preserving on networks. Artif. Intell. Rev. **47**(3), 341–366 (2017)
10. D'Acquisto, G., Domingo-Ferrer, J., Kikiras, P., Torra, V., de Montjoye, Y.-A., Bourka, A.: Privacy by design in big data: an overview of privacy enhancing technologies in the era of big data analytics, ENISA Report (2015)
11. Duch, J., Arenas, A.: Community identification using extremal optimization. Phys. Rev. E **72**, 027–104 (2005)
12. Duncan, G.T., Pearson, R.W.: Enhancing access to microdata while protecting confidentiality: prospects for the future. Statist. Sci. **6**(3), 219–232 (1991)
13. Ferrer, M., Valveny, E., Serratosa, F.: Median graph: a new exact algorithm using a distance based on the maximum common subgraph. Pattern Recogn. Lett. **30**(5), 579–588 (2009)

14. Gionis, A., Mannila, H., Mielikäinen, T., Tsaparas, P.: Assessing data mining results via swap randomization. ACM Trans. Knowl. Discov. Data 1(3), 14 (2007)
15. Girvan, M., Newman, M.E.J.: Community structure in social and biological networks. Proc. Nat. Acad. Sci. US Am. 99(12), 7821–7826 (2002)
16. Gleiser, P., Danon, L.: Community structure in jazz. Adv. Complex Syst. 6, 565 (2003)
17. Guimera, R., Danon, L., Diaz-Guilera, A., Giralt, F., Arenas, A.: Self-similar community structure in a network of human interaction. Phys. Rev. E 68, 065103(R) (2003)
18. Hakimi, S.L.: On realizability of a set of integers as degrees of the vertices of a linear graph I. J. Soc. Ind. Appl. Math. 10, 496–506 (1962)
19. Havel, V.: A remark on the existence of finite graphs. Časopis Pro Pěstování Matematiky (in Czech) 80, 477–480 (1955)
20. Hay, M., Miklau, G., Jensen, D., Weis, P., Srivastava, S.: Anonymizing social networks, Technical report No. 07-19, Computer Science Department, University of Massachusetts Amherst, UMass Amherst (2007)
21. Hay, M., Miklau, G., Jensen, D., Towsley, D., Weis, P.: Resisting structural reidentification in anonymized social networks. Proc. VLDB Endow. 1(1), 102–114 (2008)
22. Jerrum, M., McKay, B.D., Sinclair, A.: When is a graphical sequence stable? In: Frieze, A., Luczak, T. (eds.) Random Graphs, vol. 2, pp. 101–115. Wiley-Interscience, Hoboken (1992)
23. Jerrum, M., Sinclair, A.: Fast uniform generation of regular graphs. Theoret. Comput. Sci. 73, 91–100 (1990)
24. Koutra, D., Shah, N., Vogelstein, J.T., Gallagher, B., Faloutsos, C.: Deltacon: principled massive-graph similarity function with attribution. ACM Trans. Knowl. Discov. Data 10(3), 28:1–28:43 (2016)
25. Leskovec, J., Kleinberg, J., Faloutsos, C.: Graph evolution: densification and shrinking diameters. ACM Trans. Knowl. Discov. Data 1(1), 1–40 (2007)
26. Miklós, I., Erdös, P.L., Soukup, L.: Towards random uniform sampling of bipartite graphs with given degree sequence. Electr. J. Comb. 20(1), 16 (2013)
27. Milo, R., Kashtan, N., Itzkovitz, S., Newman, M.E.J., Alon, U.: On the uniform generation of random graphs with prescribed degree sequences. arXiv:cond-mat/0312028v2 (2004)
28. Nettleton, D.F., Salas, J.: Approximate matching of neighborhood subgraphs - an ordered string graph levenshtein method. Int. J. Unc. Fuzz. Knowl.-Based Syst. 24(03), 411–431 (2016)
29. Nickel, C.L.M.: Random dot product graphs: a model for social networks, PhD. dissertation, Maryland (2006)
30. Robles-Kelly, A., Hancock, E.R.: String edit distance, random walks and graph matching. Int. J. Pattern Recogn. Artif. Intell. 18(03), 315–327 (2004)
31. Ros-Martín, M., Salas, J., Casas-Roma, J.: Scalable non-deterministic clustering-based k-anonymization for rich networks. Int. J. Inf. Secur. 18(2), 219–238 (2019)
32. Salas, J., Torra, V.: Graphic sequences, distances and k-degree anonymity. Discrete Appl. Math. 188, 25–31 (2015)
33. Salas, J., Torra, V.: Improving the characterization of p-stability for applications in network privacy. Discrete Appl. Math. 206, 109–114 (2016)
34. Sanfeliu, A., Fu, K.: A distance measure between attributed relational graphs for pattern recognition. IEEE Trans. Syst. Man Cybern. SMC–13(3), 353–362 (1983)
35. Stokes, K., Torra, V.: Reidentification and k-anonymity: a model for disclosure risk in graphs. Soft Comput. 16(10), 1657–1670 (2012)

36. Sun, Z., Wang, H., Wang, H., Shao, B., Li, J.: Efficient subgraph matching on billion node graphs. Proc. VLDB Endow. **5**(9), 788–799 (2012)
37. Tian, Y., Patel, J.M.: Tale: a tool for approximate large graph matching. In: 2008 IEEE 24th International Conference on Data Engineering, pp. 963–972, April (2008)
38. Torra, V.: Data Privacy: Foundations, New Developments and the Big Data Challenge. SBD, vol. 28. Springer, Cham (2017). https://doi.org/10.1007/978-3-319-57358-8
39. Torra, V., Jonsson, A., Navarro-Arribas, G., Salas, J.: Synthetic generation of spatial graphs. Int. J. Intell. Syst. **33**(12), 2364–2378 (2018)
40. Torra, V., Navarro-Arribas, G.: Big data privacy and anonymization. In: Lehmann, A., Whitehouse, D., Fischer-Hübner, S., Fritsch, L., Raab, C. (eds.) Privacy and Identity 2016. IAICT, vol. 498, pp. 15–26. Springer, Cham (2016). https://doi.org/10.1007/978-3-319-55783-0_2
41. Vento, M.: A long trip in the charming world of graphs for pattern recognition. Pattern Recogn. **48**(2), 291–301 (2015)
42. Yan, X., Han, J.: gspan: graph-based substructure pattern mining. In: 2002 IEEE International Conference on Data Mining, 2002. Proceedings., pp. 721–724, December (2002)
43. Ying, X., Pan, K., Wu, X., Guo, L.: Comparisons of randomization and k-degree anonymization schemes for privacy preserving social network publishing. In: Proceedings of the 3rd Workshop on Social Network Mining and Analysis, ser. SNA-KDD 2009, pp. 10:1–10:10. New York, NY, USA, ACM (2009)
44. Zachary, W.W.: An information flow model for conflict and fission in small groups. J. Anthropol. Res. **33**, 452–473 (1977)
45. http://vlado.fmf.uni-lj.si/pub/networks/pajek/data/gphs.htm

Towards Minimising Timestamp Usage
In Application Software
A Case Study of the Mattermost Application

Christian Burkert$^{(\boxtimes)}$ ⓘ and Hannes Federrath

University of Hamburg, Hamburg, Germany
{burkert,federrath}@informatik.uni-hamburg.de

Abstract. With digitisation, work environments are becoming more digitally integrated. As a result, work steps are digitally recorded and therefore can be analysed more easily. This is especially true for office workers that use centralised collaboration and communication software, such as cloud-based office suites and groupware. To protect employees against curious employers that mine their personal data for potentially discriminating business metrics, software designers should reduce the amount of gathered data to a necessary minimum. Finding more data-minimal designs for software is highly application-specific and requires a detailed understanding of the purposes for which a category of data is used. To the best of our knowledge, we are the first to investigate the usage of timestamps in application software regarding their potential for data minimisation. We conducted a source code analysis of Mattermost, a popular communication software for teams. We identified 47 user-related timestamps. About half of those are collected but never used and only 5 are visible to the user. For those timestamps that *are* used, we propose alternative design patterns that require significantly reduced timestamp resolutions or operate on simple enumerations. We found that more than half of the usage instances can be realised without any timestamps. Our analysis suggests that developers routinely integrate timestamps into data models without prior critical evaluation of their necessity, thereby negatively impacting user privacy. Therefore, we see the need to raise awareness and to promote more privacy-preserving design alternatives such as those presented in this paper.

Keywords: Privacy by design · Data minimisation · Timestamps

1 Introduction

The ongoing process of digitisation greatly affects the way people work: equipment, work environments, processes and habits. Office workers use centralised software systems to advance collaboration and the exchange of knowledge. Even field workers are in frequent interaction with software systems in their headquarter, e.g., package delivery personnel that reports its current position for

C. Pérez-Solà et al. (Eds.): DPM 2019/CBT 2019, LNCS 11737, pp. 138–155, 2019.
https://doi.org/10.1007/978-3-030-31500-9_9

tracking and scheduling. All these interactions of employees with software have the potential to create digital traces that document behaviour and habits.

At the same time, there is a demand to use such data about work processes and employees to benefit the company and deploy (human) resources more productively. This interest is commonly referred to as *people analytics* [6,30]. Major software vendors such as Microsoft are already integrating people analytics functionality into their enterprise software products to provide managers and employees with more analytics data [21].

On the other hand, employees might object to their personal data being analysed by their employer, e.g., for the fear of being discriminated or stigmatised [1]. In fact, the EU General Data Protection Regulation (GDPR) is also applicable in the context of employment [22]. It specifies in Article 25 that controllers, i.e., employers, shall implement technical measures, which are designed to implement data-protection principles, such as data minimisation. Thereby, controllers are generally obligated to use software that follows data minimisation principles.

While there has been a lot of work on applying data minimisation to software engineering [4], no particular focus has been given to minimise a particular kind of metadata: timestamps. We reason that timestamps are a ubiquitous type of metadata in application software that is both very privacy-sensitive and not yet understood regarding its potential for data minimisation.

In combination with location data, it is well understood that timestamps function as so-called quasi identifiers [29] and contribute to the linkability and de-anonymisation of data [32]. Take for instance the NYC taxi dataset, where prominent passengers could be re-identified from seemingly anonymised taxi logs by correlating pickup location and time with external knowledge such as paparazzi shots [24]. We reason that timestamps should be considered as similarly sensitive outside the context of location-based services as well, especially regarding user profiling through application software. In the latter case, the identifying potential of timestamps is not a prerequisite for them to be a privacy risk, because users are commonly already identified by other means (e.g. credentials). Instead, timestamping actions allows for a temporal dimension in user profiling, which gives deeper insights into behavioural patterns, as one does not only learn where users are going, but also when and in what intervals. For instance, recent work has shown that timestamps in edit logs of real-time collaboration services such as Google Docs are sufficiently detailed for impersonation attacks on typing-biometric authentication systems [20].

We use the term personally identifiable timestamp (short: *PII timestamp*) to describe a timestamp that can be directly linked to a person within the data model of an application. Regarding application software, timestamps are a basic data type and their potential applications manifold. However, to design more data-minimal alternatives to timestamps, an understanding about their current usage in application software is required.

To gain insights into possible uses and alternatives for timestamps in application software, we picked Mattermost as the target of evaluation for this case study. Mattermost presented itself as a suitable target because of its open source, centralised data management and popularity as a communication platform for teams and Slack alternative [19].

Our main contributions are: (i) We conduct a source code analysis to identify and describe timestamp usage in Mattermost server. (ii) We describe alternative design patterns for each type of usage.

The remainder of the paper is structured as follows: Sect. 2 provides background on our adversary model. Section 3 analyses the usage of timestamps in Mattermost. Section 4 presents alternative design patterns. Section 5 discusses related work and Sect. 6 concludes the paper.

2   Adversary Model

Our adversary model follows the established honest-but-curious (HBC) notion commonly used to assess communication protocols. Paverd, Martin and Brown [25] define an HBC adversary as a legitimate participant in a communication protocol, who will not deviate from the defined protocol but will attempt to learn all possible information from legitimately received messages. Adapted to the context of application software and performance monitoring, we consider an adversary to be an entity that is in full technical and organisational control of at least one component of a software system, e.g., the application server. The adversary will not deviate from default software behaviour and its predefined configuration options, but will attempt to learn all possible information about its users from the collected data. This especially means that an adversary will not modify software to collect more or different data, or employ additional software to do so. However, an adversary can access all data items that are collected and recorded by the software system irrespective of their exposure via GUIs or APIs. We reason that this adversary model fits real world scenarios, because employers lack the technical abilities to modify their software systems or are unwilling to do so to not endanger the stability of their infrastructure.

3   Application Analysis

We analysed Mattermost as an exemplary application to gather insights into developers' usage of timestamps. Our analysis uses Mattermost Server version 4.8 released in Nov. 2018 (current version in June 2019: 5.12). The source code has been retrieved from the project's public GitHub repository [17]. To determine which timestamps are presented to the user, we used Mattermost Web Client in version 5.5.1 [18].

The analysis is structured as follows: First, we identify timestamps in the source code, then we determine those timestamps that are relatable to users. Subsequently, we investigate their type, user-visibility, and programmatic use, before we discuss our findings.

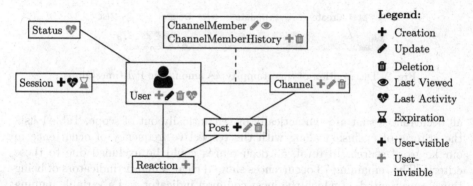

Fig. 1. Mattermost's core components with their respective timestamps, categorized according to their type and visibility.

3.1 Identification of Timestamps

Initially, we identify all timestamps that are part of Mattermost's data model code located in the dedicated directory `model`. Therein, we searched for all occurrences of the keyword `int64`, which denotes a 64-bit integer in the Go programming language. This integer type is used by Mattermost to store time values in milliseconds elapsed since January 1st, 1970. From our keyword search, we excluded all test code, which is by Go's design located in files whose filenames end in `_test.go` [9]. This initial keyword search yielded a list of 126 occurrences which not only contains timestamp-related data model declarations, but also other integer uses and occurrences of the keyword within type signatures.

Table 1. Exclusion criteria for occurrences of the keyword `int64` in Mattermost's data model source code. The top three are syntactical criteria, whereas the remaining are semantic criteria based on indicators in variable and file names.

Criterion	Description	Freq
Cast	Keyword is used to type cast a variable	12
Signature	Keyword is used within a type signature	7
Local	Keyword is used to declare a local variable	6
Counter	As indicated by the name containing `count`, `sequence` or `progress`	14
Setting	A setting as located in `config.go` or `data_retention_policy.go`	8
Identifier	Used as object identifier as indicated by the name `id`	4
Size	Used to record object sizes as indicated by the name containing `size`	2
Priority	Used as priority level as indicated by the name `priority`	1

Based on the list of 126 occurrences of the keyword `int64`, we narrowed down the candidates for timestamps in Mattermost's data model by excluding

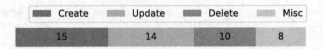

Fig. 2. Distribution of timestamp types among the PII timestamps.

all occurrences that are syntactically or semantically out of scope. Table 1 lists the criteria of exclusion along with the respective frequency of occurrence in our keyword search. In total, 53 occurrences could be excluded due to these criteria. The remaining 73 occurrences showed clear semantic indicators of being timestamp-related, of which the most common indicator was a variable naming scheme in the form a state-defining verb followed by the preposition *at*, e.g., CreateAt. This At-naming scheme occurred 64 times, followed by the naming-based indicators time and Last...Update with 7 and 2 occurrences, respectively.

3.2   Selection of PII Timestamps

As described before, we limit the scope of our analysis to PII timestamps, i.e., timestamps that mark an event which is directly or indirectly linked to a natural person. Regarding Mattermost, we consider timestamps as personal or PII, if the enclosing composition type also includes a direct or indirect reference to the user object, e.g., the creation time of a post is personal because the post object also contains a reference to the creating user.

To determine whether or not the timestamp members identified in Sect. 3.1 meet the criteria for PII timestamps, we inspected Mattermost's source code. We conclude that a timestamp is PII, if its composition type, i.e. struct, contains the User type, or any of the referenced composition types – including their references recursively – contain the User type. Out of the 73 timestamps identified in Mattermost's data model, 47 are directly or indirectly linked to a user.

3.3   Distribution of Timestamp Types

Having identified the PII timestamps, we analysed the type of these timestamps. By type, we mean the type of event that is recorded in this timestamp, e.g., the creation or deletion of an object. To conduct this analysis, we took advantage of the variable naming scheme mentioned in Sect. 3.1, which allowed us to infer the type of the timestamped event from the verb used in the variable name. For instance, we can infer from the variable name UpdateAt that this timestamp records the time when the respective object is updated. Figure 2 shows the distribution of timestamp types as inferred from their names. The most common types are create and update timestamps, that each make up almost a third of all PII timestamps. Delete timestamps are less frequent and occur only 10 times. We classified timestamps as type *create* if their name contains the word create or join, as type *update* if their name contains the word update or edit, and as type *delete* if their name contains the word `delete` or `leave`. The remaining timestamps are classified as *miscellaneous* or *misc*, and include timestamps named `LastActivityAt` (3 occurrences) and `ExpiresAt` (2).

3.4   User-Visible Timestamps

One possible use of timestamps is to inform users, e.g., about the time when a password has last been changed. To assess how many of the identified PII timestamps serve that purpose, we inspected the graphical user interface of Mattermost's Web Client in version 5.5.1 [18]. We executed a manual depth-first walk through the graphical user interface starting from the town square channel view. Alternatively, we considered using a (semi-)automatic approach of data flow analysis from the data source (REST API) to the data sink (renderer). However, we abandoned that approach as we found no way to determine all possible sinks.

During the manual GUI inspection, we clicked on and hovered over every apparent GUI element, looking for timestamps that are visible to the user. In doing so, we found the following 5 timestamps that are visible to users without special privileges:

- `Post.CreateAt`: The creation time appears next to the username above a post, or left of the post if the same user posts repeatedly without interruption. The presented time is not altered by editing the post, but remains the creation time. It is visible to all members of the respective channel.
- `Session.CreateAt` and `Session.LastActivityAt`: Both, the time of creation and the time of last activity in a session are shown in the *Active Sessions* that are reachable via the *Security* section in the account settings.
- `User.LastPasswordUpdate`: The time of the last password update is shown in the *Security* section of the account settings dialog. Each user can only see their respective timestamp.
- `User.LastPictureUpdate`: The time of the last picture update is shown in the *General* section of the account settings dialog. Each user can only see their respective timestamp.

Note that the visibility assessment only considers timestamps that are visually rendered as part of the graphical user interface and not timestamps that are readable via an API.

3.5   Programmatic Uses of Timestamps

To identify other uses of PII timestamps apart from informing users, we conducted a source code analysis of programmatic timestamp uses. In the following, we first describe the process of source code analysis which we used to locate potential programmatic uses of the identified PII timestamps. Second, we explain the process of classifying uses as programmatic. In short, we consider a use as programmatic, if the value of the timestamp has an impact on the behaviour of the application. Take for instance the usage of a post's creation timestamp that determines if a user is still allowed to edit their post.

Locating Timestamp Uses. The aim of this source code analysis is to find all uses of PII timestamps within Mattermost's server code. To locate all uses of PII timestamps, we used *gorename*, a refactoring tool that is part of the Go tools package [10]. Gorename is intended as a refactoring tool for type-safe renaming of identifiers in Go source code. We modified gorename to discover and list all occurrences of a given identifier in a type-safe manner, which allows us to automatically determine, e.g., an operation on o.UpdateAt as belonging to Session.UpdateAt and not Channel.UpdateAt solely based on the static typing of o. We used this ability to locate all occurrences of PII timestamp identifiers, e.g. User.CreateAt, within the server code base. This yielded a list of file names and line numbers referencing the found location of timestamp uses.

Table 2. Types of use of PII timestamps within Mattermost's server code. Usage types that we consider as programmatic are highlighted by a grey background.

Type of Use	Description
AutoReply	Set date of system-initiated auto-replies
Copy	Copy of object including timestamp
CurAss	Current time is assigned
Definition	Timestamp variable is defined
EditLimit	Enforce edit limit for posts
Etag	Calculate Etag for HTTP header
Expiry	Enforce the expiry of an object
Filter	Filter a sequence of objects by time
Format	Format timestamp for human readability
ImportOld	Support import of old Mattermost data
Inter	Used as intermediary in assignment of another PII timestamp
MinElapse	Ensure that a minimum amount of time has elapsed
EmailDate	Inform about post creation time in an email notification
PostNovelty	Highlight new posts
SetZero	Set timestamp to zero
Sort	Sort a sequence of objects by time
State	Track the state of an object
StateDeleted	Check if an object has been deleted
Timeout	Enforce a timeout
Valid	Validation of timestamp value

Classification of Types and Programmatic Uses. Based on the list generated by gorename, we inspected each found use of a PII timestamp identifier and conducted a bottom-up classification, by which we formed groups of uses that fulfil the same or a similar programmatic purpose. The resulting usage type classification is shown in Table 2. Following this basic classification, we assessed for each usage type whether or not it constitutes a programmatic use.

We consider a use as *programmatic* if it (a) determines the behaviour of Mattermost, and (b) is not self-referential, i.e., is used for purposes other than

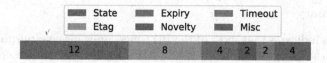

Fig. 3. Distribution of programmatic uses between the identified usage types.

maintaining timestamps. For instance, we consider `AutoReply` not as a programmatic use, because the need to derive a date for auto-reply posts only arises from having creation dates in posts in the first place (self-referential). Similarly, the assignment of the current time to (`CurAs`) and the validation (`Valid`) of timestamps are also not programmatic, because both are only necessary to prepare for other uses. On the other hand, we consider `EditLimit` a programmatic use, because it implements the policy that posts should only be editable for a certain amount of time (determines behaviour). Table 2 highlights usage types that are classified as programmatic by a grey background.

Of these 10 types of programmatic uses of PII timestamps, we found 32 instances. Figure 3 shows the frequency in which each type occurred. Note that `StateDelete` is included in `State` and is regarded as a special case of the latter from now on. The types that are summarised under miscellaneous are EditLimit, Filter, MinElapse, and Sort, each occurring once.

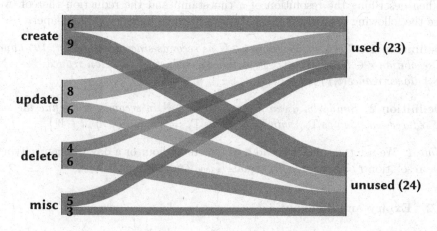

Fig. 4. Distribution of timestamps types between used and unused timestamps.

Our investigation also showed that 24 out of the 47 PII timestamps have no programmatic use at all. Figure 4 shows that the timestamp types are almost evenly distributed between the used and unused timestamps. Only 40% of create and delete timestamps are used, whereas almost 60% of update timestamps are used. The four uses of delete timestamps are all of the type `StateDeleted`, which only checks if the timestamp equals zero or not. Thus the actual time of deletion is never used programmatically.

3.6 Summary

Our analysis indicates that most of the PII timestamps have no purpose because they are neither programmatically used by the application nor presented to the user. This might suggest that developers routinely add these timestamps to a data model without reflecting their necessity. Regarding the programmatic usage of timestamps, we observe that timestamps are used to track intra-object state (ETags, State), for inter-object comparisons (PostNovelty, Sort), to measure the passage of time (EditLimit, Expiry, MinElapse, Timeout), and to allow references with an external notion of time (Filter).

4 Privacy Patterns for Timestamp Minimisation

The analysis of timestamp usage within Mattermost identified several types of timestamp usage either programmatic or informative. These types of usage are currently designed to process and present timestamps with a millisecond resolution. In the following, we will present alternative design patterns for these usage types that require less detailed timestamps or none at all.

4.1 Notation

When describing the resolution of a timestamp and the reduction thereof, we use the following definitions and notations in the remainder of this paper.

Definition 1. *Given a timestamp $t \in \mathbb{N}$ as seconds since January 1st, 1970 and a resolution $r \in \mathbb{N}$ in seconds, we define the reduction function reduce: $\mathbb{N} \to \mathbb{N}$ as follows: $reduce(t, r) := \lfloor \frac{t}{r} \rfloor r$.*

Definition 2. *Similarly, given a resolution $r \in \mathbb{N}$ in seconds, we define the set of reduced timestamps $\mathbb{T}_r$, with $\mathbb{T}_r \subseteq \mathbb{N}$, as $\mathbb{T}_r := \{ reduce(t, r) \mid t \in \mathbb{N} \}$.*

Note 1. We use the suffixes h and d to denote an hour or a day when specifying the resolution r. Therefore, $\mathbb{T}_{1h}$ equals $\mathbb{T}_{3600}$ and $\mathbb{T}_{1d}$ equals $\mathbb{T}_{24h}$.

4.2 Expiry and Timeout

Expiry or timeout mechanisms have to decide whether a given amount of time (delta) has elapsed since a reference point in time. This can be required, e.g., to check if a session has reached its maximum lifetime, or to determine whether a user's period of inactivity is long enough to set their status to absent. A naive implementation of this mechanism can either store the reference point and the delta, or the resulting point of expiry. Mattermost, for instance, uses both approaches. A more data-minimal design could reduce the resolution of the reference or expiry point, thereby recording less detailed traces of user behaviour.

4.3   Sorting

The purpose of sorting is that a sequence of objects is ordered according to a timestamp. Mattermost uses the creation timestamp of posts to present them in temporal order. Thereby, the distance between the posts' creation timestamp is irrelevant. Instead, only a timestamp's property as an ordering element is used. This functionality can also be realised by using sequence numbers that are automatically assigned to each post upon its creation.

4.4   Filtering

Mattermost allows users to filter posts by their update timestamp. A user can request Mattermost to show only posts that were last updated before, after or on a given date. Since filtering is interfacing with users, the user-provided date needs to comply with users' perception of time and allow intuitive date formats. As a consequence, sequence numbers for posts do not directly apply, because users cannot be expected to know or determine the range of sequence numbers that fits their desired date filter.

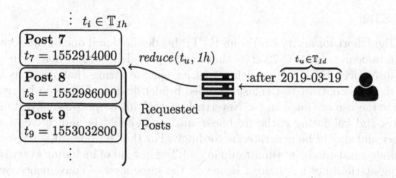

Fig. 5. Filtering can be run on timestamps with significantly reduced precision. This example illustrates filters with a one-day resolution that are applied on post timestamps with a resolution of one hour.

Instead, we propose using per-post timestamps with a reduced resolution. Figure 5 illustrates a simple filter mechanism that stores per-post timestamps with a resolution r_p of one hour. Users can specify a filter date t_u with a resolution r_u of one day. The resolution of t_u is reduced to r_p, if r_p is greater than r_u, which is necessary to ensure the discoverability of all posts within the given filter range. Note that in that case, a reduction of the resolution increases the temporal range of the given filter. As a result, a filter request might return posts that lie outside of the original filter range, thus potentially causing confusion among users, especially if $r_p \gg r_u$. We call this phenomenon *filter range extension*.

Mattermost allows to specify date filters with a resolution of one day ($r_u = 1d$). If per-post timestamps are only used for filtering, then their resolution should be equal or greater than the filter resolution ($r_p \geq r_u$). Smaller values for r_p, i.e. $r_p < r_u$, would not increase the precision of the filter mechanism, since the overall filter resolution is the minimum of r_p and r_u and thus limited by r_u. Larger values for r_p would enhance privacy protection, but cause for confusing filter range extensions on the other hand, if $r_p \gg r_u$. Therefore, $r_p = r_u$ is a sensible setting.

Note that for sequence numbers to work with filtering, a mechanism would be needed that translates human-understandable timestamp filters into sequence number filters that are comparable to sequence numbers of recorded posts. However, since posts are generally not created in equidistant time intervals, such a translation mechanism would need additional information about the distance between posts as well as at least one absolute point of reference or anchor point, to be able to map a user-given timestamp to a sequence number. Determining the respective sequence number from a given timestamp would require summing up inter-post intervals starting from the closest anchor point. Therefore, we consider sequence numbers as impractical for filtering. Also note that regarding privacy, sequence numbers in combination with anchor points provide no advantage over timestamps with reduced precision.

4.5   ETag

An ETag (short for entity-tag) is an HTTP header field and one of two forms of metadata defined in RFC 7232 [8] that can be provided to conditionally request a resource over HTTP. It is defined as an opaque string that contains arbitrary data. In contrast to the last-modified header field, an ETag can be created without the use of timestamps. Nevertheless, to fulfil its purpose of testing for updates and validating cache freshness, an ETag should be indicative of state changes and should be generated accordingly. For that purpose, it is convenient to include a last-modified timestamp in an ETag instead of including every state-defining attribute of a resource. However, the same level of convenience can be achieved by using a revision number instead of a last-modified timestamp. Such a revision counter could be added to each data model class that is requestable via HTTP. It would be incremented every time the respective object is changed. RFC 7232 [8] itself mentions in section 2.3.1 revision numbers as a way to implement ETags, alongside collision-resistant hashes of representative content and the critiqued modification timestamp.

4.6   Novelty Detection

Mattermost uses timestamps to detect and highlight unread posts. Therefore, Mattermost records the last time a user has viewed a channel. Upon revisiting a channel, Mattermost can then simply identify unread posts by comparing their creation timestamp to the channel's last visitation timestamp.

As an alternative, a sequence number could be assigned to each post. Then it would suffice to record, for each channel and user, the sequence number of the most recent post at the time of a user's last visitation. Following this design, unread posts can be identified as those posts whose sequence number exceeds the recorded sequence number of the last read post.

4.7 State Management

Mattermost uses timestamps to keep track of objects' state. An unset creation timestamp signifies that an object is not yet fully initialised and an unset deletion timestamp signifies that an object is still active and not yet deleted. However, there is no advantage in using timestamps to record the state of an object, besides saving the amount of storage that would be needed to use a boolean or integer variable in addition to a timestamp. If state management is the only purpose of a timestamp, it can be easily replaced by a boolean or state enumeration variable. In the case of Mattermost, we found that all deletion timestamps and 3 creation timestamps are only used for the purpose of managing state.

4.8 User Information

Mattermost presents some timestamps to the user in its graphical user interface (see Fig. 1). Only one of them, namely the post creation timestamp is visible to all users, whereas the other timestamps are only visible for the currently logged-in user.

The potential for minimising user-visible timestamps depends on the frequency of the timestamped event. Take, for example, the creation timestamps of posts: The higher the posting frequency the shorter the interval between posts and the more detail is required for timestamps to be distinctive. Another aspect that influences the potential for minimising user-facing timestamps is the locality of distinctiveness, i.e., the question whether users rather use timestamp information to distinguish posts in close temporal proximity to each other, or use it to gain a coarser temporal orientation.

Consider a timestamp resolution of 15 min: Posts that are created within a 15 min period would all be presented with the same one or two timestamps, potentially creating a false and confusing impression of immediate succession. Therefore, a user-facing reduction of timestamps needs to be obvious to avoid misinterpretation. This can be achieved by annotating such timestamps accordingly, e.g., by explicitly displaying an interval like *14:30–14:45*.

Besides reduction, user-visible PII timestamps can also be protected by encrypting them in a way that only authorised users can decrypt them. In case of timestamps that only concern a single user, this can be realised with a common asymmetric cryptographic system, where the secret is protected by the user's password. In case of timestamps that should be readable by multiple users, e.g., post creation timestamps, a more complex cryptographic setup is required that also has to handle churn among authorised users.

4.9 Compliance

Another reason to record timestamps that is not directly reflected in our analysis might be compliance, e.g., documentation obligations or judicial orders. In such cases, the potential for minimising PII timestamps is limited and depends on specific regulations. However, the impact of recording PII timestamps on user privacy can at least be reduced by restricting access and limiting storage periods. We suggest to store such compliance timestamps separately from production data and encrypt them using a threshold scheme [5] in order to separate the duty of decryption among multiple parties for the sake of preventing misuse.

Table 3. Overview of timestamp usage and the respectively suited design alternatives.

	EditLimit	Etag	Expiry	Filter	MinElapse	Novelty	Sort	State	Timeout	User Info
Sequence number						●	●			
Revision number		●								
Reduction	●		●	●	●				●	●
Encryption										●
Enumeration								●		

4.10   Summary and Discussion

Based on the identified timestamp usage, we were able to present more privacy-preserving alternatives to using full-resolution timestamps. The presented alternative design patterns are constructed of five technical primitives: sequence numbers, revision numbers, reduction of precision, encryption, and enumerations. Table 3 gives an overview of the design alternatives and illustrates which of the five primitives are applicable to which timestamp usage.

We find that four of the identified timestamp usage types, namely Etag, Novelty, Sort, and State, can be replaced by sequence numbers, revision numbers, and enumerations. These four usage types together make up more than half of the total number of PII timestamp usage in Mattermost (see Fig. 3). For the remaining usage types, sequence numbers are not an option because (a) values need to be comparable to a user-provided date (Filter), (b) values need to be human readable (User Information), or (c) values need to be comparable to an absolute point in time (Expiry, MinElapse, Timeout). In those cases, the privacy impact of recording timestamps can be reduced by reducing their resolution or by encrypting them.

Discussion. Article 5(1)(c) GDPR states that the extend to which the processing of personal data shall be limited is determined by the purposes for which they are processes. Hence, the principle of data minimisation does not demand absolute minimisation but minimisation relative to a given purpose. In that sense, the goal of our alternative designs and minimisation patterns in general is not to eliminate all timestamps from application software, but to replace them with less rich information wherever the latter suffices to fulfil the same purpose.

It should also be noted that the minimisation of personal data can lead to the discrimination of subjects that would not be discriminated otherwise. Take for instance a common data minimisation scheme where postal codes are collected instead of full addresses to determine service coverage. In doing so, a subject might be excluded from a service although they live very close to the serviced area. Hence, such potential negative effects should be taken into account when designing and applying data minimisation patterns.

5 Related Work

To the best of our knowledge, we are the first to investigate privacy patterns for the minimisation of timestamps based on their usage in application software. However, there is a rich body of work regarding adjacent topics. In the following, we will present privacy research that focuses on mitigating the privacy impact of timestamps, work regarding the monitoring of employees' performance, and work about the incorporation of data minimisation principles into engineering.

5.1 Timestamp Privacy

Basic redaction techniques for timestamps such as reduction and replacement with order-preserving counters have been presented by [28,34] in the context of log anonymisation. Kerschbaum [15] introduces a technique to pseudonymise timestamps in audit logs as part of a multi-party exchange of threat intelligence data. The introduced technique preserves the distance of the pseudonymised timestamps by using a grid representation. However, the distance calculation requires a third party which is generally not available for our application.

In wireless sensor networks, the term *temporal privacy* describes the effort to prevent a passive network observer from inferring the creation time of an event from the observation time of a related message or signal [12]. Countermeasures in that area include buffering to break the temporal correlation of creation and observation [12] and temporal perturbation by adding Laplace noise [33].

Since timestamps can be easily encoded as integers, building blocks from the area of privacy-preserving record linkage regarding numerical values can be applied to compare timestamps in a privacy-preserving manner [13]. However, this approach also requires a trusted third party other than the data-custodians to achieve its privacy guarantees.

5.2 Performance Monitoring and People Analytics

There are several approaches to mine (meta) data to gain insight into work processes and employees. A large body of work uses software developers' commit metadata that is publicly available on GitHub. Eyolfson, Tan and Lam [7] show that late-night contributions of software developers are statistically more buggy than contributions in the morning or during the day. Claes et al. [2] investigate working hours of developers to gain insight into work patterns that can foster stress and overload detection. Others use this data to infer developers' personality traits like neuroticism [26] or to characterise them more generally [23]. Note that these analyses have been conducted with software contribution metadata.

However, they can be applied to transactional metadata in general, including interaction data from Mattermost.

People analytics (PA) promises to optimise business processes and human resource management by gathering and analysing data about how employees work [6]. While PA is a trending topic, empirical evidence for its benefits or even concrete metrics are scarce [30]. Part of PA is the understanding of relationships and collaboration dynamics among employees, including their communication. Interaction graphs can be built based on metadata from collaboration software, which then can be analysed using established graph and network algorithms like community detection [31]. Insights from such algorithms might be used to optimise team compositions or to identify candidates for management positions.

An automated and algorithmic assessment of employees also raises legal and moral concerns. Bornstein [1] highlights the conflicts of such algorithmic assessments with anti-discrimination and anti-stereotyping regulation. Regarding data protection regulations, the GDPR also protects the personal data of employees and restricts employers' rights to analyse data [22], especially regarding automatic decision making [27].

5.3   Privacy Engineering

One of the privacy design strategies postulated by Hoepman [11] is *minimise*. The strategies are meant to guide software architects to achieve privacy by design with their software designs. The *minimise* strategy demands that only as much data is collected and processed as is appropriate and proportional to fulfil a given purpose. Whereas timestamps are certainly appropriately used in Mattermost to fulfil the purposes that we identified in Sect. 3, the more privacy-preserving design alternatives presented in Sect. 4 indicate that using full-resolution timestamps is not proportional for most purposes.

Privacy design patterns are a way to formulate actionable best practices aiming to achieve privacy by design [16]. These patterns focus on concrete and recurring software engineering decisions. While there are several patterns that detail the aforementioned *minimise* strategy, e.g., the location granularity pattern or the strip-unneeded-metadata pattern [3,14], there are – to the best of our knowledge – no patterns that focus especially on the minimisation of timestamp usage and the replacement of timestamps by less detailed alternatives.

6   Conclusion

In this case study, we analysed Mattermost as an exemplary application, to gain insight into the usage of timestamps. We found that Mattermost's data model includes 47 timestamps that are directly or indirectly linked to actions of a user. More than half of these timestamps have no programmatic use within the application and only 5 timestamps are visible to the user.

We assume that Mattermost is not a special case, but that the use of PII timestamps is commonly excessive and disproportionate. We further assume that this is no result of ill intent but a result of unconscious programming habits and a lack of awareness for privacy anti-patterns.

To find substitutes for full-resolution timestamps, we investigated the potential for data minimisation relative to the identified purpose of usage. Based on that, we presented alternative design patterns that require less precise or no timestamp information. Following these alternatives, more than half of Mattermost's timestamp usage instances can be replaced by easy-to-implement sequence or revision numbers, whereas the resolution of the remaining timestamps can at least be significantly reduced.

We suggest that future work should investigate software developers for unconscious programming habits of adding unnecessary metadata such as timestamps, and find ways to raise awareness. To further design and provide practical alternatives for user-facing timestamps, a user study about the perception of various timestamp resolutions would provide valuable information for sensible defaults.

Acknowledgements. The work is supported by the German Federal Ministry of Education and Research (BMBF) as part of the project Employee Privacy in Development and Operations (EMPRI-DEVOPS) under grant 16KIS0922K.

References

1. Bornstein, S.: Antidiscriminatory algorithms. Alabama Law Rev. **70**(2), 519 (2018)
2. Claes, M. et al.: Do programmers work at night or during the weekend? In: ICSE, pp. 705–715. ACM (2018)
3. Colesky, M. et al.: privacypatterns.org, (2019). https://privacypatterns.org. Accessed on 29 Mar 2019
4. Danezis, G. et al.: Privacy and Data Protection by Design - from policy to engineering. CoRR abs/1501.03726 (2015)
5. Desmedt, Y., Frankel, Y.: Threshold cryptosystems. In: Brassard, G. (ed.) CRYPTO 1989. LNCS, vol. 435, pp. 307–315. Springer, New York (1990). https://doi.org/10.1007/0-387-34805-0_28
6. DiClaudio, M.: People analytics and the rise of HR: how data, analytics and emerging technology can transform human resources (HR) into a profit center. Strateg. HR Rev. **18**(2), 42–46 (2019)
7. Eyolfson, J., Tan, L., Lam, P.: Do time of day and developer experience affect commit bugginess. In: Proceedings of the 8th International Working Conference on Mining Software Repositories, MSR 2011 (Co-located with ICSE), pp. 153–162. ACM (2011)
8. Fielding, R.T., Reschke, J.: Hypertext Transfer Protocol (HTTP/1.1): Conditional Requests. RFC 7232 (2014)
9. Google Inc: Go testing package, (2019). https://golang.org/pkg/testing/. Accessed on 1 Mar 2019
10. Google Inc: Go Tools gorename command (2019). https://godoc.org/golang.org/x/tools/cmd/gorename. Accessed on 4 Mar 2019
11. Hoepman, J.-H.: Privacy design strategies. In: Cuppens-Boulahia, N., Cuppens, F., Jajodia, S., Abou El Kalam, A., Sans, T. (eds.) SEC 2014. IAICT, vol. 428, pp. 446–459. Springer, Heidelberg (2014). https://doi.org/10.1007/978-3-642-55415-5_38

12. Kamat, P., et al.: Temporal privacy in wireless sensor networks. In: 27th IEEE International Conference on Distributed Computing Systems (ICDCS 2007), June 25–29, 2007, Toronto, Ontario, Canada, p. 23. IEEE Computer Society (2007)

13. Karapiperis, D., Gkoulalas-Divanis, A., Verykios, V.S.: FEDERAL: a framework for distance-aware privacy-preserving record linkage. IEEE Trans. Knowl. Data Eng. **30**(2), 292–304 (2018)

14. Kargl, F., et al.: privacypatterns.eu (2019). https://privacypatterns.eu. Accessed on 29 Mar 2019

15. Kerschbaum, F.: Distance-preserving pseudonymization for timestamps and spatial data. In: Proceedings of the 2007 ACM Workshop on Privacy in the Electronic Society, WPES 2007, Alexandria, VA, USA, October 29, 2007, pp. 68–71. ACM (2007)

16. Lenhard, J., Fritsch, L., Herold, S.: A literature study on privacy patterns research. In: 43rd Euromicro Conference on Software Engineering and Advanced Applications, SEAA 2017, Vienna, Austria, August 30 – September 1, 2017, pp. 194–201. IEEE Computer Society (2017)

17. Mattermost Inc: Mattermost Server v4.8.0, (2018). https://github.com/mattermost/mattermost-server/releases/tag/v4.8.0

18. Mattermost Inc: Mattermost Webapp v5.5.1, (2018). https://github.com/mattermost/mattermost-webapp/releases/tag/v5.5.1

19. Mattermost Inc: Mattermost Website. https://www.mattermost.org. Accessed on 30 Mar 2019

20. McCulley, S., Roussev, V.: Latent typing biometrics in online collaboration services. In: Proceedings of the 34th Annual Computer Security Applications Conference, ACSAC 2018, San Juan, PR, USA, December 03–07, 2018, pp. 66–76. ACM (2018)

21. Microsoft: Workplace Analytics. https://products.office.com/en-us/business/workplace-analytics. Accessed on 30 Mar 2019

22. Ogriseg, C.: GDPR and personal data protection in the employment context. Labour Law Issues **3**(2), 1–24 (2017)

23. Onoue, S., Hata, H., Matsumoto, K.: A study of the characteristics of developers' activities in GitHub. In: 2013 20th Asia-Pacific Software Engineering Conference (APSEC), pp. 7–12 (2013)

24. Pandurangan, V.: On taxis and rainbows. lessons from NYC's improperly anonymized taxi logs (2014). https://tech.vijayp.ca/of-taxis-and-rainbows-f6bc289679a1. Accessed on 30 Mar 2019

25. Paverd, A., Martin, A., Brown, I.: Modelling and automatically analysing privacy properties for honest-but-curious adversaries. Technical report (2014)

26. Rastogi, A., Nagappan, N.: On the personality traits of GitHub contributors. In: 27th IEEE International Symposium on Software Reliability Engineering, ISSRE 2016, Ottawa, ON, Canada, October 23–27, 2016, pp. 77–86. IEEE Computer Society (2016)

27. Roig, A.: Safeguards for the right not to be subject to a decision based solely on automated processing (Article 22 GDPR). Eur. J. Law Technol. **8**(3) (2017)

28. Slagell, A.J., Lakkaraju, K., Luo, K.: FLAIM: a multi-level anonymization framework for computer and network logs. In: Proceedings of the 20th Conference on Systems Administration (LISA 2006), Washington, DC, USA, December 3–8, 2006, pp. 63–77. USENIX (2006)

29. Sweeney, L.: k-Anonymity: a model for protecting privacy. Int. J. Uncertainty, Fuzziness Knowl.-Based Syst. **10**(5), 557–570 (2002)

30. Tursunbayeva, A., Lauro, S.D., Pagliari, C.: People analytics - a scoping review of conceptual boundaries and value propositions. Int. J. Inf. Manag. **43**, 224–247 (2018)
31. Wang, N., Katsamakas, E.: A network data science approach to people analytics. Inf. Resour. Manag. J. **32**(2), 28–51 (2019)
32. Wernke, M., et al.: A classification of location privacy attacks and approaches. Personal Ubiquit. Comput. **18**(1), 163–175 (2014)
33. Yang, X., et al.: A novel temporal perturbation based privacy-preserving scheme for real-time monitoring systems. Comput. Netw. **88**, 72–88 (2015)
34. Zhang, J., Borisov, N., Yurcik, W.: Outsourcing security analysis with anonymized logs. In: Second International Conference on Security and Privacy in Communication Networks and the Workshops, SecureComm 2006, Baltimore, MD, USA, August 2, 2006 - September 1, 2006, pp. 1–9. IEEE (2006)

Card-Based Cryptographic Protocols with the Minimum Number of Rounds Using Private Operations

Hibiki Ono and Yoshifumi Manabe[✉][iD]

Kogakuin University, Shinjuku, Tokyo 163–8677, Japan
manabe@cc.kogakuin.ac.jp

Abstract. This paper shows new card-based cryptographic protocols with the minimum number of rounds using private operations under the semi-honest model. Physical cards are used in card-based cryptographic protocols instead of computers. Operations that a player executes in a place where the other players cannot see are called private operations. Using three private operations called private random bisection cuts, private reverse cuts, and private reveals, calculations of two variable boolean functions and copy operations were realized with the minimum number of cards. Though the number of cards has been discussed, the efficiency of these protocols has not been discussed. This paper defines the number of rounds to evaluate the efficiency of the protocols using private operations. Most of the meaningful calculations using private operations need at least two rounds. This paper shows a new two-round committed-input, committed-output logical XOR protocol using four cards. Then we show new two-round committed-input, committed-output logical AND and copy protocols using six cards. This paper then shows the relationship between the number of rounds and available private operations. Even if private reveal operations are not used, logical XOR can be executed with the minimum number of cards in two rounds. On the other hand, logical AND and copy operations can be executed with the minimum number of cards in three rounds without private reveal operations. Protocols that preserves an input are also shown.

Keywords: Multi-party secure computation ·
Card-based cryptographic protocols · Private operations ·
Logical computations · Copy · Round

1 Introduction

1.1 Motivation

Card-based cryptographic protocols [12,24] were proposed in which physical cards are used instead of computers to securely calculate values. They can be used when computers cannot be used. den Boer [2] first showed a five-card protocol to securely calculate logical AND of two inputs. Since then, many protocols

© Springer Nature Switzerland AG 2019
C. Pérez-Solà et al. (Eds.): DPM 2019/CBT 2019, LNCS 11737, pp. 156–173, 2019.
https://doi.org/10.1007/978-3-030-31500-9_10

have been proposed to calculate logical functions [13,25] and specific computations such as millionaires' problem [17,27,33], voting [21,26], random permutation [7,9,10], grouping [8], matching [16], proof of knowledge of a puzzle solution [3,5,36], and so on. This paper considers calculations of logical functions and the copy operation under the semi-honest model.

There are several types of protocols regards to the inputs and outputs of the computations. The first type is committed inputs [19], where the inputs are given as committed values. The players do not know the input values. The other type is non-committed inputs [15,40], where players give their private inputs to the protocol using private input operations. The private input operations were also used in millionaires' problem [33]. Protocols with committed inputs are desirable, since they can be used for non-committed inputs: each player can give his private input value as a committed value.

Some protocols output their computation results as committed values [19]. The result is unknown to the players unless the players open the output cards. The other type of protocols [2] output the result as a non-committed value, that is, the final result is obtained only by opening cards. Protocols with committed outputs are desirable since the committed output result can be used as an input to another computation. If further calculations are unnecessary, the players just open the committed outputs and obtain the result. Thus, this paper discusses protocols with committed inputs and committed outputs.

An example of a calculation with committed inputs is a matching service between men and women. The matching service provider does not allow direct communication between clients until the matching is over. A client Anne receives information about a candidate Bruce from her agent Alice. Anne sends the reply of acceptance/rejection to Alice, but Anne does not want the matching service provider agents to know the reply. Bruce also receives information about Anne from his agent Bob. Bruce sends the reply of acceptance/rejection to Bob, but Bruce does not want the matching service provider agents to know the reply. Alice and Bob must calculate whether the matching is successful or not without knowing the inputs. In this case, a calculation with committed inputs is necessary. To prevent malicious activities by the players, Anne observes all the actions executed by Alice. Bruce observes all the actions executed by Bob. If a player executes some action that is not allowed, the observing person can point out the misbehavior. Thus Alice and Bob become semi-honest players. Note that Anne(Bruce) cannot observe Bob's(Alice's) actions. If a person observes both players' actions, the person can know a secret value.

Operations that a player executes in a place where the other players cannot see are called private operations. These operations are considered to be executed under the table or in the back so that the operations cannot be seen by the other players. Private operations are shown to be the most powerful primitives in card-based cryptographic protocols. They were first introduced to solve millionaires' problem [27] and voting [26]. Using private operations, committed-input and committed-output logical AND, logical XOR, and copy protocols can be achieved with the minimum number of cards [34]. Thus this paper considers protocols using private operations.

Table 1. Comparison of XOR protocols using private operations.

Article	# of rounds	# of cards	Preserving an input	Private reveal
[34]	3	4	No	Use
[34]	3	4	Yes	Use
[25]	2	4	No	Does not use
This paper Sect. 3	2	4	No	Use
This paper Sect. 4	2	4	No	Does not use
This paper Sect. 5	3	4	Yes	Use/Does not use

Table 2. Comparison of AND protocols using private operations.

Article	# of rounds	# of cards	Preserving an input	Private reveal
[34]	3	4	No	Use
[34]	3	6	Yes	Use
[34]	5	4	Yes	Use
[25]	2	6	No	Does not use
This paper Sect. 3	2	6	No	Use
This paper Sect. 4	3	4	No	Does not use
This paper Sect. 5	3	6	Yes	Use/Does not use
This paper Sect. 5	5	4	Yes	Does not use

The number of cards is the space complexity of the card-based protocols. Thus the time complexity must also be evaluated. Some works have been done for the protocols that do not use private operations [18]. As for the protocols using private operations, the number of rounds, defined in Sect. 2, is the most appropriate criterion to evaluate the time complexity. Roughly speaking, the number of rounds counts the number of handing cards between players. Since each private operation is relatively simple, handing cards between players and setting up so that the cards are not seen by the other players is the dominating time to execute private operations. Thus this paper discusses the number of rounds of card-based protocols using private operations.

This paper shows logical AND, logical XOR, and copy protocols with the minimum number of rounds. The summary of results are shown in Tables 1, 2, and 3. Note that the protocols in [25] needs one shuffle by each player, thus the actual execution time is larger than the ones in this paper though the number of rounds is the same. This paper then shows variations of the protocols that use a different set of private operations. In usual logical AND protocols, the input bits are lost. If one of the inputs is not lost, the input bit can be used for further computations. Such a protocol is called a protocol that preserves an input [29]. This paper shows the number of rounds of protocols that preserves an input.

In Sect. 2, basic notations, the private operations introduced in [34], and the definition of rounds are shown. Section 3 shows two round XOR, AND, and copy protocols. Section 4 shows the protocols that use a different set of private operations. Section 5 shows protocols that preserve an input. Section 6 concludes the paper.

Table 3. Comparison of COPY protocols ($m = 2$) using private operations.

Article	# of rounds	# of cards	Private reveal
[34]	3	4	Use
[25]	2	6	Does not use
This paper Sect. 3	2	6	Use
This paper Sect. 4	2	6	Does not use

1.2 Related Works

Many works have been done for calculating logical functions without private operations. den Boer [2] first showed a five-card protocol to securely calculate logical AND of two inputs. Since then, several protocols to calculate logical AND of two committed inputs have been shown [4,28,41], but they use more than six cards. Mizuki et al. [25] showed a logical AND protocol that uses six cards. It is proved that it is impossible to calculate logical AND with less than six cards when we use closed and uniform shuffles [11]. When it is allowed to use a special kind of shuffle that is not closed or uniform, the minimum number of cards of logical AND protocols is decreased to five [1,12,14,32,35]. Also, when Las Vegas protocols are allowed, logical AND protocols with five or four cards were shown [12,22,35].

For making a copy of input bit, Mizuki et al. showed a protocol with six cards [25]. A five-card protocol was shown that uses non-uniform shuffles [31].

Mizuki et al. [25] showed a logical XOR protocol that uses four cards, which is the minimum.

Several other protocols such as computations of many inputs [19,30,38], computing any boolean functions [13,29,39], two-bit output functions [6] were shown. Protocols using other types of cards were also shown [20,23,37].

2 Preliminaries

2.1 Basic Notations

This section gives the notations and basic definitions of card-based protocols. This paper is based on a two-color card model. In the two-color card model, there are two kinds of marks, ♣ and ♥ . Cards of the same marks cannot be distinguished. In addition, the back of both types of cards is ? . It is impossible to determine the mark in the back of a given card of ? .

One bit data is represented by two cards as follows: ♣♥ $= 0$ and ♥♣ $= 1$.

One pair of cards that represents one bit $x \in \{0,1\}$, whose face is down, is called a commitment of x, and denoted as $commit(x)$. It is written as $\underbrace{?\ ?}_{x}$.

Note that when these two cards are swapped, $commit(\bar{x})$ can be obtained. Thus, logical negation can be calculated without private operations.

A set of cards placed in a row is called a sequence of cards. A sequence of cards S whose length is n is denoted as $S = s_1, s_2, \ldots, s_n$, where s_i is i-th card of the sequence. $S = \underbrace{\boxed{?}\ \boxed{?}\ \boxed{?} \cdots, \boxed{?}}_{s_1\ \ s_2\ \ s_3 \qquad s_n}$. A sequence whose length is even is called an even sequence. $S_1 \| S_2$ is a concatenation of sequence S_1 and S_2.

All protocols are executed by multiple players. Throughout this paper, all players are semi-honest, that is, they obey the rule of the protocols, but try to obtain information x of $commit(x)$. There is no collusion among players executing one protocol together. No player wants any other player to obtain information on committed values.

2.2 Private Operations

We show three private operations introduced in [34]: private random bisection cuts, private reverse cuts, and private reveals.

Primitive 1 *(Private random bisection cut). A private random bisection cut is the following operation on an even sequence $S_0 = s_1, s_2, \ldots, s_{2m}$. A player selects a random bit $b \in \{0, 1\}$ and outputs*

$$S_1 = \begin{cases} S_0 & \text{if } b = 0 \\ s_{m+1}, s_{m+2}, \ldots, s_{2m}, s_1, s_2, \ldots, s_m & \text{if } b = 1 \end{cases}$$

The player executes this operation in a place where the other players cannot see. The player must not disclose the bit b.

Note that if the private random cut is executed when $m = 1$ and $S_0 = commit(x)$, given $S_0 = \underbrace{\boxed{?}\boxed{?}}_{x}$, the player's output $S_1 = \underbrace{\boxed{?}\boxed{?}}_{x \oplus b}$, which is $\underbrace{\boxed{?}\boxed{?}}_{x}$

or $\underbrace{\boxed{?}\boxed{?}}_{\bar{x}}$.

Primitive 2 *(Private reverse cut, Private reverse selection). A private reverse cut is the following operation on an even sequence $S_2 = s_1, s_2, \ldots, s_{2m}$ and a bit $b \in \{0, 1\}$. A player outputs*

$$S_3 = \begin{cases} S_2 & \text{if } b = 0 \\ s_{m+1}, s_{m+2}, \ldots, s_{2m}, s_1, s_2, \ldots, s_m & \text{if } b = 1 \end{cases}$$

The player executes this operation in a place where the other players cannot see. The player must not disclose b.

Note that the bit b is not newly selected by the player. This is the difference between the primitive in Definition 1, where a random bit must be newly selected by the player.

Note that in many protocols below, selecting left m cards is executed after a private reverse cut. The sequence of these two operations is called a private reverse selection. A private reverse selection is the following procedure on a even sequence $S_2 = s_1, s_2, \ldots, s_{2m}$ and a bit $b \in \{0, 1\}$. A player outputs

$$S_3 = \begin{cases} s_1, s_2, \ldots, s_m & \text{if } b = 0 \\ s_{m+1}, s_{m+2}, \ldots, s_{2m} & \text{if } b = 1 \end{cases}$$

Primitive 3 *(Private reveal). A player privately opens a given committed bit. The player must not disclose the obtained value.*

Using the obtained value, the player privately sets a sequence of cards.

Consider the case when Alice executes a private random bisection cut on $commit(x)$ and Bob executes a private reveal on the bit. Since the committed bit is randomized by the bit b selected by Alice, the opened bit is $x \oplus b$. Even if Bob privately opens the cards, Bob obtains no information about x if b is randomly selected and not disclosed by Alice. Bob must not disclose the obtained value. If Bob discloses the obtained value to Alice, Alice knows the value of the committed bit.

2.3   Definition of Round

The space complexity of card-based protocols is evaluated by the number of cards. We define the number of rounds as a criterion to evaluate the time complexity of card-based protocols using private operations. The first round begins from the initial state. The first round is (possibly parallel) local executions by each player using the cards initially given to each player. It ends at the instant when no further local execution is possible without receiving cards from another player. The local executions in each round include sending cards to some other players but do not include receiving cards. The result of every private execution is known to the player. For example, shuffling whose result is unknown to the player himself is not executed. Since the private operations are executed in a place where the other players cannot see, it is hard to force the player to execute such operations whose result is unknown to the player. The $i(>1)$-th round begins with receiving all the cards sent during the $(i-1)$-th round. Each player executes local executions using the received cards and the cards left to the player at the end of the $(i-1)$-th round. Each player executes local executions until no further local execution is possible without receiving cards from another player. The number of rounds of a protocol is the maximum number of rounds necessary to output the result among all possible inputs and random values.

Let us show an example of a protocol execution and the number of rounds.

Protocol 1 *(AND protocol in [34]).*

1. *Alice executes a private random bisection cut on $commit(x)$. Let the output be $commit(x')$. Alice hands $commit(x')$ and $commit(y)$ to Bob.*

2. *Bob executes a private reveal on commit(x'). Bob sets*

$$S_2 = \begin{cases} commit(y)\|commit(0) \ if \ x' = 1 \\ commit(0)\|commit(y) \ if \ x' = 0 \end{cases}$$

and hands S_2 to Alice.
3. *Alice executes a private reverse selection on S_2 using the bit b generated in the private random bisection cut. Let the obtained sequence be S_3. Alice outputs S_3.*

The first round ends at the instant when Alice sends $commit(x')$ and $commit(y)$ to Bob. The second round begins at receiving the cards by Bob. The second round ends at the instant when Bob sends S_2 to Alice. The third round begins at receiving the cards by Alice. The number of rounds of this protocol is three.

Since each operation is relatively simple, the dominating time to execute protocols with private operations is the time to handing cards between players and setting up so that the cards are not seen by the other players. Thus the number of rounds is the criterion to evaluate the time complexity of card-based protocols with private operations.

The minimum number of rounds of most protocols is two. Suppose that the number of rounds is one. Suppose that a player, say Alice, has some (or all) of the final outputs of the protocol. Since the number of rounds is one, handing cards is not executed. Thus all the operations to obtain Alice's outputs are executed by Alice. Thus Alice knows the relation between the committed inputs and Alice's outputs. If the output cards are faced up to know the results, Alice knows the private input values. Therefore most protocols need at least two rounds for the privacy of committed inputs.

2.4   Our Results

The protocols in [34] are three rounds and use four cards. This paper shows a two-round logical XOR protocol using four cards. Then we show two-round logical AND and copy protocols using six cards. Though the number of cards is increased, the number of rounds is the minimum. Another advantage of these two-round protocols is that each player does not need to remember the random bit. In the protocols in [34], a player needs to remember the random bit until the player receives the cards again in order to execute a private reverse cut. If a player replies late, the other player must remember the random bit for a very long time. If a player executes many instances of the protocols with many players in parallel, it is hard for the player to remember so many random values. In the two-round protocols, the first player can exit from the protocol after he hands the cards to the other player. Note that Alice obtains the final result by the three-round protocols in [34] but Bob obtains the final result by the two-round protocols in this paper. These protocols can be used only if this change is acceptable by both players. Note that two-round protocols with four card logical XOR and six card logical AND with private operations have been implicitly shown by [25], this paper shows another type of protocols with fewer shuffles.

The above two-round protocols do not use private reverse cuts. Thus, there is a question of whether we can obtain protocols without another type of private operations. This paper answers this question also. We show protocols that do not use private reveals. There is a worry in using this primitive. A player might make a mistake to open cards that are not allowed and obtain private values. If private reveals are not executed at all, protections such as putting each card in an envelope can be done to prove that opening cards are not executed during private operations. Thus, it would be better if all reveals are publicly executed. Even if we do not use private reveals, the number of rounds is unchanged for logical XOR and copy protocols. The number of rounds is three for the logical AND protocol without private reveals. Last, we show protocols that preserve an input.

3   XOR, AND, and Copy with the Minimum Number of Rounds

This section shows our new two-round protocols for XOR, AND, and copy.

The players are assumed to be semi-honest, that is, they honestly execute the protocol but they try to obtain secret values. The protocol is secure if the players obtain no information about input values and output values. There is no collusion between the two players executing one protocol together. The performance of protocols is evaluated by the number of cards and the number of rounds.

These protocols do not use private reverse cuts. Thus, the first player, Alice, does not need to remember the random bit b after she hands the cards to the other player.

3.1   XOR Protocol

Protocol 2 *(XOR protocol with the minimum number of rounds).*

1. *Alice executes a private random bisection cut on input $S_0 = commit(x)$ and $S_0' = commit(y)$ using the same random bit b. Let the output be $S_1 = commit(x')$ and $S_1' = commit(y')$, respectively. Note that $x' = x \oplus b$ and $y' = y \oplus b$. Alice hands S_1 and S_1' to Bob.*
2. *Bob executes a private reveal on $S_1 = commit(x')$. Bob privately sets output*

$$S_2 = \begin{cases} commit(\bar{y}') \text{ if } x' = 1 \\ commit(y') \text{ if } x' = 0 \end{cases}$$

Note that $commit(\bar{y}')$ can be obtained by swapping the two cards of $S_1' = commit(y')$.

The protocol is two rounds.

Theorem 1. *The XOR protocol is correct and secure. It uses the minimum number of cards.*

Proof. Correctness: Alice hands $commit(x \oplus b)$ and $commit(y \oplus b)$ to Bob. Bob swaps the pair of $commit(y \oplus b)$ if $x \oplus b = 1$. Thus the output S_2 is $(y \oplus b) \oplus (x \oplus b) = x \oplus y$. Therefore, the output is correct.

Alice and Bob's security: Alice sees no open cards. Thus Alice obtains no information. Bob sees $x \oplus b$. Since b is a random value that Bob does not know, Bob obtains no information about x.

The number of cards: At least four cards are necessary for any protocol to input x and y. This protocol uses no additional cards other than the input cards. $\square$

Note that though the same bit b is used to randomize x and y, it is not a security problem because $y \oplus b$ is not opened.

The number of rounds is the minimum. Mizuki et al. showed a four-card protocol with one public shuffle [25]. Since one public shuffle can be changed to two private shuffles by each player, the minimum number of rounds is achieved also by their protocol. However, the protocol needs two shuffles, thus our new protocol is simple. Comparison of committed-input, committed-output XOR protocols using private operations are shown in Table 1.

3.2 And Protocol

Protocol 3 *(AND protocol with the minimum number of rounds).*

1. *Alice executes a private random bisection cut on $S_0 = commit(x)$ and $S_0' = commit(0) \| commit(y)$ using the same random bit b. Two new cards are used to set $commit(0)$. Let the output be $S_1 = commit(x')$ and S_1', respectively. Note that*

$$S_1' = \begin{cases} commit(y) \| commit(0) \text{ if } b = 1 \\ commit(0) \| commit(y) \text{ if } b = 0 \end{cases}$$

   Alice hands S_1 and S_1' to Bob.
2. *Bob executes a private reveal on S_1. Bob executes a private reverse selection on S_1' using x'. Let the selected cards be S_2. Bob outputs S_2 as the result.*

The protocol is two rounds. The protocol uses six cards since two new cards are used to set $commit(0)$.

Theorem 2. *The AND protocol is correct and secure.*

Proof. Correctness: The desired output can be represented as follows.

$$x \wedge y = \begin{cases} y \text{ if } x = 1 \\ 0 \text{ if } x = 0 \end{cases}$$

Bob outputs $commit(y)$ as S_2 when $(x', b) = (0, 1)$ or $(1, 0)$. Since $x' = x \oplus b$, these cases equal to $x = 1$. Bob outputs $commit(0)$ as S_2 when $(x', b) = (0, 0)$ or $(1, 1)$. Since $x' = x \oplus b$, these cases equal to $x = 0$. Thus, the output is correct.

Alice and Bob's security: The same as the XOR protocol. $\square$

The number of rounds is the minimum. Mizuki et al. showed a six-card protocol with one public shuffle [25]. Since one public shuffle can be changed to two private shuffles by each player, the minimum number of rounds is achieved also by their protocol. However, the protocol needs two shuffles, thus our new protocol is simple. Comparison of committed-input, committed-output logical AND protocols using private operations are shown in Table 2.

3.3 COPY Protocol

Next, we show a new copy protocol with the minimum number of rounds.

Protocol 4 *(COPY protocol with the minimum number of rounds).*

1. *Alice executes a private random bisection cut on $S_0 = commit(x)$. Let the output be $S_1 = commit(x')$. Alice sets S_1' as m copies of $commit(b)$, where b is the bit selected in the random bisection cut. Note that $x' = x \oplus b$. Alice hands S_1 and S_1' to Bob.*
2. *Bob executes a private reveal on S_1 and obtains x'. Bob executes a private reverse cut on each pair of S_1' using x'. Let the result be S_2. Bob outputs S_2.*

The protocol is two rounds. The protocol uses $2m + 2$ cards.

Theorem 3. *The COPY protocol is correct and secure.*

Proof. Correctness: Since Bob obtains $x' = x \oplus b$, the output is $b \oplus (x \oplus b) = x$. Alice and Bob's security: The same as the XOR protocol.

Though the number of cards is increased, the number of rounds is the minimum. Comparison of COPY protocols(when $m = 2$) is shown in Table 3. Mizuki et al. showed a six-card protocol with one public shuffle [25]. Since one public shuffle can be changed to two private shuffles by each player, the minimum number of rounds is achieved also by their protocol. However, the protocol needs two shuffles, thus our new protocol is simple.

3.4 Any Two-Variable Logical Functions

Though this paper shows logical AND and logical XOR, any two-variable logical functions can also be calculated by a similar protocol. Though the protocol differs, the idea of the construction is similar to the one for the three-round protocol in [34].

Theorem 4. *Any two-variable logical function can be securely calculated in two rounds and at most six cards.*

Proof. Any two-variable logical function $f(x, y)$ can be written as

$$f(x, y) = \begin{cases} f(1, y) \text{ if } x = 1 \\ f(0, y) \text{ if } x = 0 \end{cases}$$

where $f(1, y)$ and $f(0, y)$ are y, $\bar{y}$, 0, or 1. From [34], we just need to consider the cases when one of $(f(1, y), f(0, y))$ is y or $\bar{y}$ and the other is 0 or 1. We can modify the first step of the AND protocol and Alice sets

$$S_1' = \begin{cases} commit(f(1, y)) \| commit(f(0, y)) \text{ if } b = 1 \\ commit(f(0, y)) \| commit(f(1, y)) \text{ if } b = 0 \end{cases}$$

using one $commit(y)$ and two new cards, since one of $(f(1, y), f(0, y))$ is y or $\bar{y}$ and the other is 0 or 1. Bob executes a private reveal on $S_1 = commit(x')$ and selects $commit(f(1, y))$ if $x = 1$. Bob selects $commit(f(0, y))$ if $x = 0$.

Thus, any two-variable logical function can be calculated. $\square$

3.5 n-Variable Logical Functions

Since two-variable logical functions, logical negation and a copy can be executed, any n-variable logical function can be calculated by the combination of the above protocols.

As another implementation with more number of cards, we show that any n-variable logical function can be calculated by the following protocol in two rounds, whose technique is similar to the one in [15]. Let f be any n-variable logical function.

Protocol 5 *(Protocol for any logical function with two rounds).*

1. *Alice executes a private random bisection cut on* $commit(x_i)$ $(i = 1, 2, \ldots, n)$. *Let the output be* $commit(x_i')(i = 1, 2, \ldots, n)$. $x_i' = x_i \oplus b_i (i = 1, 2, \ldots, n)$. *Note that one random bit* b_i *is selected for each* $x_i (i = 1, 2, \ldots, n)$. *Alice generates* 2^n *commitment* $S_{a_1, a_2, \ldots, a_n}$ $(a_i \in \{0, 1\}, i = 1, 2, \ldots, n)$ *as* $S_{a_1, a_2, \ldots, a_n} = commit(f(a_1 \oplus b_1, a_2 \oplus b_2, \ldots, a_n \oplus b_n))$.
   Alice hands $commit(x_i')(i = 1, 2, \ldots, n)$ *and* $S_{a_1, a_2, \ldots, a_n}$ $(a_i \in \{0, 1\}, i = 1, 2, \ldots, n)$ *to Bob.*
2. *Bob executes a private reveal on* $commit(x_i')$ $(i = 1, 2, \ldots, n)$. *Bob outputs* $S_{x_1', x_2', \ldots, x_n'}$.

Since $S_{x_1', x_2', \ldots, x_n'} = commit(f(b_1 \oplus x_1', b_2 \oplus x_2', \ldots, b_n \oplus x_n')) = commit(f(x_1, x_2, \ldots, x_n))$, the output is correct. The security is the same as the one of the XOR protocol. The protocol is two-round. The number of cards is $2^{n+1} + 2n$.

4   Protocols Without Private Reveals

This section shows that the protocols in [34] that use the minimum number of cards can be executed without the private reveal operations with a more number of steps, but the same number of rounds. Since it is hard to prevent mistakes of privately opening cards that are not allowed, it would be better all reveal operations are publicly executed.

4.1 XOR Protocol Without Private Reveals

Protocol 6 *(XOR protocol without private reveals).*

1. *Alice executes a private random bisection cut on $S_0 = commit(x)$ and $S_0' = commit(y)$ using the same random bit b. Let the output be $S_1 = commit(x')$ and $S_1' = commit(y')$, respectively. Note that $x' = x \oplus b$ and $y' = y \oplus b$. Alice hands S_1 and S_1' to Bob.*
2. *Bob executes a private random bisection cut on S_1 and S_1' using a private bit b'. Let the output be $S_2 = commit(x'')$ and $S_2' = commit(y'')$, respectively. $x'' = x \oplus b \oplus b'$ and $y'' = y \oplus b \oplus b'$ holds. Bob publicly opens S_2 and obtains value x''. Alice can see x''. Bob publicly sets*

$$S_3 = \begin{cases} commit(\bar{y''}) & \text{if } x'' = 1 \\ commit(y'') & \text{if } x'' = 0 \end{cases}$$

S_3 is the final result.

The protocol is two rounds.

Theorem 5. *The XOR protocol is correct and secure. It uses the minimum number of cards.*

Proof. Correctness: Bob obtains $commit(x \oplus b \oplus b')$ and $commit(y \oplus b \oplus b')$. Bob sets S_3 as $y'' \oplus x'' = y \oplus b \oplus b' \oplus x \oplus b \oplus b' = x \oplus y$. Thus, the result is correct.

Alice and Bob's security: After Bob executes a private random bisection cut on S_1, the obtained value $commit(x'') = commit(x \oplus b \oplus b')$. Even if this value is opened, no player can obtain the value of x, since Alice knows b and $x \oplus b \oplus b'$ and Bob knows b' and $x \oplus b \oplus b'$.

The number of cards: At least four cards are necessary for any protocol to input x and y. This protocol uses no additional cards other than the input cards.
□

4.2 AND Protocol Without Private Reveals

Protocol 7 *(AND protocol without private reveals).*

1. *Alice executes a private random bisection cut on $S_0 = commit(x)$. Let the output be $S_1 = commit(x')$. Alice hands S_1 and $S_0' = commit(y)$ to Bob.*
2. *Bob executes a private random bisection cut on S_1 using a private bit b'. Let the output be $S_1' = commit(x'')$. $x'' = x \oplus b \oplus b'$ holds. Bob publicly opens S_1' and obtains value x''. Alice can see x''. Bob publicly sets*

$$S_2 = \begin{cases} commit(y)||commit(0) & \text{if } x'' = 1 \\ commit(0)||commit(y) & \text{if } x'' = 0 \end{cases}$$

Bob then executes a private reverse cut on S_2 using the bit b' generated in the private random bisection cut. Let the output be S_3. Bob hands S_3 to Alice.

3. Alice executes a private reverse selection on S_3 using the bit b generated in the private random bisection cut. Alice outputs the obtained sequence of S_4.

Theorem 6. *The AND protocol is correct, secure, and uses the minimum number of cards.*

Proof. Correctness: The desired output can be represented as follows.

$$x \wedge y = \begin{cases} y \text{ if } x = 1 \\ 0 \text{ if } x = 0 \end{cases}$$

When Bob obtains $x'' = 1$, $commit(y)||commit(0)$ is set as S_2. When Bob obtains $x'' = 0$, $commit(0)||commit(y)$ is set as S_2. Since Bob executes a private reverse cut on S_2, $commit(y)||commit(0)$ is given to Alice when $(x'', b') = (1, 0)$ or $(0, 1)$. Since $x'' = x \oplus b \oplus b'$, these cases equal to $x \oplus b = 1$. $commit(0)||commit(y)$ is given to Alice when $(x'', b') = (1, 1)$ or $(0, 0)$. These cases equal to $x \oplus b = 0$.

Thus Alice's output is $commit(y)$ if $(x \oplus b, b) = (1, 0)$ or $(0, 1)$. These cases equal to $x = 1$. Alice's output is $commit(0)$ if $(x \oplus b, b) = (1, 1)$ or $(0, 0)$. These cases equal to $x = 0$. Therefore, the output is correct.

Alice and Bob's security: The same as the XOR protocol without private reveals.

The number of cards: Any committed input protocol needs at least four cards to input. When Bob sets S_2, the cards used for $commit(x'')$ can be re-used to set $commit(0)$. Thus, the total number of cards is four and the minimum.   □

The number of rounds is three. Using the argument in Sect. 3.4, any two-variable logical function can also be calculated by four cards and three rounds without private reveals.

4.3   COPY Protocol Without Private Reveals

Protocol 8 *(COPY protocol without private reveals).*

1. Alice executes a private random bisection cut on $S_0 = commit(x)$. Let the output be $S_1 = commit(x')$. Note that $x' = x \oplus b$. Alice sets S_1' as m copies of $commit(b)$. Alice hands S_1 and S_1' to Bob.
2. Bob executes a private random bisection cut on S_1 and each pair of S_1' using a private random bit b'. Let the output be S_2 and S_2', respectively.
   Bob publicly opens S_2 and obtains value x''. Alice can see the cards. Bob publicly swaps each pair of S_2' if $x'' = 1$. Otherwise, Bob does nothing. Let the result be S_3. Bob outputs S_3.

The protocol is two rounds. The protocol uses $2m + 2$ cards.

Theorem 7. *The COPY protocol is correct and secure.*

Proof. Correctness: Bob obtains $commit(x'')$ and $commit(b \oplus b')$, where $x'' = x \oplus b \oplus b'$. Bob opens x''. Bob publicly sets S_3 as $b \oplus b' \oplus x'' = b \oplus b' \oplus x \oplus b \oplus b' = x$. Thus, the result is correct.

Alice and Bob's security: The same as the XOR protocol without private reveals.   □

4.4 n-Variable Logical Functions Without Private Reveals

Let f be any n-variable logical function.

Protocol 9 *(Protocol for any logical function without private reveals).*

1. *The same as Step 1 in Protocol 5.*
2. *Bob executes a private random bisection cut on $commit(x_i')(i = 1, 2, \ldots, n)$.
   Note that one random bit b_i' is selected for each $x_i'(i = 1, 2, \ldots, n)$. Let
   $commit(x_i'')$ $(i = 1, 2, \ldots, n)$ be the obtained value. $x_i'' = x_i \oplus b_i \oplus b_i'(i =
   1, 2, \ldots, n)$ is satisfied. Bob privately relocates $S_{a_1, a_2, \ldots, a_n}(a_i \in \{0, 1\}, i =
   1, 2, \ldots, n)$ so that $S_{a_1, a_2, \ldots, a_n}' = S_{a_1 \oplus b_1', a_2 \oplus b_2', \ldots, a_n \oplus b_n'}(a_i \in \{0, 1\}, i =
   1, 2, \ldots, n)$. The cards satisfy that $S_{a_1, a_2, \ldots, a_n}' = commit(f(a_1 \oplus b_1 \oplus b_1', a_2 \oplus
   b_2 \oplus b_2', \ldots, a_n \oplus b_n \oplus b_n'))$.
   Bob publicly reveals $commit(x_i'')$ and obtains $x_i''(i = 1, 2, \ldots, n)$. Bob publicly
   selects $S_{x_1'', x_2'', \ldots, x_n''}'$.*

Since $S_{x_1'', x_2'', \ldots, x_n''}' = commit(f(x_1 \oplus b_1 \oplus b_1' \oplus b_1 \oplus b_1', x_2 \oplus b_2 \oplus b_2' \oplus b_2 \oplus
b_2', \ldots, x_n \oplus b_n \oplus b_n' \oplus b_n \oplus b_n')) = commit(f(x_1, x_2, \ldots, x_n))$, the output is correct.
The security is the same as the XOR protocol. The protocol is two rounds. The
number of cards is $2^{n+1} + 2n$.

All the other protocols that are shown in [34] can also be changed so as not
to use the private reveal operations. The conversion rule is as follows: When
Bob executes a private reveal and set a sequence S in the original protocol, Bob
executes a private random bisection cut to $commit(x \oplus b)$ instead. Let b' be
the random bit selected by Bob. Then Bob publicly opens the committed bit
and publicly sets a sequence S by the original rule. Bob then executes a private
reverse cut on S using the bit b' and outputs S'. Alice executes a private reverse
cut on S' and obtains the final result.

5   Protocols that Preserve an Input

In the above protocols to calculate logical functions, the input commitment values are lost. If an input is not lost, the input commitment can be used as an input
to another calculation. Thus, protocols that preserve an input are discussed [29].
For the three-round XOR and AND protocols in [34], protocols that preserve an
input were shown [34]. This paper uses the technique in [34] to obtain protocols
to preserve an input.

First, consider XOR protocols in Sects. 3 and 4.

Protocol 10 *(XOR protocol that preserves an input).*

1,2 *The same as Protocol 2 (or Protocol 6).*
   At the end of Step 2, Bob sends back $S_1 = commit(x')$ to Alice.
3 *Alice executes private reverse cut on S_1 and obtains $commit(x)$.*

In the protocol in Sect. 3, since $commit(x')$ is unnecessary after Bob's private reveal, the cards can be sent back to Alice. Alice can recover $commit(x)$. The number of rounds is increased to three.

For the XOR protocols in Sect. 4, the same technique can be applied. After Bob publicly opens S_2 and obtains $x \oplus b \oplus b'$, he can recover $S_1 = commit(x')$ since he knows b'. The protocol is three rounds and uses four cards.

Similarly, the AND type protocol in Sects. 3 and 4 can be modified to a three-round protocol to preserve an input.

Protocol 11 *(AND protocol that preserves an input).*

1,2 The same as Protocol 3 (or Protocol 7).
   At the end of Step 2, Bob sends back $S_1 = commit(x')$ to Alice.
  3 Alice executes private reverse cut on S_1 and obtains $commit(x)$ (in Protocol 7, executed at the end of the protocol).

The number of rounds is three and the number of cards is six. For the protocol in Sect. 4, two new cards are necessary to send $commit(x')$ to Alice.

The same technique can be applied to n-variable protocol and $commit(x'_i)$ can be sent back to Alice.

As for the AND type protocol, another protocol that preserves an input without additional cards can be obtained. Note that the function f satisfies that one of $(f(0,y), f(1,y))$ is y or $\bar{y}$ and the other is 0 or 1. Otherwise, we do not need to calculate f by the AND type two player protocol. When we execute the four-card AND type protocol without private reveals, two cards are selected by Alice at the final step. The remaining two cards are not used, but they also output some values. The unused two cards' value is

$$\begin{cases} f(0,y) \text{ if } x = 1 \\ f(1,y) \text{ if } x = 0 \end{cases}$$

thus the output value is $commit(\bar{x} \wedge f(1,y) \oplus x \wedge f(0,y))$. The output $f(x,y)$ can be written as $x \wedge f(1,y) \oplus \bar{x} \wedge f(0,y)$. Execute the above XOR protocol that preserves an input without private reveals for these two output values so that $f(x,y)$ is preserved. The output of XOR protocol is $\bar{x} \wedge f(1,y) \oplus x \wedge f(0,y) \oplus x \wedge f(1,y) \oplus \bar{x} \wedge f(0,y) = f(1,y) \oplus f(0,y)$. Since one of $(f(0,y), f(1,y))$ is y or $\bar{y}$ and the other is 0 or 1, the output is y or $\bar{y}$ (depending on f). Thus, input y can be recovered without additional cards.

Protocol 12 *(AND type protocol that preserves an input without private reveals).*

1,2 The same as Protocol 7.
  3 After Alice outputs S_4, let S'_4 be the cards that are not selected.
  4 Alice and Bob execute the XOR protocol that preserves an input without private reveals (Protocol 10) for S_4 and S'_4. Let the preserved input, S_4, be the result. Obtain $commit(y)$ from the XOR result.

Thus, the protocol achieves preserving an input by four cards. The AND type protocol needs three rounds and the XOR protocol that preserves an input needs three rounds. The last round of the AND type protocol and the first round of the XOR protocol are executed by Alice, thus they can be done in one round. Therefore, the total number of rounds is five.

6 Conclusion

This paper proposed round optimal card-based cryptographic protocols using private operations. Then this paper showed protocols without private reveal operations and several variant protocols. Further study includes round optimal protocols for the other fundamental problems.

References

1. Abe, Y., Hayashi, Y.I., Mizuki, T., Sone, H.: Five-card and protocol in committed format using only practical shuffles. In: Proceedings of 5th ACM on ASIA Public-Key Cryptography Workshop(APKC 2018), pp. 3–8. ACM (2018)
2. Boer, B.: More efficient match-making and satisfiability *The Five Card Trick*. In: Quisquater, J.-J., Vandewalle, J. (eds.) EUROCRYPT 1989. LNCS, vol. 434, pp. 208–217. Springer, Heidelberg (1990). https://doi.org/10.1007/3-540-46885-4_23
3. Bultel, X., et al.: Physical zero-knowledge proof for makaro. In: Izumi, T., Kuznetsov, P. (eds.) SSS 2018. LNCS, vol. 11201, pp. 111–125. Springer, Cham (2018). https://doi.org/10.1007/978-3-030-03232-6_8
4. Crépeau, C., Kilian, J.: Discreet solitary games. In: Stinson, D.R. (ed.) CRYPTO 1993. LNCS, vol. 773, pp. 319–330. Springer, Heidelberg (1994). https://doi.org/10.1007/3-540-48329-2_27
5. Dumas, J.-G., Lafourcade, P., Miyahara, D., Mizuki, T., Sasaki, T., Sone, H.: Interactive physical zero-knowledge proof for norinori. In: Du, D.-Z., Duan, Z., Tian, C. (eds.) COCOON 2019. LNCS, vol. 11653, pp. 166–177. Springer, Cham (2019). https://doi.org/10.1007/978-3-030-26176-4_14
6. Francis, D., Aljunid, S.R., Nishida, T., Hayashi, Y., Mizuki, T., Sone, H.: Necessary and sufficient numbers of cards for securely computing two-bit output functions. In: Phan, R.C.-W., Yung, M. (eds.) Mycrypt 2016. LNCS, vol. 10311, pp. 193–211. Springer, Cham (2017). https://doi.org/10.1007/978-3-319-61273-7_10
7. Hashimoto, Y., Nuida, K., Shinagawa, K., Inamura, M., Hanaoka, G.: Toward finite-runtime card-based protocol for generating hidden random permutation without fixed points. IEICE Trans. Fundam. Electron. Commun. Comput. Sci. **101–A**(9), 1503–1511 (2018)
8. Hashimoto, Y., Shinagawa, K., Nuida, K., Inamura, M., Hanaoka, G.: Secure grouping protocol using a deck of cards. IEICE Trans. Fundam. Electron. Commun. Comput. Sci. **101–A**(9), 1512–1524 (2018)
9. Ibaraki, T., Manabe, Y.: A more efficient card-based protocol for generating a random permutation without fixed points. In: Proceedings of 3rd International Conference on Mathematics and Computers in Sciences and in Industry (MCSI 2016), pp. 252–257 (2016)
10. Ishikawa, R., Chida, E., Mizuki, T.: Efficient card-based protocols for generating a hidden random permutation without fixed points. In: Calude, C.S., Dinneen, M.J. (eds.) UCNC 2015. LNCS, vol. 9252, pp. 215–226. Springer, Cham (2015). https://doi.org/10.1007/978-3-319-21819-9_16

11. Kastner, J., et al.: The minimum number of cards in practical card-based protocols. In: Takagi, T., Peyrin, T. (eds.) ASIACRYPT 2017. LNCS, vol. 10626, pp. 126–155. Springer, Cham (2017). https://doi.org/10.1007/978-3-319-70700-6_5

12. Koch, A.: The landscape of optimal card-based protocols. IACR Cryptology ePrint Archive, Report 2018/951 (2018)

13. Koch, A., Walzer, S.: Private function evaluation with cards. IACR Cryptology ePrint Archive, Report 2018/1113 (2018)

14. Koch, A., Walzer, S., Härtel, K.: Card-based cryptographic protocols using a minimal number of cards. In: Iwata, T., Cheon, J.H. (eds.) ASIACRYPT 2015. LNCS, vol. 9452, pp. 783–807. Springer, Heidelberg (2015). https://doi.org/10.1007/978-3-662-48797-6_32

15. Kurosawa, K., Shinozaki, T.: Compact card protocol. In: Proceedings of 2017 Symposium on Cryptography and Information Security (SCIS 2017), pp. 1A2–6 (2017). (in Japanese)

16. Marcedone, A., Wen, Z., Shi, E.: Secure dating with four or fewer cards. IACR Cryptology ePrint Archive, Report 2015/1031 (2015)

17. Miyahara, D., Hayashi, Y., Mizuki, T., Sone, H.: Practical and easy-to-understand card-based implementation of yao's millionaire protocol. In: Kim, D., Uma, R.N., Zelikovsky, A. (eds.) COCOA 2018. LNCS, vol. 11346, pp. 246–261. Springer, Cham (2018). https://doi.org/10.1007/978-3-030-04651-4_17

18. Miyahara, D., Ueda, I., Hayashi, Y., Mizuki, T., Sone, H.: Analyzing execution time of card-based protocols. In: Stepney, S., Verlan, S. (eds.) UCNC 2018. LNCS, vol. 10867, pp. 145–158. Springer, Cham (2018). https://doi.org/10.1007/978-3-319-92435-9_11

19. Mizuki, T.: Card-based protocols for securely computing the conjunction of multiple variables. Theor. Comput. Sci. **622**, 34–44 (2016)

20. Mizuki, T.: Efficient and secure multiparty computations using a standard deck of playing cards. In: Foresti, S., Persiano, G. (eds.) CANS 2016. LNCS, vol. 10052, pp. 484–499. Springer, Cham (2016). https://doi.org/10.1007/978-3-319-48965-0_29

21. Mizuki, T., Asiedu, I.K., Sone, H.: Voting with a logarithmic number of cards. In: Mauri, G., Dennunzio, A., Manzoni, L., Porreca, A.E. (eds.) UCNC 2013. LNCS, vol. 7956, pp. 162–173. Springer, Heidelberg (2013). https://doi.org/10.1007/978-3-642-39074-6_16

22. Mizuki, T., Kumamoto, M., Sone, H.: The five-card trick can be done with four cards. In: Wang, X., Sako, K. (eds.) ASIACRYPT 2012. LNCS, vol. 7658, pp. 598–606. Springer, Heidelberg (2012). https://doi.org/10.1007/978-3-642-34961-4_36

23. Mizuki, T., Shizuya, H.: Practical card-based cryptography. In: Ferro, A., Luccio, F., Widmayer, P. (eds.) FUN 2014. LNCS, vol. 8496, pp. 313–324. Springer, Cham (2014). https://doi.org/10.1007/978-3-319-07890-8_27

24. Mizuki, T., Shizuya, H.: Computational model of card-based cryptographic protocols and its applications. IEICE Trans. Fundam. Electron. Commun. Comput. Sci. **100–A**, 3–11 (2017)

25. Mizuki, T., Sone, H.: Six-card secure AND and four-card secure XOR. In: Deng, X., Hopcroft, J.E., Xue, J. (eds.) FAW 2009. LNCS, vol. 5598, pp. 358–369. Springer, Heidelberg (2009). https://doi.org/10.1007/978-3-642-02270-8_36

26. Nakai, T., Shirouchi, S., Iwamoto, M., Ohta, K.: Four cards are sufficient for a card-based three-input voting protocol utilizing private permutations. In: Shikata, J. (ed.) ICITS 2017. LNCS, vol. 10681, pp. 153–165. Springer, Cham (2017). https://doi.org/10.1007/978-3-319-72089-0_9

27. Nakai, T., Tokushige, Y., Misawa, Y., Iwamoto, M., Ohta, K.: Efficient card-based cryptographic protocols for millionaires' problem utilizing private permutations. In: Foresti, S., Persiano, G. (eds.) CANS 2016. LNCS, vol. 10052, pp. 500–517. Springer, Cham (2016). https://doi.org/10.1007/978-3-319-48965-0_30
28. Niemi, V., Renvall, A.: Secure multiparty computations without computers. Theor. Comput. Sci. **191**(1), 173–183 (1998)
29. Nishida, T., Hayashi, Y., Mizuki, T., Sone, H.: Card-based protocols for any boolean function. In: Jain, R., Jain, S., Stephan, F. (eds.) TAMC 2015. LNCS, vol. 9076, pp. 110–121. Springer, Cham (2015). https://doi.org/10.1007/978-3-319-17142-5_11
30. Nishida, T., Hayashi, Y., Mizuki, T., Sone, H.: Securely computing three-input functions with eight cards. IEICE Trans. Fundam. Electron. Commun. Comput. Sci. **98–A**(6), 1145–1152 (2015)
31. Nishimura, A., Nishida, T., Hayashi, Y., Mizuki, T., Sone, H.: Five-card secure computations using unequal division shuffle. In: Dediu, A.-H., Magdalena, L., Martín-Vide, C. (eds.) TPNC 2015. LNCS, vol. 9477, pp. 109–120. Springer, Cham (2015). https://doi.org/10.1007/978-3-319-26841-5_9
32. Nishimura, A., Nishida, T., Hayashi, Y., Mizuki, T., Sone, H.: Card-based protocols using unequal division shuffles. Soft Comput. **22**(2), 361–371 (2018)
33. Ono, H., Manabe, Y.: Efficient card-based cryptographic protocols for the millionaires' problem using private input operations. In: Proceedings of 13th Asia Joint Conference on Information Security (AsiaJCIS 2018), pp. 23–28 (2018)
34. Ono, H., Manabe, Y.: Card-based cryptographic protocols with the minimum number of cards using private operations. In: Zincir-Heywood, N., Bonfante, G., Debbabi, M., Garcia-Alfaro, J. (eds.) FPS 2018. LNCS, vol. 11358, pp. 193–207. Springer, Cham (2019). https://doi.org/10.1007/978-3-030-18419-3_13
35. Ruangwises, S., Itoh, T.: AND protocols using only uniform shuffles. In: van Bevern, R., Kucherov, G. (eds.) CSR 2019. LNCS, vol. 11532, pp. 349–358. Springer, Cham (2019). https://doi.org/10.1007/978-3-030-19955-5_30
36. Sasaki, T., Mizuki, T., Sone, H.: Card-based zero-knowledge proof for sudoku. In: Proceedings of 9th International Conference on Fun with Algorithms (FUN 2018), Leibniz International Proceedings in Informatics (LIPIcs), vol. 100, pp. 29:1–29:10 (2018)
37. Shinagawa, K., Mizuki, T.: Secure computation of any boolean function based on any deck of cards. In: Chen, Y., Deng, X., Lu, M. (eds.) FAW 2019. LNCS, vol. 11458, pp. 63–75. Springer, Cham (2019). https://doi.org/10.1007/978-3-030-18126-0_6
38. Shinagawa, K., Mizuki, T.: The six-card trick: secure computation of three-input equality. In: Lee, K. (ed.) ICISC 2018. LNCS, vol. 11396, pp. 123–131. Springer, Cham (2019). https://doi.org/10.1007/978-3-030-12146-4_8
39. Shinagawa, K., Nuida, K.: A single shuffle is enough for secure card-based computation of any circuit. IACR Cryptol. ePrint Arch. **2019**, 380 (2019)
40. Shirouchi, S., Nakai, T., Iwamoto, M., Ohta, K.: Efficient card-based cryptographic protocols for logic gates utilizing private permutations. In: Proceedings of 2017 Symposium on Cryptography and Information Security (SCIS 2017), pp. 1A2–2 (2017). (in Japanese)
41. Stiglic, A.: Computations with a deck of cards. Theor. Comput. Sci. **259**(1), 671–678 (2001)

CBT Workshop: Lightning Networks and Level 2

TEE-Based Distributed Watchtowers for Fraud Protection in the Lightning Network

Marc Leinweber[✉], Matthias Grundmann, Leonard Schönborn, and Hannes Hartenstein

Karlsruhe Institute of Technology, Karlsruhe, Germany
{marc.leinweber,matthias.grundmann,hannes.hartenstein}@kit.edu,
leonard.schoenborn@student.kit.edu

Abstract. The Lightning Network is a payment channel network built on top of the cryptocurrency Bitcoin. It allows Bitcoin to scale by performing transactions off-chain to reduce load on the blockchain. Malicious payment channel participants can try to commit fraud by closing channels with outdated balances. The Lightning Network allows resolving this dispute on the blockchain. However, this mechanism forces the channels' participants to watch the blockchain in regular intervals. It has been proposed to offload this monitoring duty to a third party, called a watchtower. However, existing approaches for watchtowers do not scale as they have storage requirements linear in the number of updates in a channel. In this work, we propose TEE Guard, a new architecture for watchtowers that leverages the features of Trusted Execution Environments to build watchtowers that require only constant memory and are thus able to scale. We show that TEE Guard is deployable because it can run with the existing Bitcoin and Lightning Network protocols. We also show that it is economically viable for a third party to provide watchtower services. As a watchtower needs to be trusted to be watching the blockchain, we also introduce a mechanism that allows customers to verify that a watchtower has been running continuously.

1 Introduction

On top of the Bitcoin protocol [19], the Lightning Network [21] has emerged as a second layer solution to reduce the load of transactions on the Bitcoin blockchain by implementing a payment channel network. The Lightning Network consists of interconnected bilateral payment channels. Each payment channel encodes the distribution of a fixed amount of bitcoins between its two participants. At any time, either party can close the channel by publishing a commitment transaction to the Bitcoin blockchain that pays each party the amount of coins they own in the channel. However, in payment channels, it is possible to commit a *fraud* by publishing an outdated commitment transaction to the blockchain. A fraud is profitable when the outdated commitment transaction assigns more coins to the cheating party than the most current channel state. To mitigate the fraud,

C. Pérez-Solà et al. (Eds.): DPM 2019/CBT 2019, LNCS 11737, pp. 177–194, 2019.
https://doi.org/10.1007/978-3-030-31500-9_11

the cheated party has the opportunity to react and invalidate the outdated transaction during a predefined time span.

Channels are usually configured to allow users 24 h (144 blocks) to react to such transactions with a revocation transaction that reverts the fraudulent commitment transaction. This means that users have to be online and search the blockchain at least once every 24 h to defend against fraud attempts. To avoid the obligatory activity, to allow for longer phases of inactivity and to enable the development of lightweight nodes, watchtower (or custodian) services have already been proposed. Dryja [9] proposed in 2016 a monitor that searches new blocks on the blockchain for commitment transactions concerning the channels of its users and publishes, in the case of a fraud, revocation transactions on the user's behalf. The monitor needs to store a revocation transaction for every state of the channel which obviates the need for sharing of secrets but introduces linear memory consumption and is therefore limited in its scalability. Dryja's approach was followed by several other monitoring proposals [4,16,20], whose strengths and limitations we discuss in Sect. 2.

To run such watchtower services economically, their ability to *scale* to many users is of vital importance. To keep the watchtower's provider's cost economical, the computation time and storage requirements need to be low per monitored channel and, for channels being open for a long time, the storage requirements need to be *constant with respect to the number of updates in the channel*. In this work, we present TEE Guard, a watchtower solution for the Lightning Network based on Trusted Execution Environments (TEEs) that is scalable due to its low and constant storage requirements. With TEE Guard, users of the Lightning Network can create a watchtower for their channels that is hosted by a so-called platform provider. To improve reliability, such a user can become a customer of multiple independent platforms and interconnect the watchtowers running at these different platforms. For initialisation of the watchtower service, the customer transmits, in a secure channel and unseen by the platform provider, a public key hash where revocation transactions will be paid to and references to the Lightning channels to monitor. With each channel update, the customer has to transmit the required secrets to invalidate the now outdated commitment to their watchtowers. As the watchtower provider is not trusted by its customers and might even collude with a customer's channel partner, all *secrets need to be kept private* by the watchtower service. By storing them inside the TEE, we ensure that they cannot be read by the provider. The key derivation system of the Lightning Network makes it possible to store the secrets for all past updates of a channel using constant space. The watchtower provider may not only try to learn secrets from its customers but also try to cheat them by stopping the watchtower service to save operational costs. It would be too late if a customer would notice that a watchtower stopped their service only after the watchtower did not react to an outdated commitment transaction. To allow a watchtower's customers to *verify the watchtower's availability*, TEE Guard enables the customers to interconnect multiple watchtowers forming a watchtower network and the watchtowers will monitor each other and create logs about the others' availability. These logs are signed and, in conjunction with the remote attestation feature of TEEs, the customer can thus verify that the provider has been running the watchtower continuously.

In sum, TEEs allow us to design an architecture, on top of the currently deployed Bitcoin and Lightning protocols, that is (1) scalable, (2) economically viable, and (3) in its redundancy tunable and verifiable by the user. The remainder of this work is structured as follows. The background on payment channels in general and the Lightning Network in particular as well as a definition and an overview on TEEs is given in Sect. 2. Furthermore, related work is discussed. The TEE Guard architecture is presented in detail in Sect. 3, quantitatively evaluated in Sect. 4 and its security is assessed in Sect. 5. We conclude the paper and give an outlook on future work with Sect. 6.

2 Background and Related Work

2.1 Lightning Network

The Lightning Network consists of bilateral payment channels that encode the distribution of bitcoins between their two participants. To open a new channel, two parties agree on an initial distribution, exchange payment_basepoints, and the funder of the bitcoins, who is one of the two participants, creates, signs, and publishes a transaction that spends the bitcoins to a 2-of-2 multi-signature output. The current distribution is managed in a commitment transaction which, in turn, is not published until the channel is closed. Hence, with each change of the channel's balance, a new commitment transaction is negotiated. Each commitment transaction uses keys derived from a new per_commitment_secret and includes a counter (called commitment_number) that is incremented for each update in the channel. After a new commitment transaction has been created, both parties exchange the per_commitment_secrets for the outdated transaction. In case one of the channel's participants publishes an outdated commitment transaction, the other party can spend the transaction's outputs using the corresponding per_commitment_secret and its own revocation_basepoint_secret. To allow for the revocation transaction to be published, the commitment transaction's outputs can only be spent by the other party after a fixed time span called to_self_delay (measured in number of blocks). This mechanism forces the channel's participants to check the Bitcoin blockchain in regular intervals for invalid commitments and to store all old per_commitment_secrets. The secrets can, however, be stored efficiently at constant size using a key derivation mechanism [1].

In the Lightning Network, payment channels are interconnected and the resulting network can be used for transactions over multiple hops using Hashed Time-Locked Contracts (HTLCs). The HTLC construction ensures that either all participants are able to execute the payment (using their knowledge of a secret x) or none of them are. A conditional output is added to each commitment transaction along the payment's path that includes $y = h(x)$ (with h being a collision-resistant hash function) in the condition. The output can be redeemed by the receiver by publishing x or becomes invalid after a timeout. If a commitment transaction with an HTLC output is published on chain, it can either be spent by the receiver with an HTLC success transaction providing x or after the specified timeout with an HTLC timeout transaction by the sender.

2.2 Trusted Execution Environments (TEEs)

By using trustworthy features in hardware and software that isolate processes from other user and system processes, an environment (called enclave) is established that allows the trusted execution of code. Such TEEs are implemented, for example, by Intel in the form of *Security Guard Extensions (SGX)* [2,17] or by AMD's *Memory Encryption Technology* [12]. Besides these proprietary approaches, Keystone [13] is an open-source framework for implementing TEEs based on the open hardware architecture RISC-V. These implementations have different design goals and features. Current proposals for systems using TEEs are typically based on SGX. SGX protects the confidentiality of the application's data and the integrity of code and data against malicious applications and a malicious operating system. The only trusted component is the Intel CPU whereby the required trust is reduced to the hardware designer and manufacturer.

For TEE Guard, we require a TEE to provide *isolated execution, protected memory*, and *remote attestation*. With these features, code is protected from manipulation and runtime data is guaranteed to be of integrity, confidential and fresh. The remote attestation enables the remote verification of the platform's identity (CPU secrets) and the code's identity (its instruction order). The TEE can derive encryption keys from its own identity and the code's identity to store data on disk.

2.3 Blockchain and TEEs

The work presented in this paper connects the research fields of blockchain and TEEs. Previous work in this intersection has been done with different focuses. TEEs have been used to implement alternatives to proof-of-work as used in Bitcoin, e.g. proof-of-luck [18] or proof-of-elapsed-time [11]. They have also been used to implement a trusted exchange [5], second layer architectures (e.g. Teechain [14] and Banklaves [10]), and an extension to Bitcoin to run arbitrary smart contracts [7]. In this work, we present TEE Guard, which is, to the best of our knowledge, a new approach to use TEEs in the world of cryptocurrencies by leveraging their features to implement a watchtower service for payment channel networks.

2.4 Watchtowers

Watchtowers have already been proposed very early in the development of the Lightning Network by Dryja [9]. In his approach, a customer creates a revocation transaction for every old commitment transaction, encrypts it with the old commitment transaction's ID, and sends it to the watchtower. When an old commitment transaction is published, the watchtower uses the old commitment transaction's ID to decrypt and then publish the revocation transaction. The approach preserves the channel's privacy and the watchtower does not learn any secrets that could be abused, but requires the watchtower to store $O(n)$ transactions where n is the number of updates in the channel. To incentivise the watchtower to perform this service, the customer can add an additional output to the revocation transaction that pays the watchtower provider a fee. However, this leaves the watchtower provider unpaid in case no fraud happens in a channel and in case of a fraud only the first watchtower to publish the revocation transaction earns the fee.

Ounstokun [20] proposed an updated construction of channels for the Lightning Network that requires the introduction of a new opcode to the Bitcoin protocol, but allows for constant storage for channel participants and watchtowers. However, the opcode has not been implemented since and, thus, the approach is not deployable yet.

Later, Pisa [16] was proposed, which is an approach for watchtowers (called custodians) which is constructed to work with state channels on top of Ethereum instead of payment channels as used in the Lightning Network. As the state channel construction uses a smart contract that can compare states by their commitment numbers, the custodian only needs constant sized memory. However, in the Lightning Network, as currently deployed, the state is replaced by revocation [21], which requires a specific revocation transaction for each new state. Storing all revocation transactions requires storage linear to the number of payments in a channel. As our approach uses TEEs, we can outsource the secrets needed to generate the revocation transactions which reduces the required storage of the watchtower to a constant size. Pisa requires the custodian to provide a security deposit which is being destroyed in case the custodian fails to settle a dispute. To prevent the custodian from colluding with a channel partner, the security deposit has to be high enough that the custodian would not profit from such a collusion. This necessitates high security deposits by the custodian and the required high investment makes it somewhat unrealistic that someone would assume the role of such a custodian. In our approach, we do not need such a security deposit. In return, we require that multiple custodians (or watchtowers) are used and we assume that they do not collude. We assume that it is more realistic to find non-colluding service providers for watchtowers (at least one correctly working watchtower is enough to protect the channel) than finding custodians that are willing to place a high amount of capital as security deposits.

Decker et al. proposed eltoo [8] which is a payment channel construction for Bitcoin with a different approach than the currently deployed Lightning Network. Instead of replacing old states by revocation, it adds a counter to each transaction and uses a construction that allows old transactions to be spent by transactions with higher counter values. This requires both parties to only store the transactions with the latest state and is thus also a solution to the linear storage problem. However, the concept uses so called *floating transactions* which require an opcode added to the Bitcoin protocol which has not been included yet. On a payment channel network using eltoo, watchtowers would require only constant sized storage per channel. Yet, watchtowers would be necessary to watch the blockchain for outdated commitment transactions. Our approach could still be a valuable addition on such a network because TEE Guard introduces redundancy and the mutual monitoring of watchtowers allows verification of the watchtower providers' availability.

In [4], Avarikioti et al. present DCWC, a protocol for a network of watchtowers which forward revocation transactions to other watchtowers and publish them during dispute. They also present a modified protocol DCWC*, which is implementable on the Lightning Network in contrast to DCWC. Watchtowers are incentivised to participate in the protocol by receiving a part of a channel's funds in case of dispute. The issue here is that watchtowers are not paid in case no dispute arises and therefore they might not be sufficiently incentivised. Our approach incentivises watchtowers by users paying the watchtower providers a fixed rate per month. DCWC* requires the watchtowers to store $O(n)$ messages, where n is the number of channel updates. Using TEEs to store the revocation secrets, we limit the watchtower's storage to a constant size, which makes it possible to watch a high number of channels for an unlimited time using a single machine.

Recently, with Brick [3], a new state channel construction has been proposed. Brick replaces the on-chain dispute process by a mediator committee of third parties. Therefore, watchtowers are no longer necessary in Brick. However, the channel partners have to agree on the committee members and trust that at least a third of them will be honest.

3   TEE Guard Architecture

TEE Guard implements a watchtower solution for the Lightning Network in a TEE. The TEE assures untampered and confidential execution of the watchtower code and enables remote attestation of the TEE's platform identity and the executed code identity. Watchtowers are distributed to increase reliability and implement mutual monitoring.

A Lightning Network user can become a *customer* of one or more independent TEE Guard *platform providers*. Each platform provides watchtower *instances* for an arbitrary number of customers, who send updates to the watchtower after each update in their channels. The watchtowers monitor the blockchain for outdated commitment transactions and react to such transactions with revocation transactions. By interconnecting their watchtower instances, customers can construct a distributed system that fits their needs. The instances will then monitor each other's state to increase the reliability and to detect potentially malicious or unreliable providers. On the detection of outdated information, correct states will be exchanged. The activity monitoring results are logged and provided to the customer. A provider can be, for example, a company, organisation or a private person.

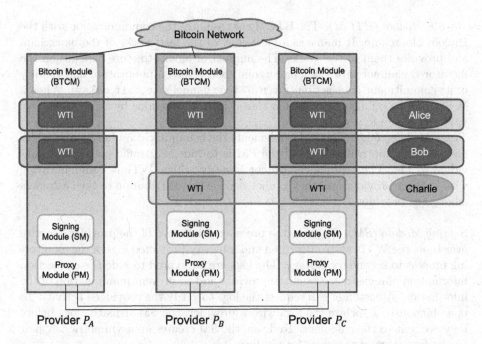

Fig. 1. Abstract overview of the TEE Guard interconnected watchtowers showing the TEE Guard software running at three watchtower providers with their different modules. Each provider hosts different watchtower instances (WTI) that are owned by a specific customer. In this scenario three users (Alice, Bob, and Charlie) operate a network of interconnected watchtowers hosted at two or three providers respectively.

Figure 1 illustrates an example of the distributed system with three provider platforms P_A, P_B and P_C and three customers Alice, Bob and Charlie. The figure only shows the TEE Guard components and abstracts away the enclave's surrounding runtime environment. Each provider platform is connected to the Bitcoin peer-to-peer network and receives the blockchain. Because each platform has a customer that is also customer of one of the two others, all platforms are interconnected. The operating principle and the rationale behind it is discussed in detail in the remainder of this section.

3.1 Provider Platform

A provider platform is implemented in one enclave and divided into modules.

Watchtower Module (WTM). The WTM implements the watchtower functionality: monitoring of channel-related transactions, creation of revocation transactions and monitoring of connected watchtowers. An instance of the WTM (called WTI) is created for each customer.

Bitcoin Module (BTCM). The BTCM encapsulates the communication with the Bitcoin blockchain. It manages a local copy of the last blocks of the blockchain and provides them to the WTIs. The number of blocks to store depends on the monitored channel with the longest time span between publication and validity of its commitment transaction (given by the channel's `to_self_delay`). When a WTI needs to publish a revocation transaction, this is done by the BTCM, too.

Proxy Module (PM). The PM implements the communication between provider systems. The motivation behind the PM is to hide the traffic patterns of individual WTIs. Thus, providers cannot identify which WTI is communicating with another provider and they cannot deduce the distribution of the customers among the providers.

Signing Module (SM). To prove the presence or absence of the provider's service over time, the WTIs write encrypted and integrity-protected logs of the monitoring process to secondary storage. The logs are encrypted to hide the connection information of a customer from the provider and to prevent manipulation of the information. A customer can request the logs to verify the providers' activity. To unambiguously associate the logs with a provider, the SM signs the logs before they are sent to the customer. To do so, the SM creates an asymmetric key pair on the first start of the provider platform.

3.2 Setup

The customer chooses a number $n \in \mathbb{N}$ of providers, sets up the n instances ("Instance Setup") and connects them ("Network Setup").

Provider Platform Setup. To allow verifiable and reproducible builds, which is essential for meaningful remote attestation, TEE Guard has to be open-source software. All a provider has to do is to download, build, and run the software. The provider is flexible in the choice of the executed software and is free to alter the code. The remote attestation feature ensures that the customer cannot be betrayed by running code they do not agree with.

Instance Setup. The provider creates a new WTI and gives access to it to the customer. During a remote attestation, the customer verifies the provider platform's identity and establishes a secure channel to the WTI. Once the customer has completed the verification, the WTI cannot be accessed by the provider anymore. This is enforced by the implementation which is protected by the TEE. The customer then has to provide the `revocation_payout_address`, a Bitcoin address for revocation payouts, that also serves as a customer identifier. Additionally, a list of trusted TEE Guard code identities for the network setup can be transmitted. During the operation of the WTI, the customer can add and remove channels to monitor.

Network Setup. As mentioned before, TEE Guard is designed to be a distributed system. The WTIs of a customer can be connected to increase reliability and enable mutual monitoring. To connect two WTIs A and B on platforms P_A and P_B, the customer tells A to establish a connection to another provider platform's proxy module. In order to (re)establish a new or broken (down) connection, one of the platforms has to be publicly reachable. Firstly, P_A's proxy module checks if it already established a connection to P_B. If not, both proxy modules perform a remote attestation and store the identity of the remote platform. A compares P_B's identity with its list of trusted TEE Guard code identities. If the list does not contain P_B's identity, A cancels the connection process. The proxy module of P_B checks if the platform hosts a WTI B with the same revocation_payout_address. Next, B checks its list of trusted identities against P_A's identity and, if P_A's identity is not trusted, B cancels the connection process. If both WTIs A and B agree on establishing a connection, they tell their platform's proxy modules to route upcoming messages between them. Finally, the proxy module of A adds P_B to A's list of connected platforms and vice versa.

3.3 Operation

A channel participant can attempt a fraud by issuing an outdated commitment transaction tx_F with commitment_number(tx_F) < commitment_number(tx_C) of the most current commitment transaction tx_C. If no revocation transaction is issued within the time span defined by the channel's to_self_delay parameter, the fraud can claim the stolen funds.

Thus, to detect a fraud, the watchtower needs to scan the last to_self_delay blocks of the blockchain because they may contain a revocable outdated commitment transaction. To identify and revoke outdated transactions, the most current commitment_number of a channel, the information to find channel-related commitment transactions on the blockchain, and the secrets to sign a valid revocation transaction are needed (see Table 1).

In sum, the customer has to contact the watchtowers on each channel transaction and the watchtower has to get active with each new block added to the blockchain.

Channel Updates. Most of the necessary parameters are fixed for the channel's life time and need to be transmitted during the setup of the channel monitoring. For each channel update the customer has to contact its WTIs and provide the per_commitment_secret for the now outdated transaction and the new commitment_number. According to BOLT#3 [1] the per_commitment_secrets can be stored efficiently by storing at most 49 (value, index) pairs at any given point in time, which can be used to derive the keys for outdated commitment transactions.

Blockchain Updates. The BTCM of each provider platform is connected to the Bitcoin network and parses incoming messages. On the reception of a new block, the BTCM verifies the Merkle hash tree, compares the transaction inputs to the

Table 1. Information required for watchtower operation that has to be stored for a channel by each watchtower

Data	Purpose
funding_txid	identification of channel-related commitment transactions
both participants' payment_basepoints	decryption of commitment_numbers in commitment transactions
most current commitment_number	identification of outdated commitment transactions
revocation_basepoint_secret, values to derive per_commitment_secrets	revocation key derivation

list of monitored channels of the platform's WTIs and stores the block together with a signature of it on the untrusted secondary storage of the provider's system. On future accesses, there is no need to verify the block again but only the signature. If a transaction input in the block is on the list of monitored channels[1], the corresponding WTI is informed. The WTI reconstructs the commitment_number using the payment_basepoints and compares it to the stored most current commitment_number.

In case the commitment transaction is outdated, the issuer of the transaction has to be identified. If the WTI succeeds in calculating the revocation key corresponding to the outdated commitment transaction, the transaction was not issued by the customer but by the customer's channel partner and the transaction is identified as fraudulent. The WTI constructs and publishes a revocation transaction that spends the fraud's output to the revocation_payout_address. If the outdated commitment transaction contains HTLC outputs, these outputs are used as inputs for the revocation transaction, too. However, the revocation transaction issued by the WTI may be preceded by an HTLC success or HTLC timeout transaction. If so, the revocation transaction would be a double spend of this transaction because both spend the same outputs. Thus, the watchtower has to monitor the blockchain until the revocation transaction becomes valid and, in case an HTLC transaction was issued by the fraud, the revocation transaction has to be reissued using the HTLC transaction's output instead of the commitment transaction's output. Channels can be removed from the list of monitored channels if a block old enough contains the revocation or a closing transaction.

With each new block, the WTI contacts the connected WTIs on other provider platforms with a New Block message. This message contains the list of monitored channels, the most currently known commitment number and the current blockchain height. If the receiving WTI identifies outdated channel information (missing channels or outdated commitment numbers), it answers with a New Block Response message. The incoming New Block messages are logged.

---

[1] More specifically, this means that the transaction id of the input equals a funding_txid contained in the list of monitored channels.

Rollback Protection. In case the freshness of the state information cannot be ensured, the temporal correctness of code execution is not guaranteed. As long as a TEE system is running, it can assure the freshness of the main memory data. This does not hold for reboots or secondary storage. Thus, state has to be identified with respect to time.

A WTI compares the correctness of its current channel information with each blockchain update. The list of connected provider platforms, however, is not transmitted. When a provider platform is rebooted, it stores its current state to the secondary storage (encrypted and signed). During initialisation, this stored state is restored to main memory. If a provider would manipulate the information on the secondary storage by replaying a previously stored image, they could remove, however not fine-grained, information on connected provider platforms. Hardware monotonic counters in current TEE implementations have been shown to be of reduced usability [15]. We use a distributed `state_counter` to circumvent hardware monotonic counters. Each time a channel is added or removed or a provider platform is connected or disconnected, the WTI where the action was taken updates the counter and broadcasts the change to the connected WTIs. When the provider platform is restarted, the WTIs ask the known connected WTIs for their current state and update if necessary. The WTIs log received and sent state information.

Verification of Availability. To verify that a provider fulfills the required availability, a user can collect and analyse the providers' logs. If the logs of watchtower A show that a message for a new block from watchtower B has been received after receiving a newer block from the blockchain, this indicates that A or B was not online for some time. Comparing the logs of all watchtowers allows distinguishing whether A or B was offline.

4   Quantitative Evaluation

In this section, we analyse TEE Guard assuming an implementation with Intel SGX. We analyse the reliability of our approach using a simulation and its scalability by analysing the computing and storage requirements. Using the results of the analysis, we estimate the cost for a watchtower provider and find that it is economically viable to provide this service for a cost below one USD cent per channel per month, provided that enough customers per provider platform use the service.

4.1   Reliability

The reliability solely rests on the availability of the provider platforms. If a system is ready to work and not hindered in its execution, it will scan the blockchain and emit revocation transactions if necessary. Availability is disrupted by outages of any kind or malicious providers. Both kinds of disruption can be proven by the logging utility, which will be discussed in Sect. 5.2.

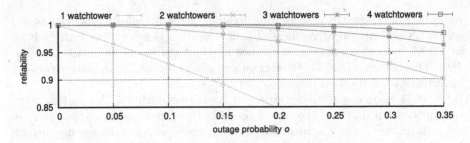

Fig. 2. Reliability study with $u = 5$ users, mean channel lifetime $c = 30$ d, mean transaction interval $t = 36$ h and simulated time $s = 20$ y set fix. The interval delimiters show the 99.9% Student's t-distribution confidence interval.

To understand the implications of outages not only for TEE Guard but distributed watchtower scenarios in general, we conducted a simulation study using a discrete-event simulator written in Java. A standard failure model cannot be applied and analytically analysed, because not only outages of watchtowers are relevant but also their state. The simulation model is simplified in contrast to Sect. 3.1 and consists of users, watchtowers (WTIs), channels, and a simplified blockchain. Channels and watchtowers are assigned to a user. Provider platforms are not modelled; each watchtower runs independently of the other watchtowers. Modelling the provider platforms would cause correlations of WTI outages because a provider platform outage causes the outage of all WTIs running on it. The blocks of the blockchain are only containing references to channels that are being closed in this block. The opening of new channels is not modelled explicitly. The parameters are the number of users u, the number of watchtowers per user w, the outage probability of a watchtower o, the mean channel life time c, the mean transaction interval for a channel transaction t and the simulated time s. The parameters c and t are mean values for exponential distributions; a new block is emitted every 10 min on average, following an exponential distribution. To get a worst case estimation, outage periods of significant length are needed. The outage length is set fix to 48 h. Thus, the watchtowers fail independently every 48 h for a period of 48 h with probability o. On startup, the users and their watchtowers are created and for each user one channel is opened. The channel transaction events as well as the channel close events are scheduled. All channels are closed with an outdated commitment number chosen uniformly at random and a new channel is immediately created for the user. After 144 blocks it is evaluated, whether the fraud was detected. The watchtowers exchange state information with every block update as discussed in Sect. 3.3.

We conducted the simulation with $u = 5$ users, mean channel lifetime $c = 30$ d, mean transaction interval $t = 36$ h and simulated time $s = 20$ y set fix. The number of users might seem low, but an increased number of users does not affect the reliability, because each user has their own watchtowers which fail independently of others. Increasing the number of users u or the simulated time s only reduces the confidence intervals at the cost of a higher simulation time.

A reduced channel transaction interval causes better reliability results because it increases the probability that at least one watchtower learns the most current commitment number before the channel partner tries to close the channel. We have chosen channels to live a month on average and to have a new transaction every 36 h on average to model the behaviour of a super market and its customer. The parameters w and o were varied in a range of 1 to 10 in steps of 1 resp. 0.00 to 1.00 in steps of 0.05. Each parameter combination was simulated 120 times with different seeds. Figure 2 shows the reliability for 1 to 4 watchtowers with an outage probability range from 0.00 to 0.35. The simulation results show that, using three watchtower instances, the mean reliability is over 96% for an unrealistically high outage probability of 35 %. The following analytical evaluation is based on these results and three watchtowers per user are assumed.

4.2 Scalability

Memory. For the provider platforms, we have to distinguish between the enclave memory that resides in main memory when the enclave is running and persistent memory that is used to store data. In Intel SGX, the enclave memory is limited to about 97 MB, but the persistent storage can be as high as the disk space available in the hosting system. In Table 2, we show which data needs to be stored in an enclave's memory. Summing it up leads to the following memory requirements, where n_U is the number of customers, n_C the number of monitored channels, n_E the number of connected provider platforms per WTI, and n_P the total number of connected provider platforms:

$$M(n_U, n_C, n_E, n_P) = n_U \cdot 125 \text{ byte} + n_C \cdot 1996 \text{ byte} + n_P \cdot 81 \text{ byte}$$
$$+ n_E \cdot n_U \cdot 4 \text{ byte} + 36 \text{ byte}$$

The provider platform also needs runtime memory for the verification of new blocks. As they can be verified consecutively, only one block needs to be stored at a time. We assume 2.5 MB memory usage for block verification which is more than enough for the currently biggest block of 2.3 MB. We estimate the memory requirement of the enclave's code to be limited by 4 kB. Assuming $n_P = 100$ different connected provider platforms, $n_E = 3$ connected provider platforms per customer, and 5 channels per customer, the enclave's memory suffices for monitoring the channels of more than 9,500 users.

Disk Space. The received blocks need to be stored on disk. However, it suffices to store the last $144 + 6$ blocks (the highest value of to_self_delay + the blocks needed for a block to be considered confirmed). Using the values from above ($n_C = 45,000$, $n_E = 3$), the logs take a maximum of 3.5 MB per block. For six months this would accumulate to 90 GB. However, this can be reduced massively by using a non-trivial storage format. To save storage cost, the watchtower providers can agree with their customers on a time after which they are allowed to remove old logging entries.

Table 2. Memory requirements of a provider platform

Data	Type and size
key material for communication between WTI and customer	ECDH public/private key, 65 byte per customer, symmetric key 32 byte per customer
revocation_payout_addresses	32 byte hash per customer
blockchain height	4 byte integer
most current block hash	32 byte hash
state counter	per customer: 4 byte integer
monitored channels	per channel: 32 byte funding_txid, 6 byte commitment_number, both participants' payment_basepoints, 2 * 32 byte 32 byte revocation_basepoint_secret, 49 * 38 byte to derive per_commitment_secrets
connected provider platforms	per provider platform: 16 byte IP address, ECDH public key 33 byte, symmetric key 32 byte
connected provider platforms per WTI	list of 4 byte indizes

Bandwidth. When connecting provider platforms, messages are exchanged during remote attestation. As this is only done once for establishing a connection, we can ignore it for estimating the required bandwidth for the continuous operation of the system. We need to download each new block that is added to the blockchain, which results in an incoming traffic of about 10 Mbit/10 min = 17 kBit/s. Assuming that, in the unrealistic worst case, a revocation transaction has to be created for every transaction in a block, we need the same outgoing bandwidth for publishing the revocation transactions. As new blocks are mined every 10 min on average in Bitcoin, the messages sent between watchtowers after receiving a new block are so rare that they can be neglected for estimating the bandwidth. For every update to a channel, a user has to send 74 bytes to the watchtowers. Therefore, we can estimate the required bandwidth for channel updates to $\beta \cdot n_c \cdot 74$ byte, where β is the update rate per channel and n_c is the number of channels a user has. For a high channel update rate of $\beta = 10/s$ for five channels per user, this results in a bandwidth requirement of 28.9 kBit/s. With 9,000 users on a system updating their five channels with such a high rate, this results in an incoming bandwidth of 254 MBit/s, which is still realistic for a professionally hosted system.

Computation Time. Channel updates do not require much computation and searching the transactions inside a block could take a few minutes without a problem, because new blocks arrive only every 10 min on average. Therefore, the computation time is not a bottleneck.

4.3 Deployability

Estimating that 100 GB of disk space will be more than enough for the persistent storage, we used the Amazon AWS Calculator[2] to estimate the operational cost of running a watchtower system. We require an SGX-enabled CPU, 256 GB outgoing traffic per month and 100 GB of disk space. This results in a monthly cost of about 80 USD. Assuming a system working to capacity with 9,000 users and 5 channels per user, the system is viable if each user pays at least 0.18 cents per channel per month. Assuming only 1000 users and 5 channels per user, each user would have to pay at least 1.6 cents per channel per month to make the system viable. As self hosted systems might be cheaper than using AWS, this shows that providers are incentivised to host watchtower systems and offer them to their customers for reasonable prices. Payment channels can be used for the monthly payments to keep transaction fees low.

5 Security Assessment

5.1 Confidentiality

Based on the assumptions of TEEs, all secrets that are stored in the watchtower are protected from observation by any third party. The secrets are stored securely in the enclave's memory resp. encrypted on hard disk. To send secrets to the provider platform, customers use an encrypted authenticated channel which is established during remote attestation. This ensures that secrets cannot be leaked when uploading them to the provider platform and that they are stored only at attested and trusted platforms.

5.2 Verification of Availability

To fulfill the requirements considering verification of availability, it is necessary to recognise the outage of enclaves and prove it to others. TEE Guard creates logs of the availability of connected enclaves. To establish trust in these logs, the messages are signed with the providers signing key. The logs are stored in an encrypted and integrity-protected form. To disguise a time of unavailability in the past, a provider would have to manipulate the logs stored at their competitors' instances. The only way to manipulate the log in order to make a competitor's instance look unresponsive is by suppressing messages sent from and to that instance. However, this also leads to the provider's own instance being seen as unresponsive from the competitor's instance, which is not in the provider's interest. If the provider would delete the logs before the user retrieves them, the manipulation would be detected by retrieving the logs from the connected platforms. The availability of other enclaves is verified using logs of all received New Block messages which only allows determining the availability once per block interval. Since a WTI searches the transactions inside a block for fraud before sending the New Block message, this suffices to verify the availability required for a watchtower.

---

[2] https://calculator.aws/#/configureEc2, June 2019.

5.3   Attacks on TEEs

It has been shown that practical implementations of TEEs are not free of bugs [6] and experience tells us that any implementation might have flaws and thus potential bugs have to be considered. Therefore, we lastly analyse how the security of TEE Guard is affected when an attacker breaks the security assumptions of TEEs. If an attacker extracts the hardware secrets from the enclave, they might be able to impersonate an enclave and fool customers into sending them the data instead of a legitimate watchtower. In this case, the attacker could use the revocation keys to create revocation transactions that do not send the revoked funds to the customer but an address owned by the attacker. However, this also requires an outdated commitment transaction to be published on the blockchain. To run this attack, an active fraud attempt is needed. If the customer's channel partner does not attempt a fraud, the attack is only possible when the provider colludes with the customer's channel partner. The maximum profit of such an attack run by a watchtower provider, who broke the TEE and colludes with the channel partner of the customer, would be the difference between the customer's balance in the last commitment transaction and the commitment transaction with the lowest balance for the customer. In the worst case, this could be all funds of the channel. It is not possible for an attacker to create new commitment transactions because the `basepoint_secret` is not sent to the watchtower and thus cannot be stolen even by an attacker breaking the TEE.

6   Conclusion

In this work, we presented TEE Guard, a proposal for a watchtower solution for the Lightning Network. The use of TEEs led to a concept that scales, because TEEs allow secrets to be securely outsourced to the watchtower and, thus, revocation transactions to be created directly by the watchtower instead of the customer. The decentralised nature of TEE Guard ensures a high reliability of the watchtower service. We have shown with a simulation that, even at unrealistically high outage rates per platform provider, the whole system already has a reliability of 96% with only three watchtowers. As customers do not trust the platform providers to keep their watchtowers running, we presented the logging mechanism that allows the providers to prove to their customers that they have continuously been running the service. Even a very strong attacker, who is able to break the TEE, does only get access to the revocation secrets, which cannot be abused as long as there is no dispute in the channel or the watchtower provider colludes with the customer's channel partner. We quantitatively analysed the concept showing that the system is deployable and running a provider platform is financially profitable for the provider even with very low prices for the customers.

For future work, we plan to analyse the requirements for watchtowers for other payment channel networks and the security trade-offs implied by adapting TEE Guard to these networks. We also want to generalise the concept by transferring the ideas of TEE Guard to state channels.

References

1. BOLT 3: Bitcoin Transaction and Script Formats (2018). https://github.com/lightningnetwork/lightning-rfc/blob/914ebab9080ccccb0ff176/03-transactions.md
2. Anati, I., Gueron, S., Johnson, S., Scarlata, V.: Innovative technology for CPU based attestation and sealing. In: Proceedings of the 2nd International Workshop on Hardware and Architectural Support for Security and Privacy. HASP 2013, ACM, New York (2013)
3. Avarikioti, G., Kogias, E.K., Wattenhofer, R.: Brick: asynchronous state channels. arXiv preprint arXiv:1905.11360, May 2019
4. Avarikioti, G., Laufenberg, F., Sliwinski, J., Wang, Y., Wattenhofer, R.: Towards secure and efficient payment channels. arXiv preprint arXiv:1811.12740 (2018)
5. Bentov, I., et al.: Tesseract: Real-Time Cryptocurrency Exchange using Trusted Hardware. IACR Cryptology ePrint Archive **2017**, 1153 (2017)
6. Bulck, J.V., et al.: Foreshadow: extracting the Keys to the Intel SGX Kingdom with transient out-of-order execution. In: 27th USENIX Security Symposium (USENIX Security 18). USENIX Association, Baltimore, MD (2018)
7. Das, P., et al.: FastKitten: practical smart contracts on bitcoin. In: 28th USENIX Security Symposium (USENIX Security 19), pp. 801–818. USENIX Association, Santa Clara. https://www.usenix.org/conference/usenixsecurity19/presentation/das
8. Decker, C., Russell, R., Osuntokun, O.: eltoo: a simple Layer2 protocol for Bitcoin. White paper (2018). https://blockstream.com/eltoo.pdf
9. Dryja, T.: Unlinkable Outsourced Channel Monitoring (10 2016), talk at Scaling Bitcoin, Milano (2016)
10. Grundmann, M., Leinweber, M., Hartenstein, H.: Banklaves: concept for a trustworthy decentralized payment service for Bitcoin. In: 2019 IEEE International Conference on Blockchain and Cryptocurrency (ICBC), pp. 268–276, May 2019. https://doi.org/10.1109/BLOC.2019.8751394, https://publikationen.bibliothek.kit.edu/1000092459
11. Intel: PoET 1.0 Specification (2015). https://sawtooth.hyperledger.org/docs/core/releases/latest/architecture/poet.html
12. Kaplan, D., Powell, J., Woller, T.: AMD Memory Encryption (2016). http://developer.amd.com/wordpress/media/2013/12/AMD_Memory_Encryption_Whitepaper_v7-Public.pdf
13. Lee, D., Kohlbrenner, D., Shinde, S., Song, D., Asanović, K.: Keystone: A Framework for Architecting TEEs. arXiv preprint arXiv:1907.10119 (2019)
14. Lind, J., Eyal, I., Kelbert, F., Naor, O., Pietzuch, P.R., Sirer, E.G.: Teechain: Scalable Blockchain Payments using Trusted Execution Environments (2017). http://arxiv.org/abs/1707.05454
15. Matetic, S., et al.: ROTE: Rollback Protection for Trusted Execution, pp. 1289–1306, August 2017. https://www.usenix.org/conference/usenixsecurity17/technical-sessions/presentation/matetic
16. McCorry, P., Bakshi, S., Bentov, I., Miller, A., Meiklejohn, S.: Pisa: Arbitration Outsourcing for State Channels. IACR Cryptology ePrint Archive **2018**, 582 (2018)
17. McKeen, F., et al.: Innovative instructions and software model for isolated execution. In: Proceedings of the 2nd International Workshop on Hardware and Architectural Support for Security and Privacy. HASP 2013. ACM, New York (2013)

18. Milutinovic, M., He, W., Wu, H., Kanwal, M.: Proof of luck: an efficient blockchain consensus protocol. In: Proceedings of the 1st Workshop on System Software for Trusted Execution. SysTEX 2016, pp. 2:1–2:6. ACM, New York (2016). https://doi.org/10.1145/3007788.3007790
19. Nakamoto, S.: Bitcoin: A Peer-to-Peer Electronic Cash System (2008). https://bitcoin.org/bitcoin.pdf
20. Osuntokun, O.: Hardening Lightning (01 2018), talk at Blockchain Protocol Analysis and Security Engineering (2018)
21. Poon, J., Dryja, T.: The Bitcoin Lightning Network: Scalable Off-Chain Instant Payments (2016). https://lightning.network/lightning-network-paper.pdf

Payment Networks as Creation Games

Georgia Avarikioti[(✉)], Rolf Scheuner, and Roger Wattenhofer

ETH Zurich, Zürich, Switzerland
{zetavar,schrolf,wattenhofer}@ethz.ch

Abstract. Payment networks were introduced to address the limitation on the transaction throughput of popular blockchains. To open a payment channel one has to publish a transaction on-chain and pay the appropriate transaction fee. A transaction can be routed in the network, as long as there is a path of channels with the necessary capital. The intermediate nodes on this path can ask for a fee to forward the transaction. Hence, opening channels, although costly, can benefit a party, both by reducing the cost of the party for sending a transaction and by collecting the fees from forwarding transactions of other parties.

This trade-off spawns a network creation game between the channel parties. In this work, we introduce the first game theoretic model for analyzing the network creation game on blockchain payment channels. Further, we examine various network structures (path, star, complete bipartite graph and clique) and determine for each one of them the constraints (fee value) under which they constitute a Nash equilibrium, given a fixed fee policy. Last, we show that the star is a Nash equilibrium when each channel party can freely decide the channel fee. On the other hand, we prove the complete bipartite graph can never be a Nash equilibrium, given a free fee policy.

Keywords: Blockchain · Payment channels · Layer 2 ·
Creation game · Network design · Nash equilibrium · Payment hubs

1 Introduction

Distributed ledgers that employ the Nakamoto [13] or similar consensus mechanisms suffer from major scalability problems [8]. In essence, the security of the consensus is based on the ability of each node to verify and store a replica of the entire blockchain history. Prominent solutions to this problem are payment channels [10,14,17]. Payment channels are constructions that allow participants of the blockchain to execute the transactions off-chain while maintaining the security guarantees of the blockchain. The parties that enter a payment channel open a joint account with a specific capital and update the distribution of this capital every time they exchange a transaction. This way the parties can execute an unlimited number of transactions as long as they all agree on the current distribution of capital. In case of a dispute, the blockchain acts as a judge, ensuring that the latest agreed capital distribution is enforced.

© Springer Nature Switzerland AG 2019
C. Pérez-Solà et al. (Eds.): DPM 2019/CBT 2019, LNCS 11737, pp. 195–210, 2019.
https://doi.org/10.1007/978-3-030-31500-9_12

Multiple payment channels create a payment network, where a transaction can be executed even though there is no direct channel between payer and payee. Currently, this is achieved with the Bitcoin Lightning Network [14] using Hash Timelock Contracts (HTLCs) [1,10]. When a sender wants to route a transaction through the network, a path of channels is discovered that has enough capital on each edge to route the transactions to the receiver [12,15,16]. The intermediate nodes that move their capital act as service providers for the sender, and ask for a fee for their service. Therefore, operating a payment channel can offer revenue to the owner of the channel apart from reducing the cost of executing transactions on-chain. On the other hand, creating a payment channel is costly, since the opening and closing of the channel must occur on-chain, which requires to pay the regular blockchain transaction fee to the miner.

In this work, we study this trade-off. We investigate possible strategies for a participant. Assuming a constant blockchain fee for each transaction, when does it make sense to open a channel? If a path from sender to receiver already exists, does it make sense to create a cheaper path (or even a direct channel) to claim the fees for forwarding transactions? Our goal is to understand under which constraints specific network structures are Nash equilibiria. In other words, when can the participants of the network increase their profit (or decrease their costs) by changing the network structure? Further, we ask which network structures are stable under a free fee policy where each node on the network that operates a channel can set its own fee on the channel. To the best of our knowledge, our work is the first to analyze payment channels in a formal game theoretic model.

Our Contributions. First, we introduce a formal game-theoretic model which we later use to study the potential strategies for network creation for the nodes of the payment network. The model encapsulates a network creation game on payment channels: each node of the network can create multiple channels to other nodes to collect fees from the transactions that are routed through their channel. However, each channel creation costs the blockchain fee, and also each node competes with the other nodes in the network, since the sender of each transactions will choose the cheapest path to route the transaction to the receiver.

We assume there is a fixed fee a sender has to pay to each intermediate node to route the transaction to the receiver. Further, we assume nodes have unlimited temporary capital, and that each pair of nodes are sender and receiver to an equal amount of transactions. Under these assumptions, we explore various network structures, namely the path, the star, the complete bipartite graph and the clique. We find that each network structure constitutes a Nash equilibrium for a specific fee value. In particular, we show that the path is a weak Nash equilibrium only if the network fee is zero. Then, we show an upper bound for value of the fee on the star graph. The bound depends on the number of transactions, number of nodes and the value of the blockchain fee. This means that creating a hub is a Nash equilibrium as long as the fee is very low (compared to the blockchain fee). We observe that if the fee is above the upper bound, a two-stars structure emerges as a Nash equilibrium. We generalize this observation by examining complete bipartite graphs. We provide upper and lower bounds for such graphs which additionally depend on the number of centers (smaller side of the complete bipartite graph). This specific network structure defines an entire class of Nash equilibria. Finally, we consider the complete graph (clique), which is naturally a Nash equilibria when the network fee is very high.

From the plethora of network structures that constitute Nash equilibria when the fee policy is fixed, only the star is stable under a free fee policy. Specifically, we show that the star is a pure Nash equilibrium, and the nodes will set the fee almost equal to the upper bound of the fixed fee policy. Further, we prove that the complete bipartite graph can never be a Nash equilibrium when nodes chose the fee of their channels freely as part of their strategy.

2   Model

In this section, we introduce the game theoretic model. To this end, we first define the necessary notation and assumptions.

Capital & fees. We assume all participants (nodes of the network) have unlimited temporary capital and thus the capital locked in all channels is also considered unlimited. This means that the channels can never be depleted. Moreover, this leads to stable fees that do not depend on the value of the routed transaction, because the participants are only interested in the number of transactions routed through their channel since the capital movement does not cost (they have unlimited capacity).

The cost of every transaction and hence the cost of opening and closing a channel on the blockchain costs the blockchain fee. We assume a fixed blockchain fee, $F_B \in \mathbb{R}^+$, i.e., the fee is constant and stable over time. Further, we assume the fee is unilaterally paid by the node opening or closing the channel. We also assume a fee f_0 for forwarding a transaction through a channel (that is not owned by the sender of the transaction) is the same for all nodes and stable in time.

Information & time. We assume full information, i.e., every node of the system knows the complete payment scenario, and the channels created by other nodes. Although the decision of closing and opening a channel can occur at any time by any participant of the network, we assume a simultaneous game, i.e., the participants open the channels in the beginning before executing any transactions. This is reasonable because we assume a full information game, thus every node knows its optimal strategy apriori.

Notation. The set of nodes participating in the network is denoted by N, and the set of transactions to be executed (payment scenario) by P. We assume $N > 3$ and we use the terms transaction and payment interchangeably throughout the paper. We define as $R(x, p)$ the (set of) cheapest route(s) of a transaction $p \in P$ in network state x. The network state x is dependent on the strategy of the nodes, i.e. which channels the nodes have opened in the network. We note that if there is no route with cost lower than the blockchain fee, the set will return empty and the transaction will be executed on the blockchain. Thus, the cost for a transaction p for the sender of the transaction on a network state x is the number of edges of the shortest path (for constant fee f_0 the cheapest is the shortest path) times the fee f_0,

$$f(x,p) = \begin{cases} (|\boldsymbol{R}(x,p)| - 1) \cdot f_0, & if\ \boldsymbol{R}(x,p) \neq \emptyset \\ F_B,\ else \end{cases}$$

On the other hand, the revenue of a node $u \in \boldsymbol{N}$ from a transaction $p \in \boldsymbol{P}$ when executed on the network in state x is f_0 if node u is part of the cheapest route $\boldsymbol{R}(x,p)$; otherwise the revenue is zero.

Moreover, the *strategy set* of a node $u \in \boldsymbol{N}$ is the set of all strategies available to node u. It is denoted as $\boldsymbol{S}_u$. The strategy of node u is denoted as $\mu_u \in \boldsymbol{S}_u$. The strategy set in our setting represents the channels a node decides to open at the beginning of the game. A *strategy combination* is a set containing a strategy for every node. The set of all possible strategy combinations is defined as $\boldsymbol{S}^N := \prod_{u \in N} \boldsymbol{S}_u$, while a strategy combination is denoted by $\boldsymbol{\mu} \in \boldsymbol{S}^N$. For simplicity, we will abuse the notation μ and μ_u to also denote the cardinality of the set, i.e., how many channels are open in the network and how many channels node u opens in the network, respectively. Last, we define the number of on-chain payments made by a node u as b_u and the total number of on-chain payments as $b = \sum_{\forall u \in N} b_u$.

Next, we define the necessary functions for the analysis, namely the cost function, the social cost and the social optimum.

Cost Function. The cost function contains the cost for channel creation, on-chain payments, payments routed through the network and the revenue from forwarding payments. For a node u it is defined as

$$c(\boldsymbol{\mu}, \mu_u) = \mu_u \cdot F_B + b_u \cdot F_B + \sum_{p \in P: s(p)=u} f(x,p) - \sum_{p \in P: u \in R(x,p)} f_0$$

where $s(p)$ denotes the sender of transaction p. In our setting, sender and receiver of a transaction p are the only relevant pieces of information of each payment, since all fees are independent of the value of a transaction.

Social Cost. The social cost (or negative welfare) is the sum of the costs of all nodes

$$-W = \sum_{n \in N} c(\boldsymbol{\mu}, \mu_u) = (\mu + b) \cdot F_B$$

Social Optimum. The social optimum is the minimum social cost, which depends on the number of open channels and the number of payments executed on-chain. This term is minimized when all transactions are executed off-chain and the network forms a tree (connected with minimum number of channels). Hence, the social optimum is $min(-W) = (N - 1) \cdot F_B$

3 Channel Creation Game

First, we show some observations that hold generally under any set of transactions and graph structure. Then, we analyze specific structures and determine under which parameters they constitute a Nash equilibrium.

3.1 Basic Properties

Lemma 1. *In a pure Nash equilibrium, no channels are opened twice.*

Proof. (Towards contradiction.) Suppose there is a pure Nash equilibrium in which two nodes have opened a channel with each other twice. In this case, each node can reduce his cost by not opening the second channel and thus the strategy cannot be a Nash equilibrium.   □

Lemma 2. *In a pure, strict Nash equilibrium[1], none of the transactions are executed on-chain.*

Proof. (Towards contradiction.) Suppose there is a pure, strict Nash equilibrium in which a node (sender) executes a transaction on-chain. If there is a channel to the receiver of the transaction, the sender can reduce his cost by simply using the channel. The same holds if there is a path of channels with total fees less than the blockchain fee. Therefore, either there is no cheap path from sender to receiver or no path at all. In this case, if the sender has at least two transactions to send to the receiver, he would open a channel and reduce the cost. Thus, the sender has a single transaction to send to the receiver. However, the cost of opening a channel and the cost of executing the transaction on-chain is exactly the same (blockchain fee). If we assume there are no transactions routed through the channel from sender to receiver, the payoff (cost) of the sender in both strategies is the same. Hence, executing the transaction on-chain cannot be a strict Nash equilibrium. Thus, there are transactions routed through the channel from sender to receiver. Then, the payoff of the sender increases and hence the dominant strategy is to open the channel. This contradicts the assumption that executing the transaction on-chain is a Nash equilibrium.   □

Next, we analyse the channel creation game for a homogeneous payment scenario, where every node makes exactly $k \geq 1$ payments to every other node. The number of transactions is therefore $P = k \cdot N \cdot (N - 1)$. For this payment scenario, we analyze multiple strategy combinations, i.e., different graph structures such as the path, the star, a complete bipartite graph and the clique, to discover under which parameters these graph structures constitute a Nash equilibrium. We note that all graph structures that are trees (e.g. path, start, complete bipartite graph) are social optima.

---

[1] If a strategy is always strictly better than all others for all profiles of other players' strategies, then it is strictly dominant. If the strategy is strictly dominant for all players, then it is a strict Nash equilibrium.

3.2 Path

The first graph structure we investigate is the path: each node connects to the node with the next higher ID. The node with the highest ID does not create a channel, but he is connected to the network through the node with the second highest ID.

Social Cost. The social cost is $-W = (N - 1) \cdot F_B$.

Nash Equilibrium. A specific strategy is a Nash equilibrium if none of the players can increase their payoff by deviating from it. This means that the path is a NE only if all possible deviations lead to higher cost (or equivalent if it is a weak NE) for the deviating node. We observe that for $f_0 = 0$, deviating from the path structure cannot decrease the cost, thus the path is a weak NE (similarly to every other tree structure). However, for any fee $f_0 > 0$, the first node can increase the revenue from the fees, and thus decrease his cost, by connecting to a middle node on the path and allowing transactions to be routed through his channel. Thus, for any non-zero fee, the path is not a NE.

3.3 Star

The second graph structure we investigate is the star: one node creates channels to everyone else, while the other nodes do not create any channels. The strategy to create channels to $a \in [0, N - 1]$ outer nodes is denoted as (a).

Cost Functions. The cost of the center node is $c(\mu, (N - 1)) = (N - 1) \cdot F_B - (N - 1) \cdot (N - 2) \cdot k \cdot f_0$.
The cost of the outer nodes is $c(\mu, (0)) = (N - 2) \cdot k \cdot f_0$.
The social cost is $-W = (N - 1) \cdot F_B$.

Nash Equilibrium. A node can only deviate from the strategy as follows: An outer node creates channels to $a \in [1, N - 2]$ other outer nodes. This holds because in the homogeneous payment scenario, a pure Nash equilibrium demands a connected graph, else Lemma 2 is violated. This means the center node will not disconnect the graph. Moreover, from Lemma 1 no outer node will create a channel to the center node.

If an outer node creates channels to $a \in [1, N - 2]$ other outer nodes, his cost function is $c(\mu, (a)) = a \cdot F_B + (N - 2 - a) \cdot k \cdot f_0 - a \cdot (a - 1) \cdot k \cdot \frac{1}{2} \cdot f_0$.

If there is an $a \in [1, N - 2]$ for which the cost function of an outer node is decreased, then the star is not a NE. Since the second derivative with respect to a is strictly negative, we only have to check the corner cases below:

- For $a = 1$: $c(\mu, (0)) < c(\mu, (1)) \Leftrightarrow f_0 < \frac{F_B}{k}$.
- For $a = N - 2$: $c(\mu, (0)) < c(\mu, (N - 2)) \Leftrightarrow f_0 < \frac{F_B}{k} \cdot \frac{2}{N-1}$.

Thus, for $N > 3$, the star is a NE when

$$f_0 < \frac{F_B}{k} \cdot \frac{2}{N - 1}$$

3.4   Star with Two Centers

We observe that for a very low constant fee any star can be a Nash equilibrium. Further, we notice that if the fee is high enough the dominant strategy for the outer nodes is to create more channels, eventually becoming the center of a second star. We examine this exact case, where there are two center nodes, each creating channels to all outer nodes, but not to each other. The outer nodes do not create any channels. We denote the strategy to create $a \in [0,2]$ channels to center nodes and $b \in [0, N-2]$ channels to outer nodes as (a,b).

Cost Functions. The cost of the center nodes is $c(\mu, (0, N-2)) = (N-2) \cdot F_B + k \cdot f_0 - (N-2) \cdot (N-3) \cdot k \cdot \frac{1}{2} f_0$.
The cost of the outer nodes is $c(\mu, (0,0)) = (N-3) \cdot k \cdot f_0 - 2 \cdot k \cdot \frac{1}{(N-2)} f_0$.
   The social cost is $-W = 2 \cdot (N-2) \cdot F_B$.

Nash Equilibrium. The nodes can deviate as follows:

(A) A center node creates channels to $b \in [1, N-3]$ outer nodes.
(B) A center node creates channels to $b \in [0, N-2]$ outer nodes and creates a channel to the other center node.
(C) An outer node creates channels to $b \in [1, N-3]$ other outer nodes.

We analyze each case to determine the parameter space for which these deviations do not decrease the cost of a node, and thus the two center structure is a NE.

(Deviation A). If a center node created channels to only $b \in [1, N-3]$ outer nodes, his cost function would become

$$c(\mu, (0, b)) = b \cdot F_B + k \cdot f_0 + (N - 2 - b) \cdot k \cdot 2f_0 - b \cdot (b-1) \cdot k \cdot \frac{1}{2} f_0$$

This cost function must be higher than $c(\mu, (0, N-2))$ for all b. Since the second derivative w.r.t. b is strictly negative, we only have to check the corner cases:

- For $b = 1$: $c(\mu, (0, N-2)) < c(\mu, (0, 1)) \Leftrightarrow f_0 > \frac{F_B}{k} \cdot \frac{2}{N+2}$.
- For $b = N - 3$: $c(\mu, (0, N-2)) < c(\mu, (0, N-3)) \Leftrightarrow f_0 > \frac{F_B}{k} \cdot \frac{1}{N-1}$

(Deviation B). If an center node creates a channel to the other center node and channels to $b \in [0, N-2]$ outer nodes, his cost function is

$$c(\mu, (1, b) = (b+1) \cdot F_B + (N - 2 - b) \cdot k \cdot f_0 - b \cdot (b-1) \cdot k \cdot \frac{1}{2} f_0$$

This cost function must be higher than $c(\mu, (0, N-2))$ for all b. Similarly, we only have to check the corner cases:

- For $b = 0$: $c(\mu, (0, N-2)) < c(\mu, (1, 0)) \Leftrightarrow f_0 < \frac{F_B}{k} \cdot \frac{2}{N}$.
- For $b = N - 2$: $c(\mu, (0, N-2)) < c(\mu, (1, N-2)) \Leftrightarrow f_0 > \frac{F_B}{k}$.

(Deviation C). If an outer node creates channels to $b \in [1, N-2]$ other outer nodes, his cost function is

$$c(\mu, (0,b)) = b \cdot F_B + (N-3-b) \cdot k \cdot f_0 - 2 \cdot k \cdot \frac{1}{(N-2)} f_0 - b \cdot (b-1) \cdot k \cdot \frac{1}{3} f_0$$

This cost function must be higher than $c_o(\mu, (0,0))$ for all b. Similarly, we only have to check the corner cases:

- For $b = 1$: $c(\mu, (0,0)) < c(\mu, (0,1)) \Leftrightarrow f_0 < \frac{F_B}{k}$.
- For $b = N-3$: $c(\mu, (0,0)) < c(\mu, (0, N-3)) \Leftrightarrow f_0 < \frac{F_B}{k} \cdot \frac{3}{N-1}$.

Combining all the bounds from the deviating strategies, for $N > 3$, we derive the parameter space for which the two center structure is a NE. Specifically, the conditions reduce to

$$\frac{F_B}{k} \cdot \frac{2}{N} < f_0 < \frac{F_B}{k} \cdot \frac{3}{N-1}$$

3.5   Complete Bipartite Graph

Previously, we showed that stars with one or two center nodes can be a Nash equilibrium, if there is a constant fee that fulfills certain conditions. Furthermore, if f_0 is high, outer nodes decrease their cost by creating channels to other outer nodes. Intuitively, this leads to a NE that is a bipartite graph structure. We study exactly this case: $c \in [2, N/2]$ nodes build a center by creating channels to everyone else but each other. For simplicity we denote the number of outer nodes by $d := N - c$. The network structure now is a complete bipartite graph with c nodes in the smaller partition and d nodes in the larger partition. We denote the strategy to create channels to $a \in [0, c]$ center nodes and to $b \in [0, d]$ outer nodes as (a, b).

Cost Functions. The cost of the center nodes is $c(\mu, (0, d)) = d \cdot F_B + (c-1) \cdot k \cdot f_0 - d \cdot (d-1) \cdot k \cdot \frac{1}{c} f_0$.
   The cost of the outer nodes is $c(\mu, (0,0)) = (d-1) \cdot k \cdot f_0 - c \cdot (c-1) \cdot k \cdot \frac{1}{d} f_0$.
   The social cost is $-W = c \cdot (N - c) \cdot F_B$.

Nash Equilibrium. The nodes can deviate as follows:

(A) A center node creates channels to only $b \in [1, d-1]$ outer nodes.
(B) A center node creates channels to $a \in [1, c-1]$ center nodes and to $b = 0$ outer nodes.
(C) A center node creates channels to $a \in [1, c-1]$ center nodes and to $b \in [1, d]$ outer nodes.
(D) An outer node creates channels to $b \in [1, d-1]$ other outer nodes.

Next, we discover the parameter space for which the strategies above lead to the increase of the cost function of a node.

(Deviation A). If a center node created channels to only $b \in [1, d-1]$ outer nodes, his cost function would become

$$c(\mu, (0, b)) = b \cdot F_B + (c-1) \cdot k \cdot f_0 + (d-b) \cdot k \cdot 2f_0 - b \cdot (b-1) \cdot k \cdot \frac{1}{c} f_0$$

This cost function must be higher than $c(\mu, (0, d))$ for all b. Since the second derivative w.r.t. b is strictly negative, we only check the corner cases:

- For $b = 1$: $c(\mu, (0, d)) < c(\mu, (0, 1)) \Leftrightarrow \frac{F_B}{k} \cdot \frac{c}{N+c} < f_0$.
- For $b = d-1$: $c(\mu, (0, d)) < c(\mu, (0, d-1)) \Leftrightarrow \frac{F_B}{k} \cdot \frac{c}{2N-2} < f_0$.

(Deviation B). If a center node creates channels to $a \in [1, c-1]$ other center nodes and to $b \in [1, d]$ outer nodes, his cost function is

$$c(\mu, (a, b)) = (a+b) \cdot F_B + (d-b) \cdot k \cdot f_0 + (c-1-a) \cdot k \cdot f_0$$
$$-b \cdot (b-1) \cdot k \cdot \frac{1}{c} f_0 - a \cdot (a-1) \cdot k \cdot \frac{1}{d+1} f_0$$

This cost function must be higher than $c(\mu, (0, d))$. Since the second derivatives w.r.t. a and b are strictly negative, we only have to check the corner cases:

- For $a = 1, b = 1$:

$$c(\mu, (0, d)) < c(\mu, (1, 1)) \Leftrightarrow \frac{F_B}{k} \cdot \frac{cN - c^2 - 2c}{N^2 - cN + N - 3c} < f_0$$

- For $a = 1, b = d$: $c(\mu, (0, d)) < c(\mu, (1, d)) \Leftrightarrow f_0 < \frac{F_B}{k}$
- For $a = c-1, b = 1$:

$$c(\mu, (0, d)) < c(\mu, (c-1, 1)) \Leftrightarrow \frac{F_B}{k} \cdot \frac{cN^2 - 3c^2 N + cN + 2c^3 - 2c^2}{N^3 - 2cN^2 + cN - N + c^2 - c} < f_0$$

- For $a = c-1, b = d$: $c(\mu, (0, d)) < c(\mu, (c-1, d)) \Leftrightarrow f_0 < \frac{F_B}{k} \cdot \frac{N-c+1}{N-1}$

(Deviation C). If a center node creates channels to $a \in [1, c-1]$ other center nodes and to $b = 0$ outer nodes, his cost function is

$$c(\mu, (a, 0)) = a \cdot F_B + d \cdot k \cdot f_0 + (c-1-a) \cdot k \cdot 2f_0 - (a) \cdot (a-1) \cdot k \cdot \frac{1}{d+1} f_0$$

This cost function must be higher than $c(\mu, (0, d))$ for all a. Since the second derivative w.r.t. a is strictly negative, we only have to check the corner cases:

- For $a = 1$:

$$c(\mu, (0, d)) < c(\mu, (1, 0)) \Leftrightarrow \frac{F_B}{k} \cdot \frac{cN - c^2 - c}{N^2 - cN - N + c^2 - 2c} < f_0$$

– For $a = c - 1$:

$$c(\mu, (0, d)) < c(\mu, (c - 1, 0)) \Leftrightarrow \frac{F_B}{k} \cdot \frac{cN^2 - 3c^2 N + 2cN + 2c^3 - 3c^2 + c}{N^3 - 2cN^2 + 2cN - N} < f_0$$

(Deviation D). If an outer node creates channels to $b \in [1, d - 1]$ other outer nodes, his cost function is

$$c(\mu, (0, b)) = b \cdot F_B + (d - 1 - b) \cdot k \cdot f_0 - c \cdot (c - 1) \cdot k \cdot \frac{1}{d} f_0 - b \cdot (n - 1) \cdot k \cdot \frac{1}{c + 1} f_0$$

This cost function must be higher than $c_o(\mu, (0, 0))$ for all b. Since the second derivative w.r.t. b is strictly negative, we only have to check the corner cases:

– For $b = 1$: $c(\mu, (0, 0)) < c(\mu, (0, 1)) \Leftrightarrow f_0 < \frac{F_B}{k}$
– For $b = d - 1$: $c(\mu, (0, 0)) < c(\mu, (0, d - 1)) \Leftrightarrow f_0 < \frac{F_B}{k} \cdot \frac{c + 1}{N - 1}$

From the analysis on the possible deviations from the strategy, we derive multiple upper and lower bounds for the value of f_0. For $N > 3$ and $2 \leq c \leq N/2$ these conditions reduce to the following:

$$f_0 > \frac{F_B}{k} \cdot \frac{cN - c^2 - 2c}{N^2 - cN + N - 3c}$$

$$f_0 > \frac{F_B}{k} \cdot \frac{cN - c^2 - c}{N^2 - cN - N + c^2 - 2c}$$

$$f_0 < \frac{F_B}{k} \cdot \frac{c + 1}{N - 1}$$

We have defined not only one Nash equilibrium in this analysis, but a whole class of Nash equilibria. Table 1 shows the numerical values for the bounds of a complete bipartite graph as Nash equilibrium. Figure 1 shows a plot of the bounds for $N = 10^3$. The Nash equilibria lay in the thin area between the lowest red and the highest blue line.

3.6 Clique

The last graph structure we investigate is the clique, i.e., the complete graph. In this case, the i-th node opens $N - i$ channels. The strategy of creating channels to a nodes without self-loops is denoted as (a).

Cost Functions. The cost of the i-th node is $c(\mu, (N - i)) = (N - i) \cdot F_B$. The social cost is $-W = \frac{N \cdot (N - 1)}{2} \cdot F_B$.

Table 1. Numerical results for the lower and bounds for a complete bipartite graph as a Nash equilibrium.

N	c	lower bound $\left[\frac{F_B}{k}\right]$	upper bound $\left[\frac{F_B}{k}\right]$	active lb	active ub
10^3	2	$.2000000 \cdot 10^{-2}$	$.30030 \cdot 10^{-2}$	5	3
	3	$.2999991 \cdot 10^{-2}$	$.40040 \cdot 10^{-2}$	5	3
	5	$.4999925 \cdot 10^{-2}$	$.60060 \cdot 10^{-2}$	5	3
	10	$.9999192 \cdot 10^{-2}$	$.11011 \cdot 10^{-1}$	5	3
	100	$.9970024 \cdot 10^{-1}$	$.10110$	3	3
	499	$.4975016$	$.50050$	3	3
	500	$.4984984$	$.50150$	3	3
10^4	2	$.20000 \cdot 10^{-3}$	$.30003 \cdot 10^{-3}$	5	3
	3	$.29997 \cdot 10^{-3}$	$.40004 \cdot 10^{-3}$	5	3
	5	$.49995 \cdot 10^{-3}$	$.60006 \cdot 10^{-3}$	5	3
	10	$.99990 \cdot 10^{-3}$	$.11001 \cdot 10^{-2}$	5	3
	100	$.99981 \cdot 10^{-2}$	$.10101 \cdot 10^{-1}$	5	3
	1000	$.99971 \cdot 10^{-1}$	$.10011$	3	3
	4999	$.49976$	$.50005$	3	3
	5000	$.49985$	$.50015$	3	3
10^5	2	$.200000 \cdot 10^{-4}$	$.300003 \cdot 10^{-4}$	5	3
	3	$.299997 \cdot 10^{-4}$	$.400004 \cdot 10^{-4}$	5	3
	5	$.499995 \cdot 10^{-4}$	$.600006 \cdot 10^{-4}$	5	3
	10	$.999990 \cdot 10^{-4}$	$.110001 \cdot 10^{-3}$	5	3
	100	$.999990 \cdot 10^{-3}$	$.101001 \cdot 10^{-2}$	5	3
	1000	$.999971 \cdot 10^{-2}$	$.100101 \cdot 10^{-1}$	3	3
	10000	$.999971 \cdot 10^{-1}$	$.100011$	3	3
	49999	$.499976$	$.500005$	3	3
	50000	$.499985$	$.500015$	3	3

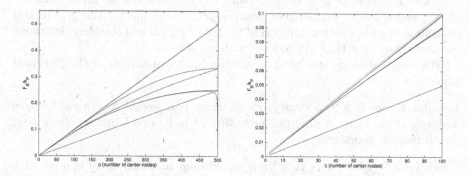

Fig. 1. Plots of the bounds for $N = 10^3$ with the upper bounds in red and the lower bounds in blue. (Color figure online)

Nash Equilibrium. The nodes can deviate as follows:

(A) The first node creates channels to only $a \in [1, N-2]$ other nodes.
(B) Node i (but not the first or last one) creates channels to only $a \in [0, N-i-1]$ nodes from the set of nodes he would originally connect to (node $i+1$ to node N).

Now, we analyze these deviation strategies to explore the parameter space for which the strategies above lead to the increase of the cost function of a node.

(Deviation A). If the first node creates channels to $a \in [1, N-2]$ other nodes his cost function is $c(\mu, (a)) = a \cdot F_B + (N-1-a) \cdot k \cdot f_0$. This cost function must be higher than $c(\mu, (N-1))$ for all a. Thus, $c(\mu, (N-1)) < c(\mu, (a)) \Leftrightarrow f_0 > \frac{F_B}{k}$.

(Deviation B). If node i (not the first or last one) creates channels to $a \in [0, N-i-1]$ nodes from the set of nodes he would originally connect to (node $i+1$ to node N), his cost functions is $c(\mu, (a)) = a \cdot F_B + (N-i-a) \cdot k \cdot f_0$. This cost function must be higher than $c(\mu, (N-i))$ for all a. Thus, $c(\mu, (N-i)) < c(\mu, (a)) \Leftrightarrow f_0 > \frac{F_B}{k}$.

To summarize, the clique is a Nash equilibrium for $f_0 > \frac{F_B}{k}$.

3.7   The Fee Game

In this subsection we investigate how the nodes set the fees, if the network structure is fixed to one of the previously found Nash equilibria. Especially, we try to find a Nash equilibrium, which also holds for the conditions discussed in the previous subsections. Therefore, we slightly change the model as follows: x is a Nash equilibrium from the previous subsections. The nodes can only set a constant fee on each of their channels. Further, the nodes cannot create new channels. We still consider simultaneous game play, i.e., the nodes must (simultaneously) choose their strategy before any transaction is executed in the network. The payment scenario is still the same (homogeneous).

Complete Bipartite Graph. We start with a complete bipartite graph with $c \in [2, N/2]$ nodes in the smaller partition creating the channels. The goal of this analysis is to gain a better intuition of the fee evolution and therefore determine the number of hubs that will be eventually established.

We make following statement about the Nash equilibrium of the described game.

Lemma 3. *For $k > 2$, a strategy combination is a (weak) Nash equilibrium if and only if there are at least two node-distinct paths which are free (zero fees) for all indirect transactions.*

Proof. $(\rightarrow)$ Suppose there is a Nash equilibrium in which there is a set of k indirect transactions (according to the payment scenario, each pair of sender and receiver executes k transactions) that can be routed only through a single path with positive fees. Then, another node will open a channel to connect the two nodes and increase his payoff, since on expectation half the k transactions will go through the new channel. Contradiction.

Suppose now, there is only a single path with zero fees to route the transaction. Then, the nodes acting as intermediaries will increase the fee almost matching the price of the blockchain fee. In such a case, we have the same effect as described above. Contradiction.

Therefore, there cannot be a single path connecting any two nodes in the network. Suppose now there are multiple node-distinct paths that connect sender and receiver, but with positive fees. Then, each node of these paths will decrease the fee in an attempt to win out the competition by being the cheapest path. Contradiction. Therefore, any strategy that is a Nash equilibrium must contain at least two node-distinct paths for each pair of sender and receiver.

($\leftarrow$) If there exist a path with zero fees for every transaction, no node stands to gain from opening a new channel. Furthermore, increasing the fee in any path will not lead to the decrease of the cost function (higher revenue) since the path containing the non-zero fee will be ignored and no transactions will be routed through such a path. Thus, no node can gain from choosing a different strategy, i.e., the strategy combination is a Nash equilibrium.   □

Theorem 1. *The complete bipartite graph is not a Nash Equilibrium when the nodes are free to chose the fees on their channels.*

Proof. Follows immediately from Lemma 3 and the lower bound established in Subsect. 3.5.   □

Star. Next, we consider the fee evolution when the network structure is a star. The reason we proceed with this specific network structure is the previous observations; when nodes can freely chose the fees they impose on their channels, having multiple paths leads to zero-fee paths. Intuitively, the star does not suffer from this problem.

Particularly, we notice that the center node is the only node that can charge fees, since all transactions are routed through the center node. However, in Subsect. 3.3, we showed that there is an upper bound on the value of the fee the center node can ask for; otherwise other nodes will deviate from the strategy combination and form a second hub.

Corollary 1. *The star is a pure Nash equilibrium when* $f_0 = \frac{F_B}{k} \cdot \frac{2}{N-1} - \epsilon$.

4   Related Work

Payment channels were originally introduced by Spilman [17]. The core idea was to use unidirectional channels with a predefined sender and receiver. Later, various constructions for bidirectional payment channels were proposed [6,9,10, 14,17]. They all use a common account for the parties and off-chain exchange of signed transactions proving the state of the channel. The creation of multiple such channels on a common blockchain network leads to the formation of channel networks, such as the Lightning network [14] on Bitcoin [13], and the Raiden network [2] on Ethereum [18]. In this work, we study different strategies for the nodes in such payment networks, independent of which payment channel construction method is used. Thus, this work applies to all payment channel solutions.

Avarikioti et al. [5,7] formulated a similar problem to the one studied in this paper. Their goal was to find an optimal strategy for a central coordinator, a so-called payment service provider. In contrast to [5,7], our work studies a situation with *multiple* players. In other words, our work is rooted in the area of game theory, whereas [5,7] was using optimization methods. Despite these differences, we found that the near-optimal solution of [5,7] (the star as network structure) is also a Nash equilibrium in an uncoordinated situation. So we get a similar result despite two completely different approaches. This is a strong indication that the Lightning network (and similar others) will eventually develop into a more centralized network structure.

Network creation games, originally introduced in by Fabrikant et al. [11], are used to model distributed networks with rational players. Each player wants to maximize/minimize a profit/cost function which represents the cost of creating and using the network. Fabrikant et al. [11] modeled the Internet using Network Creation Games. They introduced a cost function containing the network creation cost and the sum of the distances to the other nodes. For their model, they proved upper and lower bounds for the Price of Anarchy(PoA). They also conjectured that the Nash equilibria in this game are trees, however this was disproved by Albers et al. [3]. Alon et al. [4] aimed for stronger bounds on PoA of the Network Creation Game (sum and local-diameter version). Both these works, however, use simple cost functions, where the creation cost and the usage cost of an node are independent. In contrast, in this work the cost function of a node on the payment network contains both the revenue from the fees of the channel when the node is an intermediate node in a multihop transaction as well as the fees paid by the node when he is sending a multihop transaction. Hence, the cost function depends on the state of the network which itself contains the individual fee policies of the nodes. Overall, the channel creation game is probably more complex than previous work in this domain.

5   Conclusion

We introduced the first game-theoretic model that encapsulates the payment channel creation game on blockchain networks. First, we explored various network structures and determined the parameter space for which they constitute a Nash equilibrium. For the analysis, we initially assumed a fixed fee policy where each node benefits the same for each transaction routed through any of his channels. Then, we briefly considered a free fee policy, where the fee of each channel is part of the strategy of the node.

Particularly, for the fixed fee policy, we observed that the path is a Nash equilibrium only when the fee is zero. Otherwise, for a small positive fee we noticed the formation of a star. Furthermore, we showed that beyond an upper bound the star ceased to be a Nash equilibrium and multiple star structures emerged. This observation lead to the investigation of the complete bipartite graph which defined a class of Nash equilibria dependent on the correlation between the sizes of the two independent sets of nodes. Finally, the complete graph was proven to be a Nash equilibrium when the fee is relatively high, as expected.

More importantly, we showed that even in a free fee policy, the star with uniform fees almost equal to the upper bound discussed above is a pure Nash equilibrium. On the contrary, the complete bipartite graph was proven unstable under this fee policy; we proved that a complete bipartite graph can never be a Nash equilibrium. We note, that these observations indicate the stability of a star structure, even though its centralized nature is opposed to the philosophy of decentralized and distributed payment networks.

References

1. Bitcoin Wiki: Hashed Time-Lock Contracts. https://en.bitcoin.it/wiki/Hashed_Timelock_Contracts. Accessed 16 May 2018
2. Raiden network (2017)
3. Albers, S., Eilts, S., Even-Dar, E., Mansour, Y., Roditty, L.: On nash equilibria for a network creation game. ACM Trans. Econ. Comput. **2**(1), 2 (2014)
4. Alon, N., Demaine, E.D., Hajiaghayi, M., Leighton, T.: Basic network creation games. SIAM J. Discr. Math. **27**, 106–113 (2010). https://doi.org/10.1145/1810479.1810502
5. Avarikioti, G., Janssen, G., Wang, Y., Wattenhofer, R.: Payment Network Design with Fees. In: Garcia-Alfaro, J., Herrera-Joancomartí, J., Livraga, G., Rios, R. (eds.) DPM/CBT -2018. LNCS, vol. 11025, pp. 76–84. Springer, Cham (2018). https://doi.org/10.1007/978-3-030-00305-0_6
6. Avarikioti, G., Kogias, E.K., Wattenhofer, R.: Brick: asynchronous state channels. arXiv preprint: arXiv:1905.11360 (2019)
7. Avarikioti, G., Wang, Y., Wattenhofer, R.: Algorithmic channel design. In: 29th International Symposium on Algorithms and Computation (ISAAC), Jiaoxi, Yilan County, Taiwan, December 2018
8. Croman, K., et al.: On scaling decentralized blockchains. In: Clark, J., Meiklejohn, S., Ryan, P.Y.A., Wallach, D., Brenner, M., Rohloff, K. (eds.) FC 2016. LNCS, vol. 9604, pp. 106–125. Springer, Heidelberg (2016). https://doi.org/10.1007/978-3-662-53357-4_8
9. Decker, C., Russell, R., Osuntokun, O.: eltoo: a simple layer2 protocol for bitcoin (2018)
10. Decker, C., Wattenhofer, R.: A fast and scalable payment network with bitcoin duplex micropayment channels. In: Pelc, A., Schwarzmann, A.A. (eds.) Stabilization, Safety, and Security of Distributed Systems, pp. 3–18. Springer International Publishing, Cham (2015). https://doi.org/10.1007/978-3-319-21741-3_1
11. Fabrikant, A., Luthra, A., Maneva, E., Papadimitriou, C.H., Shenker, S.: On a network creation game. In: Proceedings of the Twenty-Second Annual Symposium on Principles of Distributed Computing, pp. 347–351. ACM (2003)

12. Moreno-Sanchez, P., Kate, A., Maffei, M.: Silentwhispers: enforcing security and privacy in decentralized credit networks (2017)
13. Nakamoto, S.: Bitcoin: a peer-to-peer electronic cash system (2008)
14. Poon, J., Dryja, T.: The bitcoin lightning network: scalable off-chain instant payments (2015)
15. Prihodko, P., Zhigulin, S., Sahno, M., Ostrovskiy, A., Osuntokun, O.: Flare: an approach to routing in lightning network white paper (2016)
16. Roos, S., Moreno-Sanchez, P., Kate, A., Goldberg, I.: Settling payments fast and private: efficient decentralized routing for path-based transactions. arXiv preprint arXiv:1709.05748 (2017)
17. Spilman, J.: Anti dos for TX replacement. https://lists.linuxfoundation.org/pipermail/bitcoin-dev/2013-April/002433.html. Accessed 17 Apr 2019
18. Wood, G., et al.: Ethereum: a secure decentralised generalised transaction ledger (2014)

An Efficient Micropayment Channel
on Ethereum

Hisham S. Galal[(✉)], Muhammad ElSheikh, and Amr M. Youssef

Concordia Institute for Information Systems Engineering,
Concordia University, Montréal, QC, Canada
h_galal@encs.concordia.ca

Abstract. Blockchain protocols for cryptocurrencies offer secure payment transactions, yet their throughput pales in comparison to centralized payment systems such as VISA. Moreover, transactions incur fees
that relatively hinder the adoption of cryptocurrencies for simple daily
payments. Micropayment channels are second layer protocols that allow
efficient and nearly unlimited number of payments between parties at
the cost of only two transactions, one to initiate it and the other one
to close it. Typically, the de-facto approach for micropayment channels
on Ethereum is to utilize digital signatures which incur a constant gas
cost but still relatively high due to expensive elliptic curve operations.
Recently, ElSheikh *et al.* have proposed a protocol that utilizes hash
chain which scales linearly with the channel capacity and has a lower
cost compared to the digital signature based channel up to a capacity of
1000 micropayments. In this paper, we improve even more and propose a
protocol that scales logarithmically with the channel capacity. Furthermore, by utilizing a variant of Merkle tree, our protocol does not require
the payer to lock the entire balance at the channel creation which is an
intrinsic limitation with the current alternatives. To assess the efficiency
of our protocol, we carried out a number of experiments, and the results
prove a positive efficiency and an overall low cost. Finally, we release
the source code for prototype on Github (https://github.com/HSG88/
Payment-Channel).

Keywords: Micropayment channel · Ethereum · Merkle tree

1 Introduction

Cryptocurrencies such as Bitcoin [8] and Etheruem [11] enable secure payment
transactions between parties using blockchain technology. The major innovation
that made the success of Bitcoin was the Nakamoto consensus [8] that utilizes
Proof of Work (PoW) to enable distrusting peers to reach consensus. However,
this level of security comes at the cost of limited throughput and delayed confirmation. For example, the transaction throughput in Bitcoin and Ethereum
are roughly 5 and 20 transactions per second [1], respectively. Furthermore, it
was shown [5] that blockchain protocols based on PoW can hardly scale beyond

© Springer Nature Switzerland AG 2019
C. Pérez-Solà et al. (Eds.): DPM 2019/CBT 2019, LNCS 11737, pp. 211–218, 2019.
https://doi.org/10.1007/978-3-030-31500-9_13

60 transactions per second without considerably weakening their security. Additionally, the transaction fees are not constant and they vary significantly based on the current price of the underlying coin, and also whether the network faces high traffic of transactions. These limitations make applications such as micropayments expensive to realize directly without further modification to the consensus protocol.

A payment channel [3,6,7,9] is a protocol between two parties to send nearly an unlimited number of payments interactions off-chain. To establish it, only two transactions are required, one to open the channel (i.e., creating a smart contract and funding it), and the second one to close it (i.e., reclaiming the funds). Furthermore, a channel can have certain properties such as being a unidirectional and monotonically increasing which allows us to have entirely off-line channel. In other words, none of parties have to stay on-line to monitor changes on the smart contract. All they have to watch for is the channel timeout. Furthermore, a unidirectional and monotonically increasing payment channel is convenient for a merchant and buyer scenario.

The de-facto standard payment channel protocols on Ethereum depend heavily on digital signature schemes which incur a constant cost, yet a relatively high one due to the cost of elliptic curve operations to verify the signature. Recently, ElSheikh *et al.* [4] have proposed EthWord, a protocol that utilizes hash chain to create an efficient payment channel as it scales linearly with the channel capacity (i.e., number of unit payments), and it has a lower gas cost up to roughly a channel capacity of 1000 units of payments compared to the digital signature based channels.

In this work, we improve even more on the efficiency and gas cost of the above protocols and propose a scheme which scales logarithmically with the channel capacity. Furthermore, the payer does not have to lock up the entire amount in the construction of the payment channel, which is an intrinsic limitation in the EthWord and digital signature based channels. Furthermore, as part of our contribution, we also provide an open-source prototype on Github (https://github.com/HSG88/Payment-Channel) for the community to review it.

The rest of this paper is organized as follows. Section 2 provides a brief review of a digital signature based channel construction referred to as Pay50, and the hash chain based channel EthWord. In Sect. 3, we present the cryptographic primitives utilized in the proposed scheme. Then, in Sect. 4, we provide the design for our approach and compare its efficiency and cost to the other constructions. Finally, we present our conclusions in Sect. 5.

2   Related Work

Pay50. To argue the simplicity of building payment channels on Ethereum, Di Ferrante [2] showed how to construct one using only '50 lines of code' Solidity implementation of a uni-directional, monotonic channel. Simply, the payer initializes a smart contract with the payee's address, a timeout value, and deposits some balance. Once it is deployed, the smart contract constructor stores the payer's address in order to verify the digital signatures of off-chain payments via calling `ecrecover` (an op-code in Ethereum that returns the signer's address). After the deployment, the payer can send digitally signed payments messages to the payee off-chain. The payee verifies the digital signature, and on success, the payee provides the service or the item to the payer. At a later point of time but before the channel timeout, the payee sends the last signed payment message to the smart contract which releases the amount on successful verification.

EthWord. Elsheikh et al [4] proposed EthWord as a deployment for PayWord [10] on Ethereum. It depends mainly on a hash chain which is constructed by iteratively applying a public one-way hash function H on a secret random number s. More precisely, assume that $H_i(s) = H(H_{i-1}(s))$ for $i \in [1, n]$ (i.e, the length of the hash chain is n). The last hash value in the chain $H_n(s)$ is referred to as the *tip*. Furthermore, since the utilized hash function H is assumed to have the preimage resistance property, then it is computationally infeasible for an adversary to find any preimage in the chain given its *tip*. In the context of payments, the payer creates a hash chain of length n which represents the maximum number of possible payments. Then, the payer creates a channel by deploying a smart contract on Ethereum, passing the *tip*, the payee address, and timeout value as parameters. Later, the payer can send units of payments by revealing a preimage value deeper in the chain. For example, to pay i units, the payer sends $H_{n-i}(s)$ to the payee. The payee can verify it off-chain by iteratively hashing $H_{n-i}(s)$ i times, and see if the result equals the *tip*. Before the channel timeout, the payee sends the latest H_{n-i} to reclaim i units once the smart contract has verified it. Similar to `Pay50`, after the timeout, the smart contract sends the remaining balance back to the payer.

3 Preliminaries

Merkle Tree is a core component of the blockchain protocols. In Ethereum [11], Merkle trees aggregate the hashes of transactions, states, and receipts in a particular block so that the root becomes a binding commitment to all these values in that block. Technically speaking, a Merkle tree is a balanced binary tree in which the leaf nodes hold values and each non-leaf node stores $H(LeftChild\|RightChild)$, where H is a collision-resistant hash function. Proving the membership of a value in the tree can be achieved with a logarithmic size proof (in terms of the number of leaves), known as a *Merkle proof*. For example, given a Merkle tree M with a root r, to prove that a value $x \in M$, the prover creates a Merkle proof π by retrieving the siblings of each node on the path from x to r. The verifier iterates over the nodes in π_i to construct a root r' and accepts the proof if $r' = r$. It is worth noting that since M is a binary tree, then the proof size $|\pi| = log_2(n)$ as shown in Fig. 1.

The extension to our scheme that enables the payer to add funds to the channel depends on a variant of Merkle tree which is not strictly balanced. The objective is to append new values to the tree while at the same time maintain

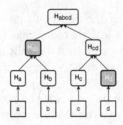

Fig. 1. An example illustrating the Merkle proof for element $c \in M$ which consists of the nodes H_d and H_{ab}

correct Merkle proofs for the old values. More precisely, to append a new set of m values to an existing Merkle tree M_n with n leaves and root r_n. First, we generate a Merkle tree M_m with a root r_m from the m values. Then, we combine the Merkle trees M_n and M_m to generate a new tree M' that contains the roots r_n and r_m as the child nodes of its root r'. Note that we still preserve the correctness of Merkle proofs for values in M_n by augmenting any valid Merkle proof π for elements in M_n with the root r_m to be valid with respect to M'.

4   Protocol Design and Implementation

Assume that Alice runs an online service for which she accepts (ether) currency on a micro-level (e.g., a very low fraction of **ether** that costs less than one dollar). Bob is interested in that service and he wants to utilize it. However, sending transactions to the Ethereum blockchain has a minimum cost of 21000 gas [11] which becomes too expensive for Bob as the number of interactions between him and Alice increases. To make it efficient and also cheap, Alice and Bob can utilize an off-chain payment channel. There are three phases in our protocol from the initiation to completion.

4.1   Channel Setup

This phase starts with Bob generating a secret random number s_0. Suppose that he estimates the number of maximum units of payments (i.e., channel capacity) he is willing to make is n. Note that we explain later how he can increase this number in case he wants to utilize the service more than he expected. Then, Bob uses a pseudo random number generator with the seed s_0 to create a sequence of random numbers $(s_1, ..., s_n)$. After that, he creates a Merkle tree M to bind the elements $((1||s_1), ..., (i||s_i), ..., (n||s_n))$ where every element is the concatenation of a value $i \in [1, n]$ and a random number s_i. We also assume that there is a minimum unit of payment u, (e.g., $u = \texttt{GWei} = 1 \times 10^{-9} \texttt{ ether}$).

At this moment, Bob deploys a smart contract on Ethereum to act as a trusted third party that holds Bob's balance and settles the final payment transaction to Alice. To deploy it, Bob passes the following parameters that control the payment channel between him and Alice to the smart contract constructor:

1. Alice's address A_{adr} on Ethereum.
2. A timeout value T_{out} before the channel is closed.
3. The root r of the Merkle tree M.

Additionally, Bob has to deposit an amount $balance = n \times u$ ether in the smart contract to be held in escrow, and pay Alice when she submits a valid Merkle proof.

4.2   Off-Chain Payments

Every time Bob wants to utilize Alice's service, he sends her a new Merkle proof π_i for an amount of i units. To generate the proof π_i, Bob runs Algorithm 1 which takes a Merkle tree M, an amount i, and the seed value s_0 as parameters. We start by generating the corresponding random value s_i. Then, s_i is concatenated to the amount i before feeding the result to the hash function H. Doing so prevents Alice from guessing the pre-images of the leaves in MT. In our implementation, we utilize Keccak256 as the hash function H due to its built-in support in the Ethereum virtual machine as an op-code. Then, we append the neighbour node of each node in the path from the leaf to the root r to the proof π_i.

Algorithm 1. Create Merkle Proof π_i for an amount i and random seed s_0

```
 1: function CREATEMERKLEPROOF(M, i, s₀)
 2:     πᵢ ← []
 3:     sᵢ ← PRNG(s₀, i)
 4:     node = H(i‖sᵢ)
 5:     while node ≠ M.root do
 6:         neighbour ← M.GetNeighbour(node)
 7:         πᵢ.Append(neighbour)
 8:         node ← M.GetParent(node)
 9:     end while
10:     return πᵢ
11: end function
```

After receiving the proof π_i, Alice has to verify it before providing the new service to Bob. So she calls Algorithm 2, and she decides to accept or reject based on the output. Note that, Algorithm 2 is also one of the functions in the smart contract that settles the payment to Alice. Essentially, Alice has to only keep track of the latest π_i (i.e., the proof for largest i amount). Once, Alice and Bob agree that there is no more payment interactions going between them and before the channel timeout, Alice sends π_i to the smart contract which will verify it and releases the payment to Alice.

4.3   Channel Termination

At this phase, Alice cannot issue any payment request to the smart contract as the channel has reached its timeout. However, Bob can reclaim his remaining

Algorithm 2. Verify Merkle Proof π_i for a Merkle tree with root r, amount i, and a random number s_i

```
 1: function VERIFYMERKLEPROOF(π_i, i, s_i, r)
 2:     node = H(i, s_i)
 3:     for j ← 1 to size(π_i) do
 4:         if node < π_{i,j} then
 5:             node ← H(node||π_{i,j})
 6:         else
 7:             node ← H(π_{i,j}||node)
 8:         end if
 9:     end for
10:     if node = r then
11:         return true
12:     else
13:         return false
14:     end if
15: end function
```

funds by calling `selfdestruct`. Typically, this will disable the smart contract from processing any further transactions. However, the smart contract can be alternatively designed to allow for reusing it for new payment channels without destructing it. Nonetheless, in our implementation, we chose a similar design to `Pay50` and `EthWord` in order to make fair comparisons.

4.4 Dynamic Refund Extension

One major advantage for our approach compared to the other alternatives is the ability to add more fund as Bob wants while the channel is open. In other words, Bob is not required to lock the full amount of balance which can be a substantial value at the time the channel construction. Toward this end, when required, Bob creates a new Merkle tree M_2 with a root r_2 that binds the additional m amounts of payments. Then, he sends r_2 to the smart contract along with the additional balance. The smart contract combines the old root r_1 with r_2 and hash the result to create a new root r'. Note that, since Ethereum is a public blockchain, then Alice can see the transaction carrying r_2. Therefore, she can still generate a valid proof for previous payments by augmenting the r_2 to the Merkle proof π_i for $i \in [1, n]$. In other words, Alice can still guarantee that she can reclaim her latest payments even if Bob acted maliciously and generated a bogus root r_2 for a fake tree M_2.

4.5 Evaluation

To assess our approach, we created experiments to check the efficiency and gas cost associated with `Pay50`, `EthWord`, and our approach. We also created off-chain clients to interact with the smart contract of each one. To our expectation, the efficiency of our approach as indicated by its gas cost far outweighs the corresponding cost of Pay50 and EthWord. The results shown in Fig. 2 indicate that our approach is better than Pay50 virtually on all practical channel capacities, however, it is slightly behind EthWord when the channel capacity is lower than 256, due to the added cost of Merkle proof size and the code logic to verify it. However, our approach scales much better after that level as the gas cost of EthWord increases at a much faster pace, and stays behind Pay50 once the channel capacity is above 1000. Interestingly, invoking the dynamic refund extension in our solution costs 28,941 gas which is much cheaper than the cost to deploy a new EthWord payment channel (318,953 gas).

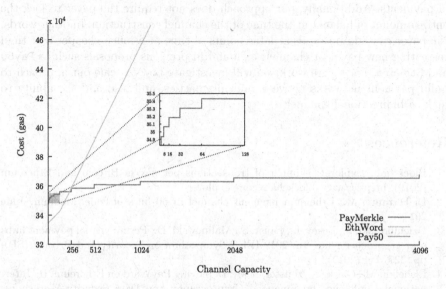

Fig. 2. The total gas cost of payment channel creation and settlement for different channel capacities.

It is also worth noting that the size of the Merkle proof is taken into account by the gas cost of the channel termination transaction. Hence, by achieving an overall lower gas cost, this implies that the cost for the increased parameters size in our protocol compared to the constant single parameter in EthWord and Pay50 is paid off by the efficiency of verifying the Merkle proofs compared to processing the hash chain in EthWord and verifying digital signature in Pay50. Concretely, the largest non-realistic capacity is 2^{256}, which requires a Merkle proof of 256 hashes. Therefore, the Merkle proof size in bytes is 256×32 bytes = 8-Kbytes. From the yellow paper of Ethereum, the fee for every non-zero byte is 68 gas. As a result, the maximum possible Merkle proof size will incur a total gas cost of 557,056 gas which is approximately 8% of the current block gas limit.

5 Conclusion

One way to improve the throughput of PoW based blockchains is by utilizing second layer improvements such as payment channels. Furthermore, payment channels increase the adoption of cryptocurrencies for simple payments as it requires only two transactions to be committed to the blockchain while allowing an unlimited number of transactions off-chain. The de-facto standard for payment channels is to utilize digital signatures which incur a constant cost, yet relatively high one compared to hash-based solutions such as EthWord. In this work, we further improved upon the efficiency of these schemes and proposed a solution, based on Merkle trees, that scales logarithmically with the number of payments. Additionally, our approach does not require the payer to lock the entire amount of balance at the time of the channel construction. In other words, the payer can add up funds at later points of time at a much cheaper cost than recreating new payment channels as found in previous proposals such as Pay50 and EthWord. For future work, we will investigate how to scale our approach to build payment networks between multiple parties, and also add the ability to make a bidirectional channel.

References

1. Blockchain Explorer: Number of transactions per day in Bitcoin and Ethereum (2019). https://www.blockchain.com/explorer
2. Di Ferrante, M.: Ethereum payment channel in 50 lines of code. Medium, June 2017
3. Dziembowski, S., Eckey, L., Faust, S., Malinowski, D.: Perun: virtual payment hubs over cryptocurrencies. In: 2019 IEEE Symposium on Security and Privacy (SP), pp. 327–344 (2019)
4. Elsheikh, M., Clark, J., Youssef, A.M.: Deploying PayWord on Ethereum. In: International Conference on Financial Cryptography and Data Security Workshops, BITCOIN, VOTING, and WTSC. Springer (2019, to appear)
5. Gervais, A., Karame, G.O., Wüst, K., Glykantzis, V., Ritzdorf, H., Capkun, S.: On the security and performance of proof of work blockchains. In: Proceedings of the 2016 ACM SIGSAC Conference on Computer and Communications security, pp. 3–16. ACM (2016)
6. Gudgeon, L., Moreno-Sanchez, P., Roos, S., McCorry, P., Gervais, A.: SoK: off the chain transactions. IACR Cryptology ePrint Archive 2019:360 (2019)
7. Miller, A., Bentov, I., Kumaresan, R., Cordi, C., McCorry, P.: Sprites and state channels: payment networks that go faster than lightning. arXiv preprint arXiv:1702.05812 (2017)
8. Nakamoto, S., et al.: Bitcoin: a peer-to-peer electronic cash system (2008)
9. Poon, J., Dryja, T.: The Bitcoin lightning network: scalable off-chain instant payments (2016)
10. Rivest, R.L., Shamir, A.: PayWord and MicroMint: two simple micropayment schemes. In: Lomas, M. (ed.) Security Protocols 1996. LNCS, vol. 1189, pp. 69–87. Springer, Heidelberg (1997). https://doi.org/10.1007/3-540-62494-5_6
11. Wood, G.: Ethereum: A secure decentralised generalised transaction ledger. Ethereum Project Yellow Paper 151, pp. 1–32 (2014)

Extending Atomic Cross-Chain Swaps

Jean-Yves Zie[1,2(✉)], Jean-Christophe Deneuville[1,3], Jérémy Briffaut[1],
and Benjamin Nguyen[1]

[1] Systems and Data Security Team (SDS), Laboratoire d'Informatique Fondamentale
d'Orléans (LIFO, EA 4022), INSA Centre Val de Loire, Bourges, France
[2] Orange Labs, Caen, France
jeanyves.ziediali@orange.com
[3] École Nationale de l'Aviation Civile, Federal University of Toulouse,
Toulouse, France

Abstract. Cryptocurrencies enable users to send and receive value in
a trust-less manner. Unfortunately trading the associated assets usually
happens on centralized exchange which becomes a trusted third party.
This defeats the purpose of a trust-less system. Atomic cross-chain swaps
solve this problem for exchanges of value without intermediaries. But
they require both blockchains to support Hash locked and Time locked
transactions. We propose an extension to atomic swap for blockchains
that only support multi-signatures (multisig) transactions. This provides
greater capabilities for cross-chain communications without adding any
extra trust hypothesis.

1   Introduction

Blockchains enable users to send and receive money in a trust-less manner. They
offer a mean for peer-to-peer (P2P) exchanges of crypto-assets such as cryptocurrencies over the Internet, which was not possible without intermediaries before.

Blockchains are a very young ecosystem, where good practices, controls, cer-
tifications and regulations still need to be defined. For instance trading the associated assets (located on different blockchains) usually takes place on centralized market places, called exchanges, which become trusted third parties. This
exposes users to loss of their funds, through hacks of those platforms or straight-
up scams when owners disappear with the funds. It emphasizes a common wis-
dom in this ecosystem: *Not your keys, Not your funds*. If we do not control the
private key of the account with the assets, we do not actually own the funds.

Atomic cross-chain swaps solve this problem for exchanges of assets without
intermediaries. Those swaps are atomic in the sense that either both parties
receive the other crypto-assets, or they both keep their own. They are used in
two ways. First, as explained earlier, they can serve as a basis for non custodial
exchanges. Using these decentralized marketplaces, users keep control of their
funds while being able to trade them.

Second, they enable Layer 2 scaling solutions, like [8, 20], to become a network
of payment channels and not just P2P channels, thus increasing the scalability

C. Pérez-Solà et al. (Eds.): DPM 2019/CBT 2019, LNCS 11737, pp. 219–229, 2019.
https://doi.org/10.1007/978-3-030-31500-9_14

of blockchains. The principle of layered scalability for blockchains is to conduct some transactions off the blockchain (*off-chain* or Layer 2) and notarize the state on the Layer 1 (*on-chain*), the blockchain, when necessary. The payment channel works like a tab between two users, with many transactions happening *off-chain* and only two transactions happening *on-chain*, one to open the channel and another one to close it. The limitation is obviously that Alice has to open a channel first with Bob before they can exchange. Atomic swaps come into play for cross-channels payments. If Alice has a payment channel with Bob and Bob has one with Charlie, Alice can atomically send a payment to Charlie with a swap between the two channels, provided there are enough funds in both channels.

The first proposal for atomic cross-chain swaps came from an online forum, *bitcointalk* [22]. It was between Bitcoin [18] and forks of Bitcoin, called altcoins for *alternative coins*. It made use of the scripting capabilities of those protocols, particularly conditional release of the coins, hashed locked and time locked transactions. This restricts the use cases to blockchains with scripting or smart contract capabilities. There are various types of cryptocurrencies with different goals and features and not all of them support Hash Time Lock Contracts (HTLC for short). In this work, we propose an extension of atomic cross-chain swaps to blockchains that do not have such capabilities with the same assumptions as HTLC atomic swaps. We construct our protocol for a blockchain with smart contracts and one that only supports multi-signatures (*multisig*) transactions. Those transactions are controlled by multiple private keys and require, to be valid, a certain number of signatures. Each signature is generated by a different private key and all those signatures need to be explicitly attached to the transaction. Using multisig transactions, we provide greater capabilities for cross-chain communications without adding any extra trust hypothesis.

The rest of this paper is organized as follows: the properties of atomic cross-chain swaps and their construction using HTLCs are recalled in Sect. 2. We show how to extend them to blockchains with multisig in Sect. 3, then discuss the relative advantages and drawbacks of our proposal in Sect. 4. A comparison to related works is proposed in Sect. 5 before concluding.

2   Hash Time Locked Contracts

2.1   Properties

Atomic swaps have two properties:

1. If all parties behave, i.e follow the protocol, all swaps happen.
2. If any party misbehaves, everyone is refunded.

This gives the users the guarantee they will not lose their assets by participating in an exchange. As stated earlier, the first proposal for blockchains came from *bitcointalk* [22]. We note that most blockchains rely on digital signatures to track the ownership of the coins. Thus, the proposed protocols, so far, make use of some scripting or smart contracts capabilities.

- Hash lock: to release the funds in the contract, one needs to provide the secret preimage s that gives $H(s)$, the lock in the contract,
- Time lock: if nothing happens before some time t on a chain, refund the user on that chain.

Those swaps are currently possible between Bitcoin and altcoins such as Litecoin [17] and on smart contract platforms like Ethereum [5] or EOS [15].

2.2 HTLC Based Atomic Swaps

We now present the atomic swap based on HTLCs. Let Alice be a user of the blockchain $\mathcal{A}$ that wants to exchange X coins with Bob a user of $\mathcal{B}$ for Y coins. We suppose that there is a common cryptographic hash function $Hash$ available on both blockchains. We also suppose that Alice and Bob agree on the timelocks T for Bob and $2T$ for Alice that are function of the blockchains involved. Specifically those timelocks are function of the Block times and confirmation times. We omit the digital signatures in the description for clarity and brevity.

Setup

1. Alice generates a secret preimage s and computes $h = Hash(s)$.
2. She also generates a HTLC on $\mathcal{A}$ with the hash lock h and the time lock $2T$.
3. Alice sends h and the HTLC address to Bob.
4. Bob can verify that the HTLC on $\mathcal{A}$ is properly constructed.
5. If it is valid, then Bob can generate a HTLC on $\mathcal{B}$ with the same hash lock h but the time lock T.
6. Alice can verify that the HTLC on $\mathcal{B}$ is properly constructed.
7. If it is valid, then Alice funds the HTLC on $\mathcal{A}$ and Bob does the same on $\mathcal{B}$.

We stress that the time locks are different on $\mathcal{A}$ and $\mathcal{B}$ and that Bob does not know s at this point.

Swap. The swap works as follows:

1. Alice presents the secret preimage s to the HTLC in $\mathcal{B}$ and receives the Y coins. She does so sufficiently early, with respect to T and the block time and confirmation time of $\mathcal{B}$.
2. Bob, who was monitoring the HTLC on $\mathcal{B}$ learns s.
3. Bob can then present s to the HTLC on $\mathcal{A}$ and receives the X coins. He does so sufficiently early, with respect to $2T$ and the block time and confirmation time of $\mathcal{A}$.

Refunds. Each user can refund herself by waiting for her time lock to expire and call the relevant HTLC. Obviously this is only possible if the funds have not been spent, in which case the other user is able to spend hers also.

Discussions. Alice has the advantage of being able to initiate the swap. This means that she can wait as long as possible to see if the trade gets more or less interesting with prices fluctuations. She can also propose a swap just to lock Bob's funds. We do not address these issues in this paper.

3   Extending Atomic Cross-Chain Swaps

The previous protocol is designed for blockchains that support Hashed Timelock Contract. But scripting or smart contract capabilities are not supported by all blockchains, including commonly used ones such as Steem [16]. We propose a protocol to conduct an atomic swap between a blockchain that supports scripting or smart contracts and one that supports multisig instead of HTLCs. Figure 1 presents a high level overview of the actors of the protocol. The basic idea is to transfer all the time locking part on the first blockchain, since the second does not support it. For example Steem supports multisig accounts and uses ECDSA signatures [13]. On Ethereum, the smart contract [19] can be used as an implementation basis. We leave the implementation of the full protocol for future work.

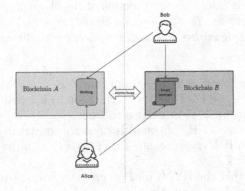

Fig. 1. Atomic Cross-Chain Swaps using multisig and smart contract

3.1   Assumptions

Our protocol enables two users, Alice and Bob, of different blockchains to securely swap their coins. As such we rely on the security model of those blockchains. We assume that each of them has a majority ($>2/3$) of consensus participants, miners for Proof-of-Work and validators for Proof-of-Stake, that are honest [4,6,9,10,14,18].

We also assume that Alice and Bob have a secure communication channel to do price discovery and exchange the swap details. We suppose that they agree on the timelocks T for Bob and $2T$ for Alice that are function of the blockchains involved (see Discussion Sect. 4).

We further assume that one blockchain supports multi-signature(*multisig*) transactions and that the second blockchain supports smart contracts and can verify the signature of a transaction issued in the first blockchain.

3.2   Notations

Let $\mathcal{A}$ be the blockchain with only multisig and $\mathcal{B}$ be the smart chain. Let $\mathsf{sk}\mathcal{A}$ be the private key for the signature scheme on a blockchain $\mathcal{A}$ and $\mathsf{pk}\mathcal{A}$ the corresponding public key. We note $\mathsf{sk}\mathcal{A}_A$ and $\mathsf{pk}\mathcal{A}_A$ if these are Alice's private and public keys and $\mathsf{sk}\mathcal{A}_B$ and $\mathsf{pk}\mathcal{A}_B$ for Bob's. We call SC the smart contract for the atomic swap on $\mathcal{B}$. We call M the multisig account on $\mathcal{A}$ that helps make the atomic swap. This can be understood as a shared account between Alice and Bob and we note $M = (\mathsf{pk}\mathcal{A}_A, \mathsf{pk}\mathcal{A}_B)$ to express that this account requires both signatures, meaning Alice and Bob need to cooperate to transact on the *behalf* of M.

Let the refund operations be $R(\mathcal{A}, \mathsf{pk}\mathcal{A}_A)$ and $R(\mathcal{B}, \mathsf{pk}\mathcal{B}_B)$, which means that Alice (resp. Bob) makes a transaction so that the public key $\mathsf{pk}\mathcal{A}_A$ (resp. $\mathsf{pk}\mathcal{B}_B$) has sole control of the funds on the chain $\mathcal{A}$ (resp. $\mathcal{B}$) and Alice (resp. Bob) is thus refunded. Let the swap operations be $S(\mathcal{B}, \mathsf{pk}\mathcal{B}_A)$ and $S(\mathcal{A}, \mathsf{pk}\mathcal{A}_B)$, which means that Alice (resp. Bob) makes a transaction so that the public key $\mathsf{pk}\mathcal{B}_A$ (resp. $\mathsf{pk}\mathcal{A}_B$) has sole control of the funds on the chain $\mathcal{B}$ (resp. $\mathcal{A}$) and Alice (resp. Bob) has completed the swap.

3.3   Setup, Transactions and Operations

Setup

1. Alice generates $\mathsf{sk}\mathcal{A}_A$, $\mathsf{pk}\mathcal{A}_A$, $\mathsf{sk}\mathcal{B}_A$, $\mathsf{pk}\mathcal{B}_A$ and sends $\mathsf{pk}\mathcal{A}_A$ and $\mathsf{pk}\mathcal{B}_A$ to Bob.
2. Bob generates $\mathsf{sk}\mathcal{A}_B$, $\mathsf{pk}\mathcal{A}_B$, $\mathsf{sk}\mathcal{B}_B$, $\mathsf{pk}\mathcal{B}_B$ and sends $\mathsf{pk}\mathcal{A}_B$ and $\mathsf{pk}\mathcal{B}_B$ to Alice.
3. Alice creates $M = (\mathsf{pk}\mathcal{A}_A, \mathsf{pk}\mathcal{A}_B)$ on $\mathcal{A}$ and sends its address to Bob for verification. She also creates and sends $tx_1 = R(\mathcal{A}, \mathsf{pk}\mathcal{A}_A)$ and $tx_2 = S(\mathcal{A}, \mathsf{pk}\mathcal{A}_B)$.
4. After verification, Bob computes (σ_1, tx_1), where $\sigma_1 = Sign_{\mathsf{sk}\mathcal{A}_B}(tx_1)$. He can then create SC, fund this contract and send its address and (σ_1, tx_1).
5. Alice verifies that everything is in order and funds the account M.

The Smart Contract SC. It consists of four procedures SC1, SC2, SC3 and SC4 respectively presented in Algorithm 1.

Algorithm 1. The smart contract SC.

1: $unlock \leftarrow$ false
2: **procedure** SC1
3:     **if** $t \geq T$ **and** $unlock ==$ false **then**
4:         R($\mathcal{B}$, pk$\mathcal{B}_B$)
5: **procedure** SC2(σ_1', tx_1)
6:     **if** Verify$_{\mathsf{pk}\mathcal{A}_A}$ (σ_1', tx_1) **then**
7:         R($\mathcal{B}$, pk$\mathcal{B}_B$)
8: **procedure** SC3(σ_2', tx_2)
9:     **if** Verify$_{\mathsf{pk}\mathcal{A}_A}$ (σ_2', tx_2) **then**
10:         $unlock \leftarrow$ true
11: **procedure** SC4
12:     **if** $t \geq 2T$ **and** $unlock ==$ true **then**
13:         S($\mathcal{B}$, pk$\mathcal{B}_A$)

Operations. The list of possible operations is summarized in Table 1. We use OPX if the operation X succeeds and will mention explicit failure if it does not.

Table 1. List of operations.

OP1	Alice creates $\sigma_1' = Sign_{\mathsf{sk}\mathcal{A}_A}(tx_1)$ and broadcasts $(\sigma_1', \sigma_1, tx_1)$
OP2	Bob waits $t \geq T$ and calls SC1
OP3	Bob creates $\sigma_2 = Sign_{\mathsf{sk}\mathcal{A}_B}(tx_2)$ and sends (σ_2, tx_2) to Alice
OP4	Alice creates $\sigma_2' = Sign_{\mathsf{sk}\mathcal{A}_A}(tx_2)$ and calls SC3
OP5	Bob learns σ_2' monitoring SC and broadcasts $(\sigma_2', \sigma_2, tx_2)$
OP6	Alice waits for $t \geq 2T$ and calls SC4
OP7	Bob learns σ_1' from $\mathcal{A}$ and calls SC2

3.4   Protocol Flow and Guarantees

The protocol flow is described in Fig. 2. It starts with $OP3$, an off-chain communication (black arrow). Alice then calls SC3 (blue arrow) while Bob learns σ_2' (dashed blue arrow). The following action (*) can happen before or after $t = T$ and Bob then controls the X coins (dashed purple arrow). The final step is $OP6$, where Alice waits for $t \geq 2T$ and calls SC4 (blue arrow) to receive the Y coins (dashed blue arrow).

The protocol needs to ensure that for every outcome, swap or refund, for one user, there exists a sequence of operations the second user can do or could have done to have the same outcome, complete the swap or get a refund. Additionally, each user should get refunded if the other user aborts the swap or does nothing.

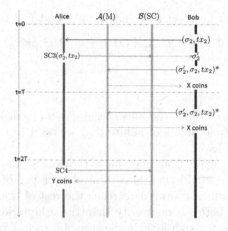

Fig. 2. Protocol flow resulting in a swap (Color figure online)

Swaps. Assuming $S(\mathcal{A}, \mathsf{pk}\mathcal{A}_B)$, i.e Bob received the X coins, we have:

$$S(\mathcal{A}, \mathsf{pk}\mathcal{A}_B) \implies OP5 \tag{1}$$

$$OP5 \implies OP4 \tag{2}$$

$$OP4 \implies unlock == True \tag{3}$$

$$\text{if } t \geq 2T \text{ and } OP6 \implies S(\mathcal{B}, \mathsf{pk}\mathcal{B}_A) \tag{4}$$

The step (3) relies on the fact that Bob can only learn σ_2' if Alice creates it, assuming $\mathcal{A}$ uses a secure signature scheme. The only way for Bob to access the Y coins is to use SC2 since $unlock == True$. He thus needs σ_1', which Alice has no reason to provide.

Assuming $S(\mathcal{B}, \mathsf{pk}\mathcal{B}_A)$, we have:

$$S(\mathcal{B}, \mathsf{pk}\mathcal{B}_A) \implies unlock == True \text{ from SC4} \tag{5}$$

$$unlock == True \implies OP4 \tag{6}$$

$$OP4 \implies OP5 \tag{7}$$

$$OP5 \implies S(\mathcal{A}, \mathsf{pk}\mathcal{A}_B) \tag{8}$$

Thus Bob can complete the swap. There is however one setting in which his part of the swap can fail: Alice has already taken the funds out of M, which is described in Fig. 3. This is covered by the refund case.

Refunds. If Alice does no operation, Bob has to wait for $t \geq T$ to be refunded using $OP2$.

Alice, on the other end, can be refunded any time by using $OP1$. The problem is that Alice can refund herself and then try to get the funds on $\mathcal{B}$ using $OP2$ after $OP1$ while $t \leq T$, as described in Fig. 3. This means that:

$$\text{Since } unlock == True \implies OP2 \text{ will fail when calling SC1} \tag{9}$$

But Alice does not have yet the funds from $\mathcal{B}$ and $T < t < 2T$.

$$\text{if } T < t < 2T \text{ and } OP7 \implies R(\mathcal{B}, \mathsf{pk}\mathcal{B}_B) \tag{10}$$

4   Discussion

Cross-chain exchanges inherit the adversarial environment of each blockchain. This introduces multiple points of failure that we need to take into account to make the swaps atomic.

For example, Alice or Bob may have the means to launch an Eclipse attack [11]. One user would then have control over the other's network connections and decide which transactions reach the rest of the network. One user can also just have a better connectivity than the other and be sure his or her transactions would go through first, in the case of a race. We can not solve such problems in the protocol without further assumptions thus we rely on the time locks the users set. They represent the intervals of time necessary for the users, in comparison to the block times, the confirmation times, to create, broadcast the relevant transactions and have them confirmed. In practice, they should also consider what is the current state of both blockchains as to not use them when they are more vulnerable. This could be the case if the mining hashing rate in Proof-of-Work, or the price of the coin in Proof-of-Stake crashes dramatically.

Another question is the pre-generation of the transactions that are in the smart contract C. This is relevant for security. It is better to require the signature of a *particular* message under a private key than *any* message under that key. This is also relevant for the fees of the transactions. Some blockchains do not take fees (Steem) or have very low fees (Litecoin) while other have a volatile fee market (Bitcoin, Ethereum). Alice can take care of this on $\mathcal{A}$ while Bob pays on $\mathcal{B}$. Still, if the fees were too low at creation, there is the risk that the transactions would take too long to appear in the blockchain(s).

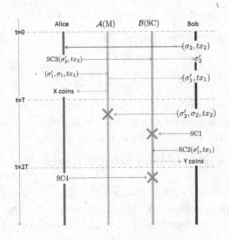

Fig. 3. Alice attack scenario

5   Related Works

The first atomic swap protocol emerged on a Bitcoin online forum and is based on HTLCs [22]. It has since been standardized on Bitcoin development resources [1,2]. This protocol is briefly presented in Sect. 2.

To our knowledge, Herlihy [12] presented the first formal study of the underlying theory of atomic cross-chain swaps. It focuses on a swap of on-chain and off-chain assets between three parties. Using graph theory and the HTLCs, they explore the time locks constraints for the atomicity in that particular setting.

Many protocols focus on cross-chain communications and exchanges. Interledger [21] uses Hashed-Timelock Agreements (HTLAs) [7], which generalized the idea of HTLCs for payments systems with or without a private or public ledger. Those agreements are used to create secure multi-hop payments using conditional transfers. The parties involved decide where their trust lies, for example a HTLC between blockchains or a legal contract.

Cosmos [3,4], Polkadot [23] and earlier versions of Interledger [21] rely on a set of validators to ensure cross-chain communication. Each round, a subset of those validators decides which cross-chain information to notarize on their chain, provided only a small protocol-defined portion of those validators is Byzantine. The problem is to guarantee that those systems are decentralized enough to be consider censorship resistant and secure while remaining scalable.

XClaim [24] is a framework for achieving trustless cross-chain exchanges using cryptocurrency-backed assets based on a smart contract blockchain and an existing link between the other blockchains like a relay. It requires vault providers to guarantee the liquidity and ownership on the original chain and to have collateral on the smart contract blockchain. Bad behaviour is discouraged using slashing or proof-of-punishment on the smart contract chain. With these assumptions, XClaim builds a trustless and non interactive protocol for issuance, redemption and swapping of tokenized assets.

Our construction does not require collateral (because there is no vault to provide) or an existing link between the blockchains. We argue that our protocol uses simpler assumptions to enable trustless exchange between users, in the same manner as HTLC based atomic cross-chain swaps.

6   Conclusion

In this work, we propose a new protocol for atomic cross-chain swaps. We extend the support of atomic swaps to blockchains without hash lock and time lock capabilities without additional trust requirements. Users can now trade in a P2P way between blockchains with smart contracts and blockchain with only mutli-signatures transactions. We leave the full implementation and live experimentation of this protocol for future work.

References

1. bitcoinwiki: Atomic swap (2019). https://en.bitcoin.it/wiki/Atomic_swap. Accessed 03 July 2019

2. Bowe, S., Hopwood, D.: Hashed time-locked contract transactions, March 2017. https://github.com/bitcoin/bips/blob/master/bip-0199.mediawiki. Accessed 03 July 2019

3. Buchman, E., Kwon, J.: Cosmos: a network of distributed ledgers. https://github.com/cosmos/cosmos/blob/master/WHITEPAPER.md. Accessed 03 July 2019

4. Buchman, E., Kwon, J., Milosevic, Z.: The latest gossip on BFT consensus. CoRR abs/1807.04938 (2018)

5. Buterin, V.: Ethereum: a next-generation smart contract and decentralized application platform. https://github.com/ethereum/wiki/wiki/White-Paper. Accessed 03 July 2019

6. Daian, P., Pass, R., Shi, E.: Snow white: robustly reconfigurable consensus and applications to provably secure proofs of stake. In: IACR, pp. 1–64 (2017)

7. de Jong, M., Schwartz, E.: Hashed-timelock agreements, Novemebr 2017. https://github.com/interledger/rfcs/blob/master/0022-hashed-timelock-agreements/0022-hashed-timelock-agreements.md. Accessed 03 July 2019

8. Decker, C., Wattenhofer, R.: A fast and scalable payment network with Bitcoin duplex micropayment channels. In: Pelc, A., Schwarzmann, A.A. (eds.) SSS 2015. LNCS, vol. 9212, pp. 3–18. Springer, Cham (2015). https://doi.org/10.1007/978-3-319-21741-3_1

9. Eyal, I., Sirer, E.G.: Majority is not enough: Bitcoin mining is vulnerable. Commun. ACM **61**(7), 95–102 (2018)

10. Gervais, A., Karame, G.O., Wüst, K., Glykantzis, V., Ritzdorf, H., Capkun, S.: On the security and performance of proof of work blockchains. In: Proceedings of the 2016 ACM SIGSAC Conference on Computer and Communications Security, CCS 2016, pp. 3–16. ACM, New York (2016)

11. Heilman, E., Kendler, A., Zohar, A., Goldberg, S.: Eclipse attacks on Bitcoin's peer-to-peer network. In: 24th USENIX Security Symposium (USENIX Security 2015), pp. 129–144 (2015)

12. Herlihy, M.: Atomic cross-chain swaps. In: Newport, C., Keidar, I., (eds.) Proceedings of the 2018 ACM Symposium on Principles of Distributed Computing, PODC 2018, Egham, UK, 23–27 July 2018, pp. 245–254. ACM (2018)

13. Johnson, D., Menezes, A., Vanstone, S.: The elliptic curve digital signature algorithm (ECDSA). Int. J. Inf. Secur. **1**(1), 36–63 (2001)

14. Kiayias, A., Russell, A., David, B., Oliynykov, R.: Ouroboros: a provably secure proof-of-stake blockchain protocol. In: Katz, J., Shacham, H. (eds.) CRYPTO 2017. LNCS, vol. 10401, pp. 357–388. Springer, Cham (2017). https://doi.org/10.1007/978-3-319-63688-7_12

15. Larimer, D.: EOS.IO technical white paper V2. https://github.com/EOSIO/Documentation/blob/master/TechnicalWhitePaper.md. Accessed 03 July 2019

16. Larimer, D., Scott, N., Zavgorodnev, V., Johnson, B., Calfee, J., Vandeberg, M.: Steem: an incentivized, blockchain-based social media platform. Self-published, March 2016

17. Lee, C.: Litecoin (2011)

18. Nakamoto, S.: Bitcoin: a peer-to-peer electronic cash system (2008). http://bitcoin.org/bitcoin.pdf

19. OpenZeppelin: ECDSA solidity library. https://github.com/OpenZeppelin/openzeppelin-contracts/blob/master/contracts/cryptography/ECDSA.sol. Accessed 03 July 2019

20. Poon, J., Dryja, T.: The Bitcoin lightning network (2016). https://lightning.network/lightning-network-paper.pdf. Accessed 03 July 2019

21. Thomas, S., Schwartz, E.: A protocol for interledger payments (2015). https://interledger.org/interledger.pdf
22. TierNolan: bitcointalk: Alt chains and atomic transfers, May 2013. https://bitcointalk.org/index.php?topic=193281.msg2224949#msg2224949. Accessed 03 July 2019
23. Wood, G.: Polkadot: vision for a heterogeneous multi-chain framework draft 1. https://polkadot.network/PolkaDotPaper.pdf. Accessed 03 July 2019
24. Zamyatin, A., Harz, D., Lind, J., Panayiotou, P., Gervais, A., J. Knottenbelt, W.: XCLAIM: trustless, interoperable, cryptocurrency-backed assets, March 2019

CBT Workshop: Smart Contracts and Applications

A Minimal Core Calculus for Solidity Contracts

Massimo Bartoletti[1]([⊠]), Letterio Galletta[2], and Maurizio Murgia[1,3]

[1] Università Degli Studi di Cagliari, Cagliari, Italy
bart@unica.it
[2] IMT Lucca, Lucca, Italy
[3] Università di Trento, Trento, Italy

Abstract. The Ethereum platform supports the decentralized execution of *smart contracts*, i.e. computer programs that transfer digital assets between users. The most common language used to develop these contracts is Solidity, a Javascript-like language which compiles into EVM bytecode, the language actually executed by Ethereum nodes. While much research has addressed the formalisation of the semantics of EVM bytecode, relatively little attention has been devoted to that of Solidity. In this paper we propose a minimal calculus for Solidity contracts, which extends an imperative core with a single primitive to transfer currency and invoke contract procedures. We build upon this formalisation to give semantics to the Ethereum blockchain. We show our calculus expressive enough to reason about some typical quirks of Solidity, like e.g. re-entrancy.

Keywords: Ethereum · Smart contracts · Solidity

1 Introduction

A paradigmatic feature of blockchain platforms is the ability to execute "smart" contracts, i.e. computer programs that transfer digital assets between users, without relying on a trusted authority. In Ethereum [5]—the most prominent smart contracts platform so far—contracts can be seen as concurrent objects [15]: they have an internal mutable state, and a set of procedures to manipulate it, which can be concurrently called by multiple users. Additionally, each contract controls an amount of crypto-currency, that it can exchange with other users and contracts. Users interact with contracts by sending transactions, which represent procedure calls, and may possibly involve transfers of crypto-currency from the caller to the callee. The sequence of transactions on the blockchain determines the state of each contract, and the balance of each user.

Ethereum supports contracts written in a Turing-complete language, called EVM bytecode [8]. Since programming at the bytecode level is inconvenient, developers seldom use EVM bytecode directly, but instead write contracts in higher-level languages which compile into bytecode. The most common of these

© Springer Nature Switzerland AG 2019
C. Pérez-Solà et al. (Eds.): DPM 2019/CBT 2019, LNCS 11737, pp. 233–243, 2019.
https://doi.org/10.1007/978-3-030-31500-9_15

languages is Solidity [1], a Javascript-like language supported by the Ethereum Foundation. There is a growing literature on the formalization of Solidity, which roughly can be partitioned in two approaches, according to the distance from the formal model to the actual language. One approach is to include as large a subset of Solidity as possible, while the other is to devise a core calculus that is as small as possible, capturing just the features of Solidity that are relevant to some specific task. In general, the first approach has more direct practical applications: for instance, a formal semantics very close to that of the actual Solidity can be the basis for developing precise analysis and verification tools. Although diverse in nature, the motivations underlying the second approach are not less strong. The main benefit of omitting almost all the features of the full language is that by doing so we simplify rigorous reasoning. This simplification is essential for the development of new proof techniques (e.g., axiomatic semantics), static analysis techniques (e.g., data and control flow analysis, type systems), as well as for the study of language expressiveness (e.g., rigorous comparisons and encodings to/from other languages). The co-existence of these two approaches is common in computer science: for instance, the formalization of Java gave rise of a lot of research since the mid 90s, producing Featherweight Java [9] as the most notable witness of the "minimalistic" approach.

Contribution. In this paper we pursue the minimalistic approach, by introducing a core calculus for smart contracts. Our calculus, called TinySol (for "Tiny Solidity"), features an imperative core, which we extend with a single construct to call contracts and transfer currency. This construct, inspired by Solidity "external" calls, captures the most paradigmatic aspect of smart contracts, i.e. the exchange of digital assets according to programmable rules. Slightly diverging from canonical presentations of imperative languages, we use key-value stores to represent the contract state, so abstracting and generalising Solidity state variables. We formalise the semantics of TinySol in Sect. 2, using a big-step operational style. In Sect. 3 we show TinySol expressive enough to reproduce *reentrancy attacks*, one of the typical quirks of Solidity; the succinctness of these proofs witnesses an advantage of our minimalistic approach. In Sect. 4 we refine our formalization, by giving semantics to transactions and blockchains. Aiming at minimality, TinySol makes several simplifications w.r.t. Solidity: in Sect. 5 we discuss the main differences between the two languages. Because of space constraints, we refer to [3] for more elaborated TinySol examples.

Related Work. Besides ours, the only other Solidity-inspired minimal core calculus we are aware of is Featherweight Solidity (FS) [6]. Similarly to our TinySol, FS focusses on the most basic features of Solidity, i.e. transfers of cryptocurrency and external calls, while omitting other language features, like e.g. internal and delegate calls, function modifiers, and the gas mechanism. The main difference between TinySol and FS is stylistic: while our design choice was to start from a basic imperative language, and extend it with a single contract-oriented primitive (external calls), FS follows the style of Featherweight Java, modelling function bodies as expressions. Compared to our calculus, FS also includes the dynamic creation of contracts, and a type system which detects some run-time errors. A further difference is that FS models blockchains as functions from contract identifiers to states; instead, we represent a blockchain as a sequence of transactions, and then we reconstruct the state by giving a semantics to this sequence. In this way we can reason e.g. about re-orderings of transactions, forks, etc.

A few papers pursue the approach of formalising large fragments of Solidity. The work [19] proposes a big-step operational semantics for several Solidity constructs, including e.g. access to memory and storage, inheritance, internal and external calls, and function modifiers. The formalization also deals with complex data types, like structs, arrays and mappings. The works [10,18] propose *tour-de-force* formalizations of larger fragments of Solidity, also including a gas mechanism. Both [18,19] mechanize their semantics in the Coq proof assistant, while [10] uses the K-Framework [14]. The work [13] extends the semantics of [10] to encompass also exceptions and return values.

2   TinySol Syntax and Semantics

We assume a set **Val** of *values* $v, k, \ldots$, a set **Const** of *constant names* $x, y, \ldots$, a set of *procedure names* $f, g, \ldots$ and a set **Addr** of *addresses* $\mathcal{X}, \mathcal{Y}, \ldots$, partitioned into *account addresses* $\mathcal{A}, \mathcal{B}, \ldots$ and *contract addresses* $\mathcal{C}, \mathcal{D}, \ldots$. We write sequences in bold, e.g. $\boldsymbol{v}$ is a sequence of values; ϵ is the empty sequence. We use $n, n', \ldots$ to range over $\mathbb{N}$, and $b, b', \ldots$ to range over boolean values.

A *contract* is a finite set of terms of the form $f(\boldsymbol{x})\{S\}$, where S is a *statement*, with syntax in Fig. 1. Intuitively, each term $f(\boldsymbol{x})\{S\}$ represents a contract procedure, where f is the procedure name, $\boldsymbol{x}$ are its formal parameters (omitted when empty), and S is the procedure body. Each contract has a key-value store, which we model as a partial function from keys $k \in \textbf{Val}$ to values $v \in \textbf{Val}$.

Statements extend those of a basic imperative language with three constructs:

- **throw** raises an uncatchable exception, rolling-back the state;
- $k := E$ updates the key-value store, binding the key k to the value denoted by the expression E;
- $\mathcal{X} : f(\boldsymbol{v})\$n$ calls the procedure f (with actual parameters $\boldsymbol{v}$) of the contract at address $\mathcal{X}$, transferring n units of currency to $\mathcal{X}$.

The expressions used within statements (Fig. 1, right) can be constants (e.g., integers, booleans, strings), addresses, and operations between expressions. We assume that all the usual arithmetic, logic and cryptographic operators are provided (since their definition is standard, we will not detail them). The expression $!k$ evaluates to *true* if the key k is bound in the contract store, otherwise it evaluates to *false*. The expression $?k$ denotes the value bound to the key k in the contract store. The expression $\mathcal{X} : E$ evaluates E in the context of the address $\mathcal{X}$. For instance, $\mathcal{X} : ?k$ denotes the value bound to k in the store of $\mathcal{X}$.

We assume a mapping Γ from addresses to contracts, such that $\Gamma(\mathcal{A}) = \{f_{\text{skip}}()\{\text{skip}\}\}$ for all account addresses $\mathcal{A}$. This allows for a uniform treatment of account and contract addresses: indeed, calling a procedure on an account address $\mathcal{A}$ can only result in a pure currency transfer to $\mathcal{A}$, since the procedure can only perform a skip. We further postulate that: (i) expressions and statements are well-typed: e.g., guards in conditionals and in loops have type bool; (ii) the procedures in $\Gamma(\mathcal{C})$ have distinct names; (iii) the key balance cannot stay at the left of an assignment; (iv) the constant names sender and value cannot stay in the formal parameters of a procedure.

We use the following syntactic sugar. For a call $\mathcal{X} : f(\boldsymbol{v})\$n$, when there is no money transfer (i.e., $n = 0$) we just write it as $\mathcal{X} : f(\boldsymbol{v})$; when the target is an account address $\mathcal{A}$ (so, the call is to the procedure f_{skip}), we write it as $\mathcal{A}\$n$. We write if E then S for if E then S else skip.

$S ::=$	statement	$E ::=$	expression
skip	skip	v	value
| throw	exception	| x	const name
| $E := E'$	store update	| $\mathcal{X}$	address
| $S ; S'$	sequence	| op E	operator
| if E then S else S'	conditional	| $?E$	key lookup
| while E do S	loop	| $!E$	key bound?
| $E_0 : f(\boldsymbol{E_1})\$E_2$	call	| $\mathcal{X} : E$	context

Fig. 1. Syntax of TinySol.

The semantics of contracts is given in terms of a function from states to states. A *state* $\sigma : \textbf{Addr} \to (\textbf{Val} \rightharpoonup \textbf{Val})$ maps each address to a key-value store, i.e. a partial function from values (keys) to values. We use the standard brackets notation for representing finite maps: for instance, $\{v_1/x_1, \cdots, v_n/x_n\}$ maps x_i to v_i, for $i \in 1..n$. When a key k is not bound to any value in $\sigma \mathcal{X}$, we write $\sigma \mathcal{X} k = \bot$. We postulate that dom $\sigma \mathcal{A} = \{\text{balance}\}$ for all account addresses $\mathcal{A}$, and dom $\sigma \mathcal{C} \supseteq \{\text{balance}\}$ for all contract addresses $\mathcal{C}$. A *qualified key* is a term of the form $\mathcal{X}.k$. We write $\sigma(\mathcal{X}.k)$ for $\sigma \mathcal{X} k$.

A *state update* $\pi : \textbf{Addr} \rightharpoonup (\textbf{Val} \rightharpoonup \textbf{Val})$ is a substitution from qualified keys to values; we denote with $\{v/\mathcal{X}.k\}$ the state update which maps $\mathcal{X}.k$ to v. We define keys(π) as the set of qualified keys $\mathcal{X}.k$ such that $\mathcal{X} \in \text{dom } \pi$ and $k \in \text{dom } \pi \mathcal{X}$. We apply updates to states as follows:

$$(\sigma \pi)\mathcal{X} = \delta_{\mathcal{X}} \quad \text{where} \quad \delta_{\mathcal{X}} k = \begin{cases} \pi \mathcal{X} k & \text{if } \mathcal{X}.k \in \text{keys}(\pi) \\ \sigma \mathcal{X} k & \text{otherwise} \end{cases}$$

We define the auxiliary operators $\sigma + \mathcal{X} : n$ and $\sigma - \mathcal{X} : n$ on states, which, respectively, increase/decrease the balance of $\mathcal{X}$ of n currency units:

$$\sigma \circ \mathcal{X} : n = \sigma\{(\sigma \mathcal{X} \text{balance}) \circ n/\mathcal{X}.\text{balance}\} \qquad (\circ \in \{+, -\})$$

Example 1. Let σ be a state which maps $\mathcal{X}$ to the store $\delta_{\mathcal{X}} = \{0/k_0, 1/k_1\}$. Let $\pi = \{2/\mathcal{X}.k_0\}$ and $\pi' = \{3/\mathcal{X}.k_2\}$ be state updates. We have $(\sigma\pi)\mathcal{X} = \{2/k_0, 1/k_1\}$, $(\sigma\pi')\mathcal{X} = \{0/k_0, 1/k_1, 3/k_2\}$, and $(\sigma\pi)\mathcal{Y} = (\sigma\pi')\mathcal{Y} = \sigma\mathcal{Y}$ for all $\mathcal{Y} \neq \mathcal{X}$.

We give the operational semantics of statements in a big-step style. The semantics of a statement S is parameterised over a state σ, an address $\mathcal{X}$ (the contract wherein S is evaluated), and an *environment* $\rho : \mathbf{Const} \rightharpoonup \mathbf{Val}$, used to evaluate the formal parameters and the special names $\mathtt{sender}$ and $\mathtt{value}$.

$$\llbracket v \rrbracket^{\mathcal{X}}_{\sigma,\rho} = v \qquad \llbracket x \rrbracket^{\mathcal{X}}_{\sigma,\rho} = \rho\, x \qquad \llbracket \mathcal{Y} \rrbracket^{\mathcal{X}}_{\sigma,\rho} = \mathcal{Y} \qquad \llbracket \mathbf{op} E \rrbracket^{\mathcal{X}}_{\sigma,\rho} = \mathrm{op}\, \llbracket E \rrbracket^{\mathcal{X}}_{\sigma,\rho} \qquad \llbracket \mathcal{Y} : E \rrbracket^{\mathcal{X}}_{\sigma,\rho} = \llbracket E \rrbracket^{\mathcal{Y}}_{\sigma,\rho}$$

$$\llbracket ?E \rrbracket^{\mathcal{X}}_{\sigma,\rho} = \sigma\,\mathcal{X}\,(\llbracket E \rrbracket^{\mathcal{X}}_{\sigma,\rho}) \qquad \llbracket !E \rrbracket^{\mathcal{X}}_{\sigma,\rho} = \begin{cases} \mathit{true} & \text{if } \llbracket E \rrbracket^{\mathcal{X}}_{\sigma,\rho} \neq \bot \text{ and } \sigma\,\mathcal{X}\,(\llbracket E \rrbracket^{\mathcal{X}}_{\sigma,\rho}) \neq \bot \\ \mathit{false} & \text{if } \llbracket E \rrbracket^{\mathcal{X}}_{\sigma,\rho} \neq \bot \text{ and } \sigma\,\mathcal{X}\,(\llbracket E \rrbracket^{\mathcal{X}}_{\sigma,\rho}) = \bot \end{cases}$$

$$\frac{}{\llbracket \mathbf{skip} \rrbracket^{\mathcal{X}}_{\sigma,\rho} = \sigma} \qquad \frac{\llbracket E \rrbracket^{\mathcal{X}}_{\sigma,\rho} = k \quad \llbracket E' \rrbracket^{\mathcal{X}}_{\sigma,\rho} = v}{\llbracket E := E' \rrbracket^{\mathcal{X}}_{\sigma,\rho} = \sigma\{v/\mathcal{X}.k\}} \qquad \frac{\llbracket E \rrbracket^{\mathcal{X}}_{\sigma,\rho} = b \in \{\mathit{true}, \mathit{false}\}}{\llbracket \mathbf{if}\ E\ \mathbf{then}\ S_{\mathit{true}}\ \mathbf{else}\ S_{\mathit{false}} \rrbracket^{\mathcal{X}}_{\sigma,\rho} = \llbracket S_b \rrbracket^{\mathcal{X}}_{\sigma,\rho}}$$

$$\frac{\llbracket S_0 \rrbracket^{\mathcal{X}}_{\sigma,\rho} = \sigma'}{\llbracket S_0; S_1 \rrbracket^{\mathcal{X}}_{\sigma,\rho} = \llbracket S_1 \rrbracket^{\mathcal{X}}_{\sigma',\rho}} \qquad \frac{\llbracket E \rrbracket^{\mathcal{X}}_{\sigma,\rho} = \mathit{false}}{\llbracket \mathbf{while}\ E\ \mathbf{do}\ S \rrbracket^{\mathcal{X}}_{\sigma,\rho} = \sigma} \qquad \frac{\llbracket E \rrbracket^{\mathcal{X}}_{\sigma,\rho} = \mathit{true} \quad \llbracket S \rrbracket^{\mathcal{X}}_{\sigma,\rho} = \sigma'}{\llbracket \mathbf{while}\ E\ \mathbf{do}\ S \rrbracket^{\mathcal{X}}_{\sigma,\rho} = \llbracket \mathbf{while}\ E\ \mathbf{do}\ S \rrbracket^{\mathcal{X}}_{\sigma',\rho}}$$

$$\frac{\begin{array}{ll} \llbracket E_0 \rrbracket^{\mathcal{X}}_{\sigma,\rho} = \mathcal{Y} & \llbracket E_2 \rrbracket^{\mathcal{X}}_{\sigma,\rho} = n \leq \sigma\,\mathcal{X}\,\mathtt{balance} \quad \sigma' = \sigma - \mathcal{X} : n + \mathcal{Y} : n \\ \llbracket E_1 \rrbracket^{\mathcal{X}}_{\sigma,\rho} = v_1 \cdots v_h & \mathtt{f}(x_1 \cdots x_h)\{S\} \in \Gamma(\mathcal{Y}) \quad \rho' = \{\mathcal{X}/\mathtt{sender}, n/\mathtt{value}, v_1/x_1, \cdots, v_h/x_h\} \end{array}}{\llbracket E_0 : \mathtt{f}(E_1)\$E_2 \rrbracket^{\mathcal{X}}_{\sigma,\rho} = \llbracket S \rrbracket^{\mathcal{Y}}_{\sigma',\rho'}}$$

Fig. 2. Semantics of statements and expressions.

Executing S may affect both the store of $\mathcal{X}$ and, in case of procedure calls, also the store of other contracts. Instead, the semantics of an expression is a value; so, expressions have no side effects. We assume that all the semantic operators are *strict*, i.e. their result is $\bot$ if some operand is $\bot$. We denote by $\llbracket S \rrbracket^{\mathcal{X}}_{\sigma,\rho}$ the semantics of a statement S in a given state σ, environment ρ, and address $\mathcal{X}$, where the partial function $\llbracket \cdot \rrbracket^{\mathcal{X}}_{\sigma,\rho}$ is defined by the inference rules in Fig. 2. We write $\llbracket S \rrbracket^{\mathcal{X}}_{\sigma,\rho} = \bot$ when the semantics of S is not defined.

The semantics of expressions is straightforward; note that we use $\mathbf{op}$ to denote syntactic operators, and op for their semantic counterpart. The environment ρ is used to evaluate constant names x, while the state σ is used to evaluate $!E$ and $?E$. The semantics of statements is mostly standard, except for the last rule. A procedure call $E_0 : \mathtt{f}(E_1)\$E_2$ within $\mathcal{X}$ has a defined semantics iff: (i) E_0 evaluates to an address $\mathcal{Y}$; (ii) E_2 evaluates to a non-negative number n, not exceeding the balance of $\mathcal{X}$; (iii) the contract at $\mathcal{Y}$ has a procedure named $\mathtt{f}$ with formal parameters $x_1 \cdots x_h$; (iv) E_1 evaluates to a sequence of values of length h. If all these conditions hold, then the procedure body S is executed in a state where $\mathcal{X}$'s balance is decreased by n, $\mathcal{Y}$'s balance is increased by n, and in an environment where the formal parameters are bound to the actual ones, and the special names $\mathtt{sender}$ and $\mathtt{value}$ are bound, respectively, to $\mathcal{X}$ (the caller) and n (the value transferred to $\mathcal{Y}$).

Example 2. Consider the following statements, to be evaluated within a contract $\mathcal{C}$ in a store σ where $\sigma \mathcal{C} k = \bot$:

$$?k:=1 \qquad k:=?k \qquad \text{if } !\,?k \text{ then } k:=0 \text{ else } k:=1$$

$$\text{throw} \qquad ?k:=1; \text{ skip} \qquad \text{while true do skip}$$

We have that: (a) $?k:=1$ evaluates to $\bot$ because the first premise of the assignment rule is not satisfied, as the lhs of the assignment evaluates to $\bot$; (b) similarly, $k:=?k$ evaluates to $\bot$ because the second premise is not satisfied, as the rhs evaluates to $\bot$; (c) if $!\,?k$ then $k:=0$ else $k:=1$ evaluates to $\bot$, because the semantics of the guard is $\bot$; (d) since there are no semantic rules for throw, implicitly this means that its semantics is undefined; (e) $?k:=1$; skip is a sequence of two commands, where the first command evaluates to $\bot$. The rule for sequences requires that the first command evaluates to some state σ', while this is not the case for $?k:=1$. Therefore, the premise of the rule does not hold, and so the overall command evaluates to $\bot$; (f) finally, while true do skip evaluates to $\bot$, because there exists no (finite) derivation tree which infers $[\![\text{while true do skip}]\!]_{\sigma,\rho}^{\mathcal{C}} = \sigma$. Summing up, all the statements above have an undefined semantics. In practice, the semantic rules for transactions (see Sect. 4) ensure that the effects of any transaction whose statement evaluates to $\bot$ will be reverted (see e.g. Example 6).

Example 3 (Wallet). Consider the following procedures of the contract at $\mathcal{C}$:

$$\mathtt{f}()\,\{\text{if sender} = A \text{ then skip else throw}\}$$

$$\mathtt{g}(x,y)\,\{\text{if sender} = A \,\&\&\, \text{value} = 0 \,\&\&\, ?\text{balance} \geq x \text{ then } y\$x \text{ else throw}\}$$

The procedure $\mathtt{f}$ allows A to deposit funds to the contract; dually, $\mathtt{g}$ allows A to transfer funds to other addresses. The guard sender $= A$ ensures that only A can invoke the procedures of $\mathcal{C}$; calls from other addresses result in a throw, which leaves the state of $\mathcal{C}$ unchanged (in particular, throw reverts the currency transfer from sender to $\mathcal{C}$). The procedure $\mathtt{g}$ also checks that no currency is transferred along with the contract call (value $= 0$), and that the balance of $\mathcal{C}$ is enough (?balance $\geq x$). Let $S_{\mathtt{g}}$ be the body of $\mathtt{g}$, let σ be such that $\sigma\mathcal{C}\text{balance} = 3$, and let $\rho = \{A/\text{sender}, 0/\text{value}, 2/x, B/y\}$. We have:

$$[\![S_{\mathtt{g}}]\!]_{\sigma,\rho}^{\mathcal{C}} = [\![y\$x]\!]_{\sigma,\rho}^{\mathcal{C}} = [\![y : \mathtt{f}_{\text{skip}}()\$x]\!]_{\sigma,\rho}^{\mathcal{C}} = [\![\text{skip}]\!]_{\sigma - \mathcal{C}:2 + B:2, \{\mathcal{C}/\text{sender}, 1/\text{value}\}}^{B}$$

$$= \sigma - \mathcal{C} : 2 + B : 2$$

Note that $[\![S_{\mathtt{g}}]\!]_{\sigma,\rho}^{\mathcal{C}} = \bot$ if $\sigma\mathcal{C}\text{balance} < 2$, or $\rho\text{sender} \neq A$, or $\rho\text{value} \neq 0$.

3 Digression: Modelling Re-Entrancy

We now show how to express in TinySol *re-entrancy*, a subtle features of Solidity which was exploited in the famous "DAO Attack" [2,12].

Example 4 (Harmless re-entrancy). Consider the following procedures:

$$\mathtt{f}(x,b)\{\mathbf{if}\ b\ \mathbf{then}\ \{\mathcal{D}:\mathtt{g}();\ x\$\mathtt{value}\}\} \qquad \in \Gamma(\mathcal{C})$$

$$\mathtt{g}()\{\mathbf{sender}:\mathtt{f}(\mathcal{B},\mathtt{false})\} \qquad \in \Gamma(\mathcal{D})$$

Intuitively, $\mathtt{f}$ first calls $\mathtt{g}$, and then transfers $\mathtt{value}$ units of currency to the address x. The procedure $\mathtt{g}$ attempts to change the currency recipient by calling back $\mathtt{f}$, setting the parameter x to $\mathcal{B}$. We prove that this attack fails. Let $S = \mathcal{C}:\mathtt{f}(\mathcal{A},\mathtt{true})\$1$. For all σ and ρ such that $\sigma\mathcal{C}\mathtt{balance}=1$, we have:

$$\llbracket S \rrbracket_{\sigma,\rho}^{\mathcal{X}} = \llbracket \mathbf{if}\ b\ \mathbf{then}\ \{\mathcal{D}:\mathtt{g}();\ x\$\mathtt{value}\} \rrbracket_{\sigma,\rho'}^{\mathcal{C}} \qquad (\rho' = \{^{\mathcal{C}}/_{\mathtt{sender}},\ ^{1}/_{\mathtt{value}},\ ^{true}/_{b},\ ^{\mathcal{A}}/_{x}\})$$

$$= \llbracket \mathcal{D}:\mathtt{g}();\ x\$\mathtt{value} \rrbracket_{\sigma,\rho'}^{\mathcal{C}}$$

$$= \llbracket x\$\mathtt{value} \rrbracket_{\sigma',\rho'}^{\mathcal{C}} \qquad (\sigma' = \llbracket \mathcal{D}:\mathtt{g}() \rrbracket_{\sigma,\rho'}^{\mathcal{C}})$$

$$= \sigma' - \mathcal{C}:1+\mathcal{A}:1$$

where $\sigma' = \llbracket \mathcal{D}:\mathtt{g}() \rrbracket_{\sigma,\rho'}^{\mathcal{C}}$

$$= \llbracket \mathbf{sender}:\mathtt{f}(\mathcal{B},\mathtt{false}) \rrbracket_{\sigma,\{^{\mathcal{C}}/_{\mathtt{sender}},\,^{0}/_{\mathtt{value}}\}}^{\mathcal{D}}$$

$$= \llbracket \mathbf{if}\ b\ \mathbf{then}\ \{\mathcal{D}:\mathtt{g}();\ x\$\mathtt{value}\} \rrbracket_{\sigma,\rho''}^{\mathcal{C}} \qquad (\rho'' = \{^{\mathcal{D}}/_{\mathtt{sender}},\ ^{0}/_{\mathtt{value}},\ ^{false}/_{b},\ ^{\mathcal{B}}/_{x}\})$$

$$= \llbracket \mathbf{skip} \rrbracket_{\sigma,\rho''}^{\mathcal{C}} = \sigma$$

Since $\sigma' = \sigma$, we conclude that $\llbracket S \rrbracket_{\sigma,\rho}^{\mathcal{C}} = \sigma - \mathcal{C}:1+\mathcal{A}:1$. So, $\mathtt{g}$ has failed its attempt to divert the currency transfer to $\mathcal{B}$. □

Example 5 (Vicious re-entrancy). Consider the following procedures:

$$\mathtt{f}()\{\mathbf{if}\ \mathbf{not}\,!k\ \&\&\ ?\mathtt{balance}\geq 1\ \mathbf{then}\ \{\mathcal{D}:\mathtt{g}()\$1;\ k:=\mathbf{true}\}\,\} \qquad \in \Gamma(\mathcal{C})$$

$$\mathtt{g}()\{\mathcal{C}:\mathtt{f}()\} \qquad \in \Gamma(\mathcal{D})$$

Intuitively, $\mathtt{f}$ would like to transfer 1 ether to $\mathcal{D}$, by calling $\mathtt{g}$. The guard $\mathbf{not}\,!k$ is intended to ensure that the transfer happens at most once. Let σ be such that $\sigma\mathcal{C}\mathtt{balance}=n\geq 1$ and $\sigma\mathcal{C}k=\bot$, and let $\rho=\{^{\mathcal{D}}/_{\mathtt{sender}},\,^{0}/_{\mathtt{value}}\}$, $\rho'=\{^{\mathcal{C}}/_{\mathtt{sender}},\,^{1}/_{\mathtt{value}}\}$. Let $S_{\mathtt{f}}$ and $S_{\mathtt{g}}$ be the bodies of $\mathtt{f}$ and $\mathtt{g}$. We have:

$$\llbracket S_{\mathtt{f}} \rrbracket_{\sigma,\rho}^{\mathcal{C}} = \llbracket \mathcal{D}:\mathtt{g}()\$1;\ k:=\mathbf{true} \rrbracket_{\sigma,\rho}^{\mathcal{C}} = \llbracket k:=\mathbf{true} \rrbracket_{\sigma_1,\rho}^{\mathcal{C}}$$

$$\sigma_1 = \llbracket \mathcal{D}:\mathtt{g}()\$1 \rrbracket_{\sigma,\rho}^{\mathcal{C}} = \llbracket S_{\mathtt{g}} \rrbracket_{\sigma-\mathcal{C}:1+\mathcal{D}:1,\rho'}^{\mathcal{D}} = \llbracket S_{\mathtt{f}} \rrbracket_{\sigma-\mathcal{C}:1+\mathcal{D}:1,\rho}^{\mathcal{C}}$$

$$= \llbracket \mathcal{D}:\mathtt{g}()\$1;\ k:=\mathbf{true} \rrbracket_{\sigma-\mathcal{C}:1+\mathcal{D}:1,\rho}^{\mathcal{C}}$$

$$= \llbracket k:=\mathbf{true} \rrbracket_{\sigma_2,\rho}^{\mathcal{C}}$$

$$\sigma_2 = \llbracket \mathcal{D}:\mathtt{g}()\$1 \rrbracket_{\sigma-\mathcal{C}:1+\mathcal{D}:1,\rho}^{\mathcal{C}} = \llbracket S_{\mathtt{g}} \rrbracket_{\sigma-\mathcal{C}:2+\mathcal{D}:2,\rho'}^{\mathcal{D}} = \llbracket k:=\mathbf{true} \rrbracket_{\sigma_3,\rho}^{\mathcal{C}}$$

$$\sigma_i = \llbracket k:=\mathbf{true} \rrbracket_{\sigma_{i+1},\rho}^{\mathcal{C}} \qquad (\text{for } i \in 3\ldots n-1)$$

$$\sigma_n = \llbracket \mathbf{skip} \rrbracket_{\sigma-\mathcal{C}:n+\mathcal{D}:n,\rho}^{\mathcal{C}} = \sigma - \mathcal{C}:n+\mathcal{D}:n$$

Summing up, $\llbracket S_{\mathtt{f}} \rrbracket_{\sigma,\rho}^{\mathcal{C}} = (\sigma - \mathcal{C}:n+\mathcal{D}:n)\{^{true}/k\}$, i.e. $\mathcal{D}$ has drained all the currency from $\mathcal{C}$. □

4 Transactions and Blockchains

A *transaction* T is a term of the form $\mathcal{A} \xrightarrow{n} \mathcal{C} : \mathtt{f}(\boldsymbol{v})$, where $\mathcal{A}$ is the address of the caller, $\mathcal{C}$ is the address of the called contract, $\mathtt{f}$ is the called procedure, n is the value transferred from $\mathcal{A}$ to $\mathcal{C}$, and $\boldsymbol{v}$ is the sequence of actual parameters. The semantics of T in a given state σ, is a new state $\sigma' = [\![\mathsf{T}]\!]_\sigma$. The function $[\![\cdot]\!]_\sigma$ is defined by the following rules:

$$\frac{\mathtt{f}(\boldsymbol{x})\{S\} \in \Gamma(\mathcal{C}) \qquad \sigma\mathcal{A}\,\mathbf{balance} \geq n \qquad [\![S]\!]^{\mathcal{C}}_{\sigma-\mathcal{A}:n+\mathcal{C}:n,\,\{\mathcal{A}/\mathtt{sender},n/\mathtt{value},v/x\}} = \sigma'}{[\![\mathcal{A} \xrightarrow{n} \mathcal{C} : \mathtt{f}(\boldsymbol{v})]\!]_\sigma = \sigma'} \text{[Tx1]}$$

$$\frac{\mathtt{f}(\boldsymbol{x})\{S\} \in \Gamma(\mathcal{C}) \quad (\sigma\mathcal{A}\,\mathbf{balance} < n \quad \text{or} \quad [\![S]\!]^{\mathcal{C}}_{\sigma-\mathcal{A}:n+\mathcal{C}:n,\,\{\mathcal{A}/\mathtt{sender},n/\mathtt{value},v/x\}} = \bot)}{[\![\mathcal{A} \xrightarrow{n} \mathcal{C} : \mathtt{f}(\boldsymbol{v})]\!]_\sigma = \sigma} \text{[Tx2]}$$

Rule [Tx1] handles the case where the transaction is successful: this happens when $\mathcal{A}$'s balance is at least n, and the procedure call terminates in a non-error state. Note that n units of currency are transferred to $\mathcal{C}$ *before* starting to execute $\mathtt{f}$, and that the names $\mathbf{sender}$ and $\mathbf{value}$ are set, respectively, to $\mathcal{A}$ and n. Instead, [Tx2] applies either when $\mathcal{A}$'s balance is not enough, or the execution of $\mathtt{f}$ fails (this also covers the case when $\mathtt{f}$ does not terminate). In these cases, T does not alter the state, i.e. $\sigma' = \sigma$.

A *blockchain* $\mathbf{B}$ is a finite sequence of transactions. The semantics of $\mathbf{B}$ is obtained by folding the semantics of its transactions:

$$[\![\epsilon]\!]_\sigma = \sigma \qquad [\![\mathsf{T}\mathbf{B}]\!]_\sigma = [\![\mathbf{B}]\!]_{[\![\mathsf{T}]\!]_\sigma}$$

Note that erroneous transactions occuring in a blockchain have no effect on its semantics (as rule [Tx2] makes them identities w.r.t. the append operation).

Example 6. Recall the contract $\mathcal{C}$ from Example 3, and let $\mathbf{B} = \mathsf{T}_0\mathsf{T}_1\mathsf{T}_0$, where:

$$\mathsf{T}_0 = \mathcal{A} \xrightarrow{3} \mathcal{C} : \mathtt{f}() \qquad \mathsf{T}_1 = \mathcal{A} \xrightarrow{0} \mathcal{C} : \mathtt{g}(2, \mathcal{B})$$

Let $S_\mathtt{f}$ and $S_\mathtt{g}$ be the bodies of $\mathtt{f}$ and $\mathtt{g}$, respectively. $\sigma\mathcal{A}\,\mathbf{balance} = 5$ and $\sigma\mathcal{C}\,\mathbf{balance} = 0$. By rule [Tx1] we have that:

$$[\![\mathsf{T}_0]\!]_\sigma = [\![S_\mathtt{f}]\!]^{\mathcal{C}}_{\sigma-\mathcal{A}:3+\mathcal{C}:3,\{\mathcal{A}/\mathtt{sender},3/\mathtt{value}\}} = [\![\mathtt{skip}]\!]^{\mathcal{C}}_{\sigma-\mathcal{A}:3+\mathcal{C}:3,\{\mathcal{A}/\mathtt{sender},3/\mathtt{value}\}}$$
$$= \sigma - \mathcal{A} : 3 + \mathcal{C} : 3$$

Now, let $\sigma' = \sigma - \mathcal{A} : 3 + \mathcal{C} : 3$. By rule [Tx1] we have that:

$$[\![\mathsf{T}_1]\!]_{\sigma'} = [\![S_\mathtt{g}]\!]^{\mathcal{C}}_{\sigma',\{\mathcal{A}/\mathtt{sender},0/\mathtt{value},2/x,\mathcal{B}/y\}} = [\![y\$x]\!]^{\mathcal{C}}_{\sigma',\{\mathcal{A}/\mathtt{sender},0/\mathtt{value},2/x,\mathcal{B}/y\}}$$
$$= \sigma' - \mathcal{C} : 2 + \mathcal{B} : 2$$

Let $\sigma'' = \sigma' - \mathcal{C} : 2 + \mathcal{B} : 2$. By rule [Tx2], we obtain $[\![\mathbf{B}]\!]_\sigma = [\![\mathsf{T}_0]\!]_{\sigma''} = \sigma''$.

5   Conclusions

We have introduced TinySol, a minimal core contract calculus inspired by Solidity. While our calculus is focussed on a single new construct to call contracts and transfer currency, other languages have been proposed to capture other peculiar aspects of smart contracts. Some of these language are domain-specific, e.g. for financial contracts [4,7] and for business processes [11,17], while some others are more abstract, modelling contracts as automata with guarded transitions [13,16]. Establishing the correctness of the compilation from these languages to Solidity would be one of the possible applications of a bare bone formal model, like our TinySol. Another possible application of a minimal calculus is the investigation of different styles of semantics, like e.g. denotational and axiomatic semantics. Further, the study of analysis and optimization techniques for smart contracts may take advantage of a succinct formalization like ours.

Differences Between TinySol **and Solidity.** Aiming at minimality, TinySol simplifies or neglects several features of Solidity. A first difference is that we do not model a *gas mechanism*. In Ethereum, when sending a transaction, users deposit into it some crypto-currency, to be paid to the miner which appends the transaction to the blockchain. Each computation step performed by the miner consumes part of this deposit; when the deposit reaches zero, the miner stops executing the transaction. At this point, all the effects of the transaction (except the payment to the miner) are rolled back. Although in TinySol we do not model the gas mechanism, we still ensure that non-terminating calls have an undefined semantics (see e.g. Example 2), so that they are rolled back by rule [Tx2]. The semantics of TinySol could be easily extended with an "abstract" gas model, by associating a cost to instructions and recording the gas consumption in the environment. However, note that any gas mechanism formalized at the level of abstraction of Solidity would not faithfully reflect the actual Ethereum gas mechanism, where the cost of instructions are defined at the EVM bytecode level. Indeed, compiler optimizations would make it hard to establish a correspondence between the cost of a piece of Solidity code and the cost of its compiled bytecode. Still, an abstract gas model could be useful in practice, e.g. to establish upper bounds to the gas consumption of a piece of Solidity code.

A second difference is that our model assumes the set of contracts to be fixed, while in Ethereum new contracts can be created at run-time. As a consequence, TinySol does not feature constructors that are called when the contract is created. Dynamic contract creation could be formalized by extending our model with special transactions which extends the mapping Γ with the contracts generated at run-time. Once this is done, adding constructors is standard.

In Ethereum, contracts can implement time constraints by using the block publication time, accessible via the variable block.timestamp. In TinySol we do not record timestamps in the blockchain. Still, time constraints can be implemented by using oracles, i.e. contracts which allow certain trusted parties to set their keys (e.g., timestamps), and make them accessible to other contracts (see e.g. the lottery contract in [3]).

In Ethereum, when the procedure name specified in the transaction does not match any of the procedures in the contract, a special unnamed "fallback" procedure (with no arguments) is implicitly invoked. Extending TinySol with this mechanism would be straightforward. Delegate and internal calls, which we have omitted in TinySol, would be simple to model as well.

Acknowledgements. Partially supported by Aut. Reg. Sardinia projects *Smart collaborative engineering* and *Sardcoin*, by IMT Lucca project *PAI VeriOSS*, and by MIUR PON *Distributed Ledgers for Secure Open Communities*.

References

1. Solidity documentation. solidity.readthedocs.io/en/v0.5.4/ (2019)
2. Atzei, N., Bartoletti, M., Cimoli, T.: A survey of attacks on ethereum smart contracts (SoK). In: Maffei, M., Ryan, M. (eds.) POST 2017. LNCS, vol. 10204, pp. 164–186. Springer, Heidelberg (2017). https://doi.org/10.1007/978-3-662-54455-6_8
3. Bartoletti, M., Galletta, L., Murgia, M.: A minimal core calculus for Solidity contracts. CoRR arXiv:1908.02709 (2019)
4. Biryukov, A., Khovratovich, D., Tikhomirov, S.: Findel: secure derivative contracts for ethereum. In: Brenner, M., et al. (eds.) FC 2017. LNCS, vol. 10323, pp. 453–467. Springer, Cham (2017). https://doi.org/10.1007/978-3-319-70278-0_28
5. Buterin, V.: Ethereum: a next generation smart contract and decentralized application platform. github.com/ethereum/wiki/wiki/White-Paper (2013)
6. Crafa, S., Pirro, M.D., Zucca, E.: Is solidity solid enough? In: Financial Cryptography Workshops (2019)
7. Egelund-Müller, B., Elsman, M., Henglein, F., Ross, O.: Automated execution of financial contracts on blockchains. Bus. Inf. Syst. Eng. **59**(6), 457–467 (2017)
8. Hildenbrandt, E., et al.: KEVM: a complete formal semantics of the Ethereum Virtual Machine. In: IEEE Computer Security Foundations Symposium (CSF), pp. 204–217 (2018)
9. Igarashi, A., Pierce, B.C., Wadler, P.: Featherweight Java: a minimal core calculus for Java and GJ. ACM Trans. Program. Lang. Syst. **23**(3), 396–450 (2001)
10. Jiao, J., Kan, S., Lin, S., Sanán, D., Liu, Y., Sun, J.: Executable operational semantics of Solidity. CoRR arXiv:1804.01295 (2018)
11. López-Pintado, O., García-Bañuelos, L., Dumas, M., Weber, I., Ponomarev, A.: Caterpillar: a business process execution engine on the Ethereum blockchain. Practice and Experience, Software (2019)
12. Luu, L., Chu, D.H., Olickel, H., Saxena, P., Hobor, A.: Making smart contracts smarter. In: ACM CCS, pp. 254–269 (2016)
13. Mavridou, A., Laszka, A., Stachtiari, E., Dubey, A.: VeriSolid: Correct-by-design smart contracts for Ethereum. In: Financial Cryptography and Data Security (2019)
14. Roşu, G., Şerbănuţă, T.F.: An overview of the K semantic framework. J. Logic Algebraic Program. **79**(6), 397–434 (2010)
15. Sergey, I., Hobor, A.: A concurrent perspective on smart contracts. In: Brenner, M., et al. (eds.) FC 2017. LNCS, vol. 10323, pp. 478–493. Springer, Cham (2017). https://doi.org/10.1007/978-3-319-70278-0_30

16. Sergey, I., Kumar, A., Hobor, A.: Scilla: a smart contract intermediate-level language. CoRR abs/1801.00687 (2018)
17. Tran, A.B., Lu, Q., Weber, I.: Lorikeet: A model-driven engineering tool for blockchain-based business process execution and asset management. In: BPM, pp. 56–60 (2018)
18. Yang, Z., Lei, H.: Lolisa: formal syntax and semantics for a subset of the Solidity programming language. CoRR arXiv:1803.09885 (2018)
19. Zakrzewski, J.: Towards verification of ethereum smart contracts: a formalization of core of solidity. In: Piskac, R., Rümmer, P. (eds.) VSTTE 2018. LNCS, vol. 11294, pp. 229–247. Springer, Cham (2018). https://doi.org/10.1007/978-3-030-03592-1_13

Multi-stage Contracts
in the UTXO Model

Alexander Chepurnoy and Amitabh Saxena[✉]

Ergo Platform, Bern, Switzerland
{kushti,amitabh123}@protonmail.ch

Abstract. Smart contract platforms such as Bitcoin and Ethereum allow writing programs that run on a decentralized computer. Bitcoin uses short-lived immutable data structures called UTXOs for data manipulation. Ethereum, on the other hand uses, long-lived mutable data structures called *accounts*. UTXOs are easier to handle, less error prone and scale better because the only operation we can do with them is to create or destroy (i.e., spend) them. The code inside a UTXO is executed only once, when it is spent. Additionally, this code refers to only local context (i.e., it is stateless). In Ethereum's account based system, there is a shared global context which each account can access and modify, thereby causing side affects. However, the benefit of persistent storage offered by accounts makes up for these drawbacks. In this work, we describe how to emulate persistent storage in UTXO based systems using a technique called *transaction trees*. This allows us to emulate the functionality of account-based systems such as Ethereum without the overhead of accounts. We demonstrate this via several examples which include contracts for a Rock-Paper-Scissors game, crowdfunding and an initial coin offering (ICO). The contracts are created in a UTXO based smart contract platform called Ergo that supports transaction trees.

1 Introduction

Smart contracts were envisioned in 1994 by Nick Szabo [1], a legal scholar and cryptographer. He proposed the concept of self-executing contracts written in executable code and stored in a replicated manner on distributed computers that enforced the rules written in the code. Bitcoin [2] can be seen as the first implementation of this concept using a fully decentralized ledger whose contracts primarily pertain to transfer and store of value, i.e., as a currency system. Ethereum [3] is an example of a general-purpose smart contract platform.

The limited application of Bitcoin allows optimizations focussed on long-term survivability and scalability. Firstly, all data and code is stored in short-lived immutable objects (called UTXOs [4]). A user can execute code inside a UTXO by supplying some input (which may contain additional code). A UTXO is destroyed once its code is executed (i.e., it is spent). Secondly, all computation is performed within a *local context*; any code pertaining to a UTXO can only operate on data for that UTXO and does not have access to the global state.

© Springer Nature Switzerland AG 2019
C. Pérez-Solà et al. (Eds.): DPM 2019/CBT 2019, LNCS 11737, pp. 244–254, 2019.
https://doi.org/10.1007/978-3-030-31500-9_16

In contrast, Ethereum follows a different set of design principles in which the code and data is contained in long-lived mutable objects called *accounts*. This was done because UTXOs are stateless and do not provide persistent storage. Not only can Ethereum code modify data in its own account, but also trigger execution of code in other accounts. Thus, Ethereum code operates over a *shared global context* representing all existing accounts.

The results of [5] allow UTXO-based systems to emulate Ethereum-like functionality by reducing the computation to Rule-110 [6, 7]. However, such reductions are not very efficient and a more practical solution for the same is desirable. In particular, we need higher-level abstractions (instead of Rule-110) that enable UTXO-based systems to efficiently emulate Ethereum functionality and maintain Bitcoin's scalability. In this work we describe a technique called *transaction trees* that allow writing advanced smart contracts in UTXO based systems. As proof of concept, we implemented such contracts on a UTXO-based platform called Ergo that supports transaction trees.

Context Enrichment. In Bitcoin and other existing UTXO systems, the context is just the UTXO being processed. In order for a UTXO-based system to support transaction trees, the context must be rich enough to contain at least the entire spending transaction. More formally, for any UTXO based blockchain, we can define the following levels of context, each extending the previous:

1. The current UTXO plus the blockchain height and time
2. The current spending transaction (other inputs and outputs)
3. The current block's solution.
4. The current block (other sibling transactions)

Any platform at Level 2 and above is suitable for transaction trees. In this regard, Bitcoin operates at Level 1 and Ergo at Level 3. Note that in Level 4 we cannot check validity of transactions independently of other transactions in the block. Hence it is more complex to implement Level 4.

In this work we show via examples how to create efficient Ethereum-like contracts in the UTXO model using transaction trees. The examples include a Rock-Paper-Scissors game, an Initial Coin Offering (ICO) campaign and a new primitive called *reversible addresses* for securely storing funds.

2   Ergo Overview

The Ergo platform is a Level 3 UTXO based blockchain that allows general-purpose smart contracts via a highly expressible language called ErgoScript. Since Ergo follows the UTXO based model, all data and code is stored in immutable objects called *boxes*. As in Bitcoin, a transaction in Ergo can spend (destroy) multiple boxes and create new ones. A box is made of up to ten *registers* labelled $R_0, R_1, \ldots R_9$, four of which are mandatory. R_0 contains the monetary value, R_1 contains the *guard script*, R_2 contains assets (tokens) and R_3 contains a unique identifier of 34 bytes made up of a transaction ID and an output index.

The guard script in R_1 encodes a spending condition, which must be satisfied for spending the box. Deploying a contract involves creating an unspent box with the relevant ErgoScript code in R_1 and populating other registers if necessary. The contract is executed by spending the box.

Similar to Bitcoin, an ErgoScript program also cannot access the global state and all computation must be done only using a local context. Unlike Bitcoin, this context is quite rich and allows access to the entire spending transaction [8]. In particular, an ErgoScript program defines the spending condition using predicates on the inputs and outputs of the transaction interleaved with Sigma protocols [9]. Thus, an ErgoScript program can enforce the spending transaction's structure (such as requiring that assets are transferable only to a certain address). Additionally, the program can require the spender to prove knowledge of the discrete logarithm of some public value using a protocol called proveDlog, which is based on the Schnorr identification scheme [9]. All public keys (such as alice and bob) in the following sections are of type proveDlog. Similar to Bitcoin, a Pay-to-Script-Hash (P2SH) address in Ergo contains the hash of a script that must be provided when spending from that address. The script encodes the actual spending condition. Ergo also supports Pay-to-Script (P2S) address, where the actual script is encoded in the address.

One useful feature of Ethereum is the ability to store and access a large amount of data, which Ergo also provides. However, Ergo contracts do not store the actual data in the blockchain. Rather, the data is stored off chain and a short digest is stored in the blockchain. A user wishing to access or modify this data must provide correct proofs of (non)-existence or modification for this digest, as in the ICO example of Sect. 4.3.

3   Transaction Trees

A powerful feature of ErgoScript is the ability to specify the spending transaction's structure in a fine-grained manner. Among the many things we can specify, the important ones are: (1) the number of input and output boxes, (2) the value of any box, and (3) the guarding script of any box. This allows us to create *transaction trees*, where the contract in an input box requires an output box to contain some predefined contract, thereby ensuring that only a certain sequence of contracts are possible. We will use this to convert an Ethereum-style long-lived contract into multi-stage contracts in the UTXO model, where each stage encodes data and code to be carried over to the next stage.

Transaction Chains: Before describing transaction trees, we describe a simpler primitive called transaction chains. A transaction chain is used for creating a multi-stage protocol whose code does not contain loops or 'if' statements. A transaction chain is created as follows:

1. Represent an Ethereum contract's execution using n sequential steps, where each step represents a transaction that modifies its state. The states before and after a transaction are the start and end nodes respectively of a directed

graph, with the transaction as the edge joining them. As an example, a 3-stage contract, such as the ICO example of Sect. 4.3 is represented as:

The states contain data and the code that was executed in the transaction.

2. Hardwire state n's code and data inside state $n-1$'s code. Then require the code of state $n-1$ to output a box containing state n's code and data. An example is given in the following pseudocode:

```
out.propositionBytes == state_n_code &&
out.R4[Int].get == SELF.R4[Int].get // ensure data is propagated
```

The above code uses the field `propositionBytes` of a box, which contains the binary representation of its guard script as collection of bytes.

3. Repeat Step 2 by replacing $(n, n-1)$ by $(n-1, n-2)$ while $n > 2$.

To avoid code size increase at each iteration, we should ideally work with hashes, as in `hash(out.propositionBytes) == state_n_code_hash`. However, for clarity of presentation, we will skip this optimization.

Transaction Trees: A transaction tree is an extension of transaction chains where the code can contain 'if' statements and *simple loops*, i.e., where some start and end nodes are the same. The following figure illustrates a transaction tree.

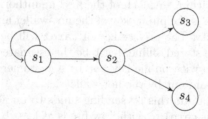

An 'if' statement is handled using the following pseudocode.

```
if (condition) { out.propositionBytes == state_3_code }
else { out.propositionBytes == state_4_code }
```

A simple loop is a special case of the 'if' statement:

```
if (condition) { out.propositionBytes == state_2_code }
else { out.propositionBytes == SELF.propositionBytes }
```

Transaction Graphs: Ergo supports a more advanced technique called *transaction graphs*, where cycles are allowed in contract references, as shown below.

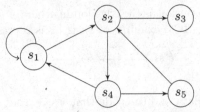

Discussion of such contracts is beyond the scope of this work and we refer the reader to [10, Section 3.3.3] for an example of such a contract. All the examples in this paper are based on transaction trees.

4 Multi-stage Contracts

4.1 Reversible Addresses

An example of multi-stage contract is a *reversible address*, which has anti-theft features in the following sense: any funds sent to a reversible address can only be spent in way that allows payments to be reversed for a certain time. To motivate this feature, consider managing the hot-wallet of an exchange or mining pool used for handling customer withdraws. A hot-wallet is an address for which the private key is stored on the server. Such addresses are necessary for facilitating automated withdrawals. Being a hot-wallet, its private key is susceptible to compromise and funds being stolen. We want to ensure that we are able to recover any stolen funds in the event of such a compromise, provided that the breach is discovered within, say, 24 h of the first unauthorized withdraw.

Assume that `alice` is the public key of the hot-wallet and `carol` is the public key of the trusted party. The private key of `carol` will be needed for reversing payments and must be stored offline. Let b be the estimated number of blocks in a 24 h period. Let Bob with public key `bob` be a customer wishing to withdraw funds, which will be paid out by the hot-wallet.

In Ethereum, we can do this by sending funds to an account having with a contract C_b that allows `carol` to withdraw funds at least b blocks and after that they can only be withdrawn by `bob`. We could use the same account (contract instance) for multiple withdraws by Bob, but the optimal way is to have a new account for each withdraw, emulating the UTXO model. The funds for this must also come from another account with a contract C_a that ensure that withdraw can only be done to a contract with the structure of C_b.

In Ergo, this is done by a two-stage protocol, where the second stage implements C_b and the first stage implements C_a. The following script called `withdrawScript` implements the second stage. This will be the guarding script of the hot-wallet's withdraw transaction paying to `bob`.

```
val bob = SELF.R4[SigmaProp].get    // public key of customer withdrawing
val bobDeadline = SELF.R5[Int].get  // max locking height
(bob && HEIGHT > bobDeadline) || (carol && HEIGHT <= bobDeadline)
```

This above script is referenced in the first stage script given next.

```
val isChange = {(b:Box) => b.propositionBytes == SELF.propositionBytes}
val isWithdraw = {(b:Box) =>
    b.R5[Int].get >= HEIGHT + blocksIn24h &&
    b.propositionBytes == withdrawScript
}
alice && OUTPUTS.forall({(b:Box) => isChange(b) || isWithdraw(b)})
```

The reversible address is the P2SH address of the above script. Any funds sent to this address are subject to the withdraw rules that we desire. In the normal case, Bob will spend the box after roughly `blocksIn24h` blocks. If an unauthorized transaction from the hot-wallet is detected, an abort procedure is triggered using the private key of `carol` and funds in any unspent boxes sent from the hot-wallet are diverted to a secure address. Note that the trusted party (`carol`) is bound to the hot-wallet address. A new address is needed for a different trusted party.

Although such addresses are designed for securing hot-wallet funds, the may have other applications. One example is for automated-release escrow payments in online shopping, where `carol` can be the public key of any mutually agreed adjudicating party.

4.2   Rock-Paper-Scissors Game

Our next example of a multi-stage contract is the Rock-Paper-Scissors game, which is often used to introduce Ethereum [11]. The game is played between two players, Alice and Bob. Each player chooses a secret independently and the game is decided after the secrets are revealed. Let $a, b \in \mathbb{Z}_3$ be the secrets of Alice, Bob respectively, with the understanding that $(0, 1, 2)$ represent (rock, paper, scissors). If $a = b$ then the game is a draw, otherwise if $a - b \in \{1, -2\}$ then Alice wins else Bob wins.

The first party to reveal the secret has a disadvantage, since the other party can adaptively choose and win. In the real world, both parties reveal their secrets simultaneously to prevent this. In the virtual world, however, this cannot be enforced. Hence this attack must be handled using *cryptographic commitments*, where the first party, Alice, does not initially reveal her secret, but rather only a commitment to that secret. The modified game using commitments is as follows:

1. Alice commits to her secret a by inputting her commitment $c = Comm(a)$.
2. Bob inputs his public value b. At this stage, Alice knows if she won or lost.
3. Alice opens her commitment and reveals a, after which the winner is decided.

This works fine assuming that Alice is *well-behaved*, i.e., she always opens her commitment irrespective of whether she won or lost. In the real world, however, we also need to consider the possibility that Alice never opens her commitment. Border cases such as these make smart contracts quite tricky, because once deployed, it is not possible to add "bug-fixes" to them. In this example, we must penalize Alice (with a loss) if she does not open her commitment within some stipulated time.

The complete game is coded in ErgoScript in two stages. In the first stage, Alice creates a *start-game* box that encodes her game rules. In the second stage, Bob spends the start-game box and creates two *end-game* boxes spendable by the winner. These new boxes indicate that the game has ended.

To start the game, Alice decides a game amount x (of Ergo's primary token), which each player must contribute. She then selects a secret s and computes a commitment

$c = H(a||s)$ to a. Finally, she locks up x tokens along with her commitment c inside the start-game box protected by the following script:

```
OUTPUTS.forall(
  {(out:Box) =>
    val b = out.R4[Byte].get
    val bobDeadline = out.R6[Int].get
    bobDeadline >= HEIGHT+30 && out.value >= SELF.value &&
    (b == 0 || b == 1 || b == 2) &&
    out.propositionBytes == outScript
  }
) && OUTPUTS.size == 2 && OUTPUTS(0).R7[SigmaProp].get == alice &&
OUTPUTS(0).R4[Byte].get == OUTPUTS(1).R4[Byte].get // same b
```

The above code requires that the spending transaction must create exactly two outputs, one paying to each player in the event of a draw or both paying to the winner otherwise. In particular, the code requires that (1) register R_7 of the first output must contain Alice's public key (for use in the draw scenario), (2) register R_4 of each output must contain Bob's choice, and (3) each output must contain at least x tokens protected by outScript, which is given below:

```
val s = getVar[Coll[Byte]](0).get  // Alice's secret byte string s
val a = getVar[Byte](1).get  // Alice's secret choice a
val b = SELF.R4[Byte].get     // Bob's public choice b
val bob = SELF.R5[SigmaProp].get // Bob's public key
val bobDeadline = SELF.R6[Int].get // after this, Bob wins by default
val drawPubKey = SELF.R7[SigmaProp].get
val valid_a = (a == 0 || a == 1 || a == 2)
val validCommitment = blake2b256(s ++ Coll(a)) == c
val validAliceChoice = valid_a && validAliceChoice
val aliceWins = (a - b) == 1 || (a - b) == -2
val receiver = if (a == b) drawPubKey else (if (aliceWins) alice else bob)
(bob && HEIGHT > bobDeadline) || (receiver && validAliceChoice)
```

The above code protects the two end-game boxes that Bob generates. The condition (bob && HEIGHT > bobDeadline) guarantees that if Alice does not open her commitment before a certain deadline, then Bob automatically wins. Note that Bob has to ensure that R_7 of the second output contains his public key. Additionally, he must ensure that R_5 of both outputs contains his public key (see below). We don't encode these conditions because if Bob doesn't follow the protocol, he will automatically lose.

4.3   Initial Coin Offering

Another popular use-case of Ethereum is an Initial Coin Offering (ICO) contract. An ICO mirrors an Initial Public Offering (IPO) and provides a mechanism for a project to collect funding in some tokens and then issue "shares" (in the form of some other tokens) to investors. Generally, an ICO comprises of 3 stages:

1. *Funding:* During this period, investors are allowed to fund the project.
2. *Issuance:* A new asset token is created and issued to investors.
3. *Withdrawal:* Investors can withdraw their newly issued tokens.

Compared to the previous examples, our ICO contract is quite complex, since it involves multiple stages and parties. The number of investors may run into thousands, and the naive solution would store this data in the contract, as in the ERC-20 standard [12]. Unlike Ethereum, Ergo does not permit storing large datasets in a contract. Rather, we store only a 40-bytes header of (a key, value) dictionary, that is authenticated like a Merkle tree [13]. To access some elements in the dictionary, or to modify it, a spending transaction should provide lookup or modification proofs. This allows a contract to authenticate large datasets using very little storage and memory.

Funding: The project initiates the ICO by creating a box with the guard script given below. The box also contains a authenticating value for an empty dictionary of (investor, balance) pairs in R_5, where investor is the hash of a script that will guard the box with the withdrawn tokens (once the funding period ends).

```
val selfIndexIsZero = INPUTS(0).id == SELF.id
val proof = getVar[Coll[Byte]](1).get
val toAdd = INPUTS.slice(1, INPUTS.size).map({(b: Box) =>
    val pk = b.R4[Coll[Byte]].get
    val value = longToByteArray(b.value)
    (pk, value)
})
val modifiedTree = SELF.R5[AvlTree].get.insert(toAdd, proof).get
val expectedTree = OUTPUTS(0).R5[AvlTree].get
val selfOutputCorrect =
  if (HEIGHT < 2000) OUTPUTS(0).propositionBytes == SELF.propositionBytes
  else OUTPUTS(0).propositionBytes == issuanceScript
val outputsCorrect = OUTPUTS.size == 1 && selfOutputCorrect

selfIndexIsZero && outputsCorrect && modifiedTree == expectedTree
```

The first funding transaction spends this box and creates a box with the same script and updated data. Further funding transactions spend the box created from the previous funding transaction. The box checks that it is first input of each funding transaction, which must have other inputs belonging to investors. The investor inputs contain a hash of the withdraw script in register R_4. The script also checks (via proofs) that hashes and monetary values of the investing inputs are correctly added to the dictionary of the new box, which must be only output with the correct amount of ergs (we ignore fee in this example). In this stage, which lasts at least till height 2,000, withdraws are not permitted and ergs can only be put into the project. The first transaction with height of 2,000 or more should keep the same data but change the output's script called **issuanceScript** described next.

Issuance: This stage requires only one transaction to get to the next stage (the withdrawal stage). The spending transactions makes the following modifications. Firstly, it changes the list of allowed operations on the dictionary from "inserts only" to "removals only". Secondly, the contract checks that the proper amount of ICO tokens are issued. In Ergo, each transaction can issue at most one new kind of token, with the (unique) identifier of the first input box. The issuance contract checks that a new token is issued with amount equal to the nano-ergs collected till now. Thirdly, the contract checks that a spending transaction is indeed re-creating the box with the guard script corresponding to the next stage, the withdrawal stage. Finally, the contract checks that the spending transaction has 2 outputs (one for the project tokens and one for the ergs withdrawn by the project). The complete script is given below.

```
val openTree = SELF.R5[AvlTree].get
val closedTree = OUTPUTS(0).R5[AvlTree].get
val correctDigest = openTree.digest == closedTree.digest
val correctKeyLength = openTree.keyLength == closedTree.keyLength
val removeOnlyTree = closedTree.enabledOperations == 4
val correctValue = openTree.valueLengthOpt == closedTree.valueLengthOpt
val tokenId: Coll[Byte] = INPUTS(0).id
val tokenIssued = OUTPUTS(0).tokens(0)._2
val correctTokenNumber = OUTPUTS(0).tokens.size == 1 &&
                         OUTPUTS(1).tokens.size == 0
val correctTokenIssued = SELF.value == tokenIssued
val correctTokenId = OUTPUTS(0).R4[Coll[Byte]].get == tokenId &&
                     OUTPUTS(0).tokens(0)._1 == tokenId
val valuePreserved = OUTPUTS.size == 2 && correctTokenNumber &&
                     correctTokenIssued && correctTokenId
val stateChanged = OUTPUTS(0).propositionBytes == withdrawScript
val treeIsCorrect = correctDigest && correctValue &&
                    correctKeyLength && removeOnlyTree


projectPubKey && treeIsCorrect && valuePreserved && stateChanged
```

Withdrawal: Investors are now allowed to withdraw ICO tokens under a guard script whose hash is stored in the dictionary. Withdraws are done in batches of N. A withdrawing transaction, thus, has $N + 1$ outputs; the first output carries over the withdrawal sub-contract and balance tokens, and the remaining N outputs have guard scripts and token values as per the dictionary. The contract requires two proofs for the dictionary elements: one proving that values to be withdrawn are indeed in the dictionary, and the second proving that the resulting dictionary does not have the withdrawn values. The complete script called `withdrawScript` is given below

```
val removeProof = getVar[Coll[Byte]](2).get
val lookupProof = getVar[Coll[Byte]](3).get
val withdrawIndexes = getVar[Coll[Int]](4).get
val tokenId: Coll[Byte] = SELF.R4[Coll[Byte]].get
val withdrawals = withdrawIndexes.map({(idx: Int) =>
    val b = OUTPUTS(idx)
    if (b.tokens(0)._1 == tokenId)
        (blake2b256(b.propositionBytes), b.tokens(0)._2)
    else
        (blake2b256(b.propositionBytes), 0L)
})
val withdrawValues = withdrawals.map({(t: (Coll[Byte], Long)) => t._2})
val total = withdrawValues.fold(0L, {(l1: Long, l2: Long) => l1 + l2 })
val toRemove = withdrawals.map({(t: (Coll[Byte], Long)) => t._1})
val initialTree = SELF.R5[AvlTree].get
val removedValues = initialTree.getMany(toRemove, lookupProof).map(
    {(o: Option[Coll[Byte]]) => byteArrayToLong(o.get)}
)
val valuesCorrect = removedValues == withdrawValues
```

```
val modifiedTree = initialTree.remove(toRemove, removeProof).get
val outTreeCorrect = OUTPUTS(0).R5[AvlTree].get == modifiedTree
val selfTokenCorrect = SELF.tokens(0)._1 == tokenId
val outTokenCorrect = OUTPUTS(0).tokens(0)._1 == tokenId
val outTokenCorrectAmt = OUTPUTS(0).tokens(0)._2 + total == SELF.tokens(0)._2
val tokenPreserved = selfTokenCorrect && outTokenCorrect && outTokenCorrectAmt
val selfOutputCorrect = OUTPUTS(0).propositionBytes == SELF.propositionBytes

outTreeCorrect && valuesCorrect && selfOutputCorrect && tokensPreserved
```

Note that the above ICO example contains many simplifications. For instance, we don't consider fee when spending the project box. Additionally, the project does not self-destruct after the withdraw stage. We refer the reader to [14] for the full example.

5 Conclusion

We gave examples demonstrating that, despite being UTXO-based, Ergo can support complex multi-stage contracts found in Ethereum. In particular, we described:

1. A Rock-Papers-Scissors game with provable fairness (Sect. 4.2).
2. Reversible Addresses having anti-theft features (Sect. 4.1).
3. A full featured ICO that accepts funding in ergs (Sect. 4.3).

The examples used the idea of *transaction trees* to emulate persistent storage by linking several UTXOs containing small pieces of code to form a large multi-stage protocol. We refer the reader to ErgoScript repository and tutorials [8,10] for additional examples of multi-stage contracts, including a Local Exchange Trading Systems (LETS), non-interactive mixing, atomic swaps and many more.

References

1. Szabo, N.: The idea of smart contracts. Nick Szabo's Papers and Concise Tutorials, vol. 6 (1997)
2. Nakamoto, S.: Bitcoin: a peer-to-peer electronic cash system (2008). https://bitcoin.org/bitcoin.pdf
3. Wood, G.: Ethereum: a secure decentralised generalised transaction ledger. Ethereum Project Yellow Paper 151, pp. 1–32 (2014)
4. Lopp, J.: Unspent transactions outputs in Bitcoin (2016). http://statoshi.info/dashboard/db/unspent-transaction-output-set. Accessed 7 Nov 2016
5. Chepurnoy, A., Kharin, V., Meshkov, D.: Self-reproducing coins as universal turing machine. In: Garcia-Alfaro, J., Herrera-Joancomartí, J., Livraga, G., Rios, R. (eds.) DPM/CBT 2018. LNCS, vol. 11025, pp. 57–64. Springer, Cham (2018). https://doi.org/10.1007/978-3-030-00305-0_4
6. Cook, M.: A concrete view of Rule 110 computation. Electron. Proc. Theor. Comput. Sci. 1, 31–55 (2009)
7. Neary, T., Woods, D.: P-completeness of cellular automaton Rule 110. In: Bugliesi, M., Preneel, B., Sassone, V., Wegener, I. (eds.) ICALP 2006. LNCS, vol. 4051, pp. 132–143. Springer, Heidelberg (2006). https://doi.org/10.1007/11786986_13
8. Ergoscript, a cryptocurrency scripting language supporting noninteractive zero-knowledge proofs, March 2019. https://docs.ergoplatform.com/ErgoScript.pdf

9. Damgård, I.: On Σ-Protocols (2010). http://www.cs.au.dk/~ivan/Sigma.pdf
10. Advanced ErgoScript tutorial, March 2019. https://docs.ergoplatform.com/sigmastate_protocols.pdf
11. Delmolino, K., Arnett, M., Kosba, A.E., Miller, A., Shi, E.: Step by step towards creating a safe smart contract: lessons and insights from a cryptocurrency lab. IACR Cryptology ePrint Archive 2015:460 (2015)
12. The Ethereum Wiki. ERC20 token standard (2018). https://theethereum.wiki/w/index.php/ERC20_Token_Standard
13. Reyzin, L., Meshkov, D., Chepurnoy, A., Ivanov, S.: Improving authenticated dynamic dictionaries, with applications to cryptocurrencies. In: Kiayias, A. (ed.) FC 2017. LNCS, vol. 10322, pp. 376–392. Springer, Cham (2017). https://doi.org/10.1007/978-3-319-70972-7_21
14. Chepurnoy, A.: An ICO example on top of Ergo (2019). https://ergoplatform.org/en/blog/2019_04_10-ico-example/

The Operational Cost of Ethereum Airdrops

Michael Fröwis[(✉)] and Rainer Böhme

Department of Computer Science, Universität Innsbruck, Innsbruck, Austria
michael.froewis@uibk.ac.at

Abstract. Efficient transfers to many recipients present a host of issues on Ethereum. First, accounts are identified by long and incompressible constants. Second, these constants have to be stored and communicated for each payment. Third, the standard interface for token transfers does not support lists of recipients, adding repeated communication to the overhead. Since Ethereum charges resource usage, even small optimizations translate to cost savings. Airdrops, a popular marketing tool used to boost coin uptake, present a relevant example for the value of optimizing bulk transfers. Therefore, we review technical solutions for airdrops of Ethereum-based tokens, discuss features and prerequisites, and compare the operational costs by simulating 35 scenarios. We find that cost savings of factor two are possible, but require specific provisions in the smart contract implementing the token system. Pull-based approaches, which use on-chain interaction with the recipients, promise moderate savings for the distributor while imposing a disproportional cost on each recipient. Total costs are broadly linear in the number of recipients independent of the technical approach. We publish the code of the simulation framework for reproducibility, to support future airdrop decisions, and to benchmark innovative bulk payment solutions.

Keywords: Airdrop · Bulk payment · ERC-20 · Token systems · Ethereum

1 Introduction

Fungible virtual assets, such as cryptocoins and tokens residing on a blockchain, are network goods: their value lies in enabling exchange. A coin is worthless if nobody else uses or accepts it. Its value grows quadratically in the number of users, according to Metcalfe's law; and still super-linear under more conservative theories [7]. As a result, new coins have to reach a critical mass until positive feedback sustains rapid growth [6].

This observation is taken to heart in the marketing of new coins. The community has adopted the term *airdrop* for the subsidized (often free) provision of new coins to selected lead users, typically holders of competing coins, with the intention to raise popularity and reach critical mass. Similar strategies are

© Springer Nature Switzerland AG 2019
C. Pérez-Solà et al. (Eds.): DPM 2019/CBT 2019, LNCS 11737, pp. 255–270, 2019.
https://doi.org/10.1007/978-3-030-31500-9_17

well understood in the economics [16] and marketing literature [14]. Whether and under which conditions airdrops are successful for cryptocoins and tokens are empirical questions that future work should tackle. Here, we study the operational costs of airdrops on Ethereum, the most popular platform for token systems .

Airdrops incur costs in the form of transactions fees paid to miners, which are shared between the initiator of the airdrop (often the developer or maintainer of a token, henceforth *distributor*) and the *recipients*, depending on the technical approach chosen by the distributor. The platform charges fees for instructions, space on the blockchain, and the size of the state information. The costs are not negligible because every recipient identifier contains cryptographic material with high entropy that must be communicated in the airdrop. Typically, the identifiers are included in the transaction payload and thus occupy space on the blockchain. Another difficulty faced by Ethereum airdrops is the lack of a bulk transaction method in the popular ERC-20 standard [1] for fungible tokens. This adds overhead due to repeated communication.

We have observed several solutions and workarounds to these problems in the wild, and synthesize our findings into the—to the best of our knowledge—first systematic overview on the technology behind airdrops on the Ethereum platform. We implement selected techniques in model smart contracts and measure their cost and resource consumption as a function of the number of recipients by executing the contracts in a simulated Ethereum node.

The rest of the paper is organized as follows. Section 2 presents technical options for carrying out airdrops on Ethereum, including a discussion of the relevant parameters and necessary prerequisites. We distinguish push and pull approaches, internal and external batching, and revisit pooled payments. Section 3 presents the cost estimates from the simulation study in units of Ethereum's internal fee model (gas), the best level of analysis for comparison between options. Section 4 interprets the main findings in units of fiat currency (USD), the level of analysis that matters for business decisions. Section 5 connects to relevant related work, before we give an outlook and conclude in Sect. 6. Technical details of the simulation framework are placed in two appendices.

2   Technical Aspects of Ethereum Airdrops

This section gives an overview about technical considerations when conducting airdrops on Ethereum. We briefly discuss parameters of an airdrop chosen by the business side, then explain shortcomings of the default token transfer interface and the resulting technical workarounds. We apply the lens of operational costs, which in case of the Ethereum platform translates to estimating transaction fees in units of gas, the most comparable metric.

2.1   Parameters of an Airdrop

Before carrying out an airdrop, a couple of parameters need to be decided. One of the first things to decide on is *who* shall receive tokens. This is often done by a simple sign-up system (using e.g., Telegram, web forms, etc.), or by defining a measure of relevance on existing addresses to select the set of recipients. Companies like *Bounty One*[1] offer matching between airdrop- distributors and recipients as a service. The rationale behind this is simple: you prefer to hand out tokens to active users participating in the ecosystem, instead of sending them to inactive accounts. For example, the banking startup *OmiseGO* conducted one of the early Ethereum airdrops [19]. They used account balance as an activity indicator and simply handed out tokens to every address holding more than 0.1 Ether (ETH) at block height 3 988 888. This airdrop serves as good running example because all code, including a documentation of the rationales behind design decisions, is publicly available [2]. The threshold of 0.1 ETH is a very simple metric. It does not consider essential aspects, such as: are those accounts still active and able to use the tokens?[2]

Two other important parameters of an airdrop are the number of tokens to be dropped and their distribution over recipients. For example, OmiseGO dropped 5% of the total supply of OMG tokens. The distribution between recipients can be uniform or depend on properties of the recipient. OmiseGO allocated tokens to recipients proportional to their ETH balances at a specified point in time.[3].

Finally, the technical approach of how to transfer the tokens to the recipients needs to be defined. This involves choosing the software implementation to distribute tokens in bulk. This decision heavily depends on the existing infrastructure of the underlying token system.

In summary, the main decisions to be taken are:

– Who receives tokens (based on what metric)?   *(recipient selection)*
– How many tokens per recipient?   *(distribution)*
– Which technical approach to use for distributing the tokens? *(implementation)*

Although recipient selection and the token distribution strategy involve technical aspects, they are mainly driven by business considerations. Those are out of the scope of this work. Here, we focus on technical implementation options and their associated cost. Our results are an important input to the business decision because they quantify the operational cost of an airdrop.

2.2   Shortcomings of Vanilla ERC-20 Airdrops

ERC-20 [1] is the most prominent standard on the Ethereum platform today. It defines an API for token systems that enables: (1) the encoding of token

---

[1] https://bountyone.io/airdrops, [Online; accessed 18 Jun 2019]. .

[2] This depends on the account type and state. For example, disabled contracts or contracts that are not programmed to transact with token systems will never be able to use the funds. This is also noted in [2].

[3] See https://github.com/omisego/airdrop/blob/master/processor.py, line 77.

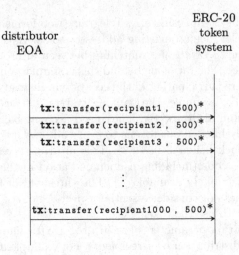

Fig. 1. Naïve push-style airdrop. One transaction per recipient from distributor to token system. The * indicates that the method is part of the ERC-20 standard.

properties,[4] (2) access to the balances of owners,[5] and (3) the transfer of tokens between accounts.[6] Moreover, ERC-20 defines logging and event notification.[7]

The first ERC-20 token systems emerged already in late 2015.[8] Airdrops, on the other hand, are a more recent phenomenon starting to gain traction in early 2018. As a consequence, the ERC-20 API does not include a batch transfer method to directly transfer tokens to *multiple* recipients in one transaction. The lack of this functionality in legacy token systems makes airdrops more expensive. The immutability of contracts, which is often a desired feature, prevents legacy token systems from adding batch capabilities after deployment.

Hence, implementing an airdrop in vanilla ERC-20 proceeds as shown in Fig. 1. The distributor uses an externally owned account (EOA) in order to send one transaction for every recipient. Each transaction invokes the `transfer` method of the token system in order to update its internal state. The fixed cost per transaction is 21 000 units of gas, which are not recoverable and add to the overhead of this approach. To avoid issuing one full transaction per airdrop recipient, the community developed several optimizations to reduce cost and avoid network congestion.

---

[4] Functions: `symbol, name, decimals, totalSupply`.

[5] Functions: `balanceOf`.

[6] Functions: `transfer, transferFrom, approve, allowance`.

[7] Events: `Approve` and `Transfer`.

[8] The first ERC-20 token. Block: 490 326, Address: `0xEff6425659825E22a3cb00d468 E769f038166ae6`.

2.3   Optimizations

We distinguish three avenues for improvements. Transaction batching helps to reduce communication costs. The pull approach shifts part of the burden to the recipient and potentially conserves tokens and fees from recipients who do not collect their share. Off-chain approval saves cost by avoiding to store the list of account identifiers on the blockchain. We discuss each of these avenues in the following subsections.

2.3.1   Transaction Batching

The easiest way to save cost is by removing the overhead of issuing one transaction for each airdrop recipient. A cheaper alternative to transactions are *message calls* (also known as *internal transactions*) invoked by contracts. The difference in fixed cost (without payload) is substantial: a transaction costs 21 000 units of gas, compared to 700 for a message call. Message calls to the same contract are yet an order of magnitude cheaper and cost about 10 units of gas. Message calls are generated in a loop over the list of recipients, which is given as an argument to a single initial transaction.

Let n be the number of recipients, then the cost savings s are

$$s_1 = n \cdot 21000 - (n \cdot 700 + 21000), \tag{1}$$

if the loop is implemented in a different contract than the token system. This method is called *external batching* and visualized in Fig. 2. To give an intuition for the source of the savings, recall that the batch is authorized by a single signature as compared to one signature per transaction in the naïve approach.

Even higher savings of

$$s_2 = n \cdot 21000 - (n \cdot 10 + 21000) \tag{2}$$

are possible if the loop is implemented directly in the token system contract (*internal batching*). The difference between external and internal batching can be explained by the penalty of fetching new code from disk, which applies n times in the case of external batching. However, changing the token systems 's contract may either be impossible because it is already deployed immutably, or not desired in the fear of introducing new bugs or losing investor trust.

The savings are upper bounds that are only achievable if all n identifiers fit into one transaction. The size of transactions is restricted by the block gas limit.[9] Larger recipient lists must be split into several batches, each incurring the fixed cost of one transaction.

2.3.2   The Pull Approach

As already mentioned, distributing tokens to inactive accounts or to recipients not interested in the token has little to no value to the distributor. This problem can be addressed during recipient selection, by

---

[9] Gas limit at the time of writing is 8 000 029 in block number 8 014 738.

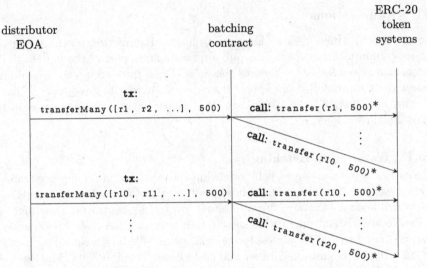

Fig. 2. External batching, push-style airdrop. The token system has **no** batching capabilities. Batching is done by an external contract (either own or service (For example, https://multisender.app/, [Online; accessed 25 Jun 2019].)) Note the calls instead of transactions. .

- evaluating appropriate technical indicators,
- collecting information provided in sign-up, or
- using the specialized services in the ecosystem who administer panels of potential recipients (see Footnote 1).

We are not aware of any literature on the effectiveness of these options and consider it out of our scope.

A more technical approach is to condition the token transfer on on-chain user interaction. This type of *pull-style airdrop* can be implemented using the `approve` function of ERC-20. Instead of directly transferring tokens to the recipients during the airdrop, the distributor gives the recipient the right to withdraw the airdropped amount. The distributor may specify a deadline for the withdrawal and reclaim the remaining tokens thereafter.

Arguably, the additional effort and cost for the recipient ensures that the distribution is more targeted.

The pull approach has a couple of downsides. First, the cost for the recipient and the distributor are significant. Both sides pay about as much as the distributor pays for a push-style airdrop. This even holds when the distributor approves in batches as described in Sect. 2.3.1 (see Fig. 3). Second, a known front-running attack against the ERC-20 `approve` logic requires the distributor to set all allowances to zero before updating them with new values [1,21]. This approximately doubles his cost. Third, many existing token systems do not implement the `approve` method and therefore cannot use the pull approach [12]. Lastly, recovering the unclaimed tokens after the deadline costs about as much as transferring them.

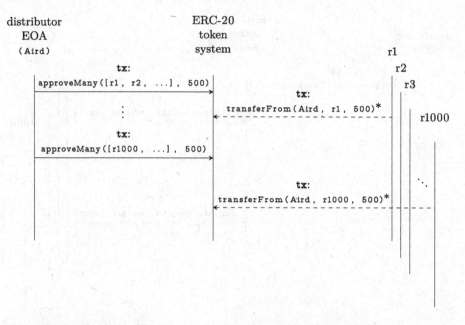

Fig. 3. Pull-style airdrop with non-standard `approveMany` for internal batching. Recipients have to interact with the ERC-20 contract to finally receive the funds. The distributor-side can be batched. Recipients must use individual transactions. Dashed lines mean recipient pays.

2.3.3 Off-Chain Approval

The Ethereum community has realized that storing every recipient address on the chain leads to network congestion as well as to high cost. One approach that does not require to store all recipient addresses on-chain are *pooled payments* [5]. They are inspired by *Merkle mine* [4], an approach developed for token systems to define the initial allocation to a large number of owners.

Pooled payments resemble pull-style airdrops, as the distributor approves the recipient to withdraw a certain amount of tokens through a transaction. But pooled payments do not store the entire approval on-chain. Instead, the distributor encodes the list of recipients with denominations in a Merkle tree [17], where leafs are concatenations of recipient addresses and amounts. The approving contract has to store the root hash of the Merkle tree only (see Fig. 4, `approveReceivers(merkle_root)`). The list of recipients is published off-chain. To claim funds, every recipient needs the list and computes the Merkle tree as well as a Merkle proof for his entry. The recipient then sends a transaction to the contract with his address (implicit by the signature on the transaction), the amount, and the Merkle proof (see Fig. 4, `claimMerkle (500, merkle_proof)`). This allows the contract to compute the hash of the leaf node from the message sender (signature) and the amount in the argument. Using the Merkle proof, it verifies the correctness of the claim, checks

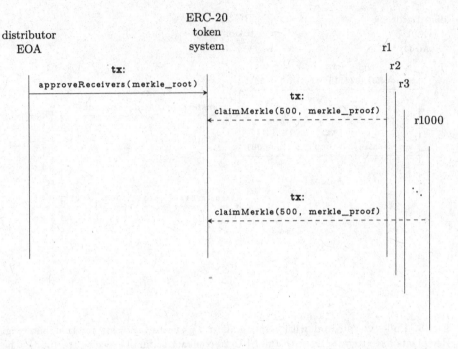

Fig. 4. Pooled payments: non-standard, internal, pull-style airdrop. Cost is constant for distributor. List of recipients is public.

its freshness, and transfers the funds. To prevent double-claiming, the contract must store a record of this withdrawal. Several methods exist to keep this as compact and cost-efficient as possible.

Unlike for normal pull payments, the distributor has constant cost independent of the number of recipients. Most of the airdrop cost is shifted to the recipient.[10]

It is also worth mentioning that pooled payments based on off-chain approvals have some usability issues that may delay their adoption for airdrops. Both recipients and distributors need tools to do off-chain computation (Merkle tree, proofs) and data retrieval (list of recipients). We are aware of one business that seems to bet on the adoption of this approach.[11]

---

[10] This accounts to: one transaction per recipient, Merkle proof verification, and storage of withdrawal record.

[11] The claim of constant distributor cost in the Coinstantine whitepaper indicates the use of pooled payments. See https://www.coinstantine.io/, [Online; accessed 22 Jun 2019].

2.4   Miscellaneous Aspects

Optimizing the airdrop strategy and the code involved in the airdrop workflow are not the only things to consider when doing airdrops.

Gas Token: Gas Token[12] provides a way to pre-acquire gas in periods when gas is cheap. Those gas tokens can be "redeemed" when the gas price is high. The gas token mechanism exploits the fact that Ethereum refunds gas when storage resources are freed. To include this mechanism in airdrops, the functionality to redeem gas tokens must be built into either the token system or the batching contract. Gas tokens are already used in other areas, such as arbitrage bots [10].

Systemic Risk: Airdrops can be a systemic risk for the Ethereum platform, if not used carefully, since they use large amounts of resources (gas). Coindesk [3] reports that the uptake of airdrops in conjunction with questionable incentives set by the exchange FCOIN led to substantial network congestion, gas price increases, and wasted resources. Reportedly, OmiseGO also considered the impact of its airdrop on the network and decided to limit their batches such that they never use more than 50% of the block gas limit [2]. We adopt the idea of such a limit for the simulations in the following.

3   Cost Estimates

In the following we compare simulated costs of different airdrop techniques. The simulation gives us valid estimates of total cost, which puts the back-of-the-envelope calculations of savings in Eqs. (1) and (2) into perspective. The simulation framework, the constants used, a complete list of scenarios can be found in the extended version of the paper.[13] The code can be found in on Github.[14] It can serve as starting point for facts-based airdrop decisions and to benchmark new solutions.

All 35 scenarios are combinations of the approaches discussed in Sect. 2.3. Table 1 resolves the labels used in the figures below.

Figure 5 presents the gas cost as a function of the number of recipients. We only show strategies viable when targeting a less than 50% block fill grade. Observe that all strategies behave broadly linear. The fixed cost per batch is negligible. NAIVE|PUSH and BASE_LINE|INTERNAL_BATCH|PUSH|UNIFORM|100 serve as upper and lower bounds and thus are benchmarks for the other strategies. The lower bound (baseline) simulates a push-style airdrop only considering communication and storage cost.[15]

Although, INTERNAL_BATCH|PULL|UNIFORM|100 appears to be the cheapest strategy, this is only true for the distributor. Recall that in pull-style airdrops recipients have to send additional transactions in order to withdraw their tokens.

---

[12] https://gastoken.io/, [Online; accessed 21 Jun 2019].

[13] https://arxiv.org/abs/1907.12383.

[14] https://github.com/soad003/TheOperationalCostOfEthereumAirdrops.

[15] This rests on the assumption that other computation cost can be optimized.

Table 1. Approach labels and descriptions.

Label	Description
NAIVE	No batching is applied. One transaction per recipient
PUSH	Push-style airdrop as discussed in Sect. 2.3
PULL	Pull-style airdrop as discussed in Sect. 2.3
EXTERNAL_BATCH	External batching as discussed in Sect. 2.3
INTERNAL_BATCH	Internal batching as discussed in Sect. 2.3
UNIFORM	One amount per batch. Otherwise n different amounts are sent
RECIPIENT_COST	Recipient cost of pull-style airdrop. All recipients claim funds
BASE_LINE	Baseline for pull style airdropsee extended version

If we sum up the costs of the recipient (PULL|RECIPIENT_COST) and distributor, the pull-style airdrop is by far the most expensive, 32% more costly than NAIVE|PUSH.

The biggest improvement for both parties compared to the NAIVE|PUSH is archived by INTERNAL_BATCH|PUSH|UNIFORM|100, saving roughly 42%. The baseline suggests that savings up to 58% are be possible. If we compare internal vs. external batching, the internal strategies are save around 8% compared to their externally batched counterparts. The uniform strategies only save about 1% compared to their counterparts. The savings might go up if larger amounts are transferred, which require more non-zero bytes to encode. However, the batching contract could support logarithmic scaling or batch-wide multipliers, which make our approximation with two non-zero bytes per amount realistic again.

Figure 6 shows a single simulation run for 1 000 recipients. Given the approximate linearity the number of recipients, this view is sufficient to compare the strategies. This time we present all strategies. We color code the minimal block fill grade in which each strategy gets feasible. The first thing to observe is that batches of more than 300 recipients are not feasible with current block gas limits. Only the pull approaches and the baseline can manage a batch size of up to 300. Note that NAIVE|PUSH as well as the RECIPIENT_COST are feasible even with a threshold of 10% block fill grade, since no batching is applied.

We did not simulate the pooled payment strategy (see Sect. 2.3.3). Since the distributor cost is constant by only storing the Merkle root, this would make it the by far cheapest option for distributors. The recipients have to withdraw the tokens in a very similar manner to the PULL|RECIPIENT_COST strategy. In addition, the recipient has to pay for the verification of the Merkle proof and the storage of the withdrawal record.

4   Discussion

The above results show differences in gas consumption. More relevant units of operational cost for the distributor are ETH, if the distributor is already invested

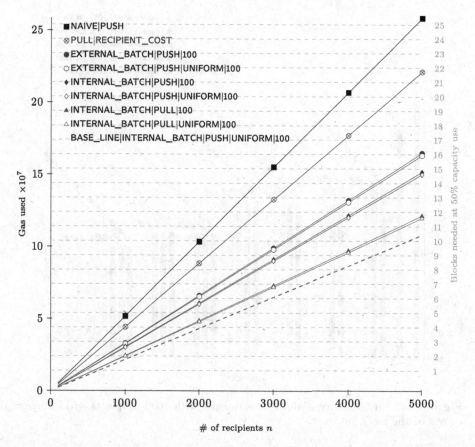

Fig. 5. Simulated cost for n recipients using different strategies. The chart shows the subset of strategies whose transactions fit into 50% of the block gas limit. Cost is not discounted, i.e., we assume all recipients are new to the token system, which makes the airdrop a bit more expensive.

in Ethereum, and fiat (USD). To put our results into perspective, we estimate the cost savings per 1000 recipients in USD. This entails two conversions with variable rates: from gas to ETH and from ETH to USD. The first conversion is governed by the fee market mechanism and the miner's transaction inclusion strategy. Since transactions compete for inclusion in the chain, the gas price (in ETH) depends on the network load. The second conversion rate is found on the cryptocurrency exchanges in the ecosystem. The price depends on demand and supply of cryptocurrency, which supposedly follow investors' economic expectations.

We do not aim to explain price formation in this work (although airdrops may affect prices in the short run), but take an empirical approach. Figure 7 shows the co-movement of both prices from January 2017 to June 2019 on a log-log scale. We calculate 60-day moving averages and represent each center day as

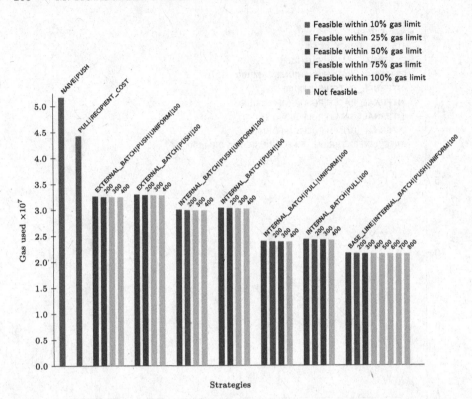

Fig. 6. Cost and feasibility of different strategies with 1000 recipients with different cutoffs on the block gas limit.

dot, color-coded by the calendar month. The dashed lines connect levels of equal gas price in USD. Observe that both prices follow different dynamics, hence are not strongly coupled by a single market mechanism. While the ETH/USD exchange rate varies over two orders of magnitude in the sample period, the gas price in ETH remains in a much narrower band. However, it exhibits more sudden jumps, which relate to extreme values (e.g., due to congestion) that enter or leave the moving window. One can also speculate if the introduction of gas tokens in spring 2018 has narrowed the band of gas price movement due to the counter-cyclical behavior of gas token investors.

To get an idea of airdrop costs in USD, the dashed line marked with 15 cents (of USD) per recipient seems a good rule of thumb. This price level was applicable for a naïve push-style airdrop in May 2017, October 2018, and in March and May 2019. More efficient strategies have costed around 7.5 cents per recipient. Given the variability of both prices (some of which is hidden by the moving average), it seems that the right timing is at least as important as the strategy.

To continue the example from above, the OmiseGO airdrop distributed tokens to 450 000 recipients, using an externally batched push-style approach.

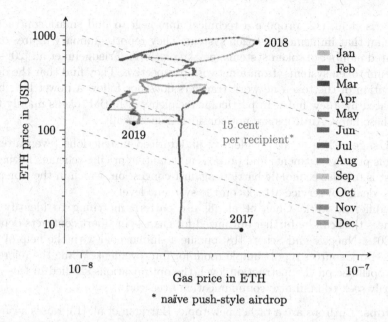

Fig. 7. Dynamics of the price of Ethereum resource use, broken down into components. The plot shows 60-day moving averages of daily prices reported by Etherscan.io. Dashed lines connect levels of equal gas price in USD.

By applying the gas-to-USD conversion rate indicated in Fig. 7, we can estimate the cost at roughly 44 523 USD. Airdrops of this size occupy 50% of the available capacity in 1440 blocks, taking at least 6 h to complete. As a consequence, early recipients have an advantage when selling tokens on an exchange immediately after receipt. This suggests that token systems should support time locks in order to enable large and *fair* airdrops.

5   Related Work

Our work connects to prior works on the systematic analysis of token systems on the Ethereum platform, gas efficiency, and one seminal publication on airdrops.

Token Systems and ICOs: Howell et al. [15] study the success factors of 440 ICOs on Ethereum based on propriety transaction data, presumably acquired from exchanges and other intermediaries, and manual labeling. Their main interest is in the relationship between issuer characteristics and indicators of success. The regression analyses find highly significant positive effects on liquidity and volume of the token for independent variables measuring the existence of a white paper, the availability of code on Github, the support by venture capitalists, the entrepreneurs' experience, the acceptance of Bitcoin, and the organization of a pre-sale. No significant effect is found for airdrops.

Fröwis et al. [12] propose a technical approach to find smart contracts on Ethereum that implement a token system. They report summary figures on the estimated number of token systems on the platform. Friedhelm et al. [20] study Ethereum token systems from a network perspective. They find that the degree distribution of the token network transfers does not follow a power law, but is dominated by a few hubs. In particular recipients of initial tokens mainly trade with these hubs. Airdrops are not considered specifically.

Gas Usage: Chen et al. [9] identify underpriced instructions (even after the 2016 gas price adjustment) and propose an adaptive pricing scheme. Their main interest is to raise economic barriers against congestion, which in the worst case enables denial of service attacks on the systemic level.

In different work, Chen et al. [8] use pattern matching to identify code sequences that can be further optimized for gas use in smart contracts deployed until 2016. Naegele and Schett [18] pursue a similar goal with the help of SMT solvers. Both sources report ample room for improvement. While the referenced works optimize on the instruction level, the optimizations studied in this paper primarily seek to minimize communication overhead.

Airdrops: Airdrops are a rather new topic. Harrigan et al. [13] raises awareness for privacy implications of airdrops when identifiers of one chain (*source chain*) are reused to distribute coins on another chain (*target chain*). Sharing identifiers between chains in general gives additional clues for address clustering.

Di Angelo et al. [11] study the temporal behavior of contracts on Ethereum. They mention that many short-lived contracts seem to harvest airdropped tokens by presenting many unique recipient addresses.

To the best of our knowledge, we are the first to review technical solutions for airdrops of tokens on the Ethereum platform and to analyze their gas efficiency.

6   Conclusion and Outlook

This work compared the efficiency of bulk transfer approaches on the Ethereum platform, a general problem that became particularly relevant with the uptake of token airdrops. It turns out that many of the approaches we systematized and reviewed are workarounds for architectural short-comings of the Ethereum platform or the popular ERC-20 standard for fungible tokens. The cost efficiency of the approaches differ roughly by a factor of two. Moreover, the most cost-efficient solutions for the distributor impose significant cost on the recipient, which might thwart the very intention of an airdrop as marketing tool. We release our simulation framework and the model contracts for reproducibility, as testbed for actual airdrops, and as benchmarking suite for new solutions.

The choice of approach is constrained by properties of the token system. This mainly relates to the penalty of repeatedly calling a method from *another* contract, which appears disproportional to the computational effort of the node. While a remote call is indeed expensive at first use, every repeated call is sped up through caching. Ethereum's gas price schedule seems to unfairly discriminate against bulk operations, an issue that designers of price schedules for future blockchain platforms should fix. On Ethereum as it stands, token issuers are best advised to reflect about airdrops before deployment of the token system contract. Planning ahead is vital in an environment where code cannot be amended easily.

While framed and motivated for the application of airdrops, our analysis generalizes to any kind of bulk operation on lists of incompressible items. Future designs of blockchain platforms should consider mitigating most of the issues discussed here by supporting a global index for constants with high entropy. Some cryptographic material, in particular public keys and commitments, must be stored on-chain in order to enable authorization of actions by the knowledge of secrets. But if they do not double as references, as in Ethereum, then every value has to be stored (and paid for) only once. Furthermore, if all contracts have access to all public information on the blockchain, sets can be reused and storage space saved. In short, the community needs a DRY (don't repeat yourself) principle for data on the blockchain.

Acknowledgments. We would like to thank Patrik Keller, Clemens Brunner and Alexandra Bertomeu-Gilles for their valuable insights and feedback.

This work has received funding from the European Union's Horizon 2020 research and innovation programme under grant agreement No. 740558.

References

1. EIP 20: ERC-20 Token Standard (2015). https://eips.ethereum.org/EIPS/eip-20. Accessed 18 June 2019
2. OmiseGO Tokens Airdrop (2017). https://github.com/omisego/airdrop. Accessed 18 June 2019
3. Ethereum's Growing Gas Crisis (And What's Being Done to Stop It) (2018). https://www.coindesk.com/ethereums-growing-gas-crisis-and-whats-being-done-to-stop-it. Accessed 18 June 2019
4. MerkleMine Specification (2018). https://github.com/livepeer/merkle-mine/blob/master/SPEC.md. Accessed 18 June 2019
5. Pooled Payments (scaling solution for one-to-many transactions) (2018). https://ethresear.ch/t/pooled-payments-scaling-solution-for-one-to-many-transactions/590. Accessed 18 June 2019
6. Böhme, R.: Internet protocol adoption: learning from Bitcoin. In: IAB Workshop on Internet Technology Adoption and Transition (ITAT), Cambridge (2013)
7. Briscoe, B., Odlyzko, A., Tilly, B.: Metcalfe's law is wrong. IEEE Spectr. **43**(7), 34–39 (2006)
8. Chen, T., Li, X., Luo, X., Zhang, X.: Under-optimized smart contracts devour your money. In: 2017 IEEE 24th International Conference on Software Analysis, Evolution and Reengineering (SANER), pp. 442–446. IEEE (2017)
9. Chen, T., et al.: An adaptive gas cost mechanism for Ethereum to defend against under-priced DoS attacks. In: Liu, J.K., Samarati, P. (eds.) ISPEC 2017. LNCS, vol. 10701, pp. 3–24. Springer, Cham (2017). https://doi.org/10.1007/978-3-319-72359-4_1
10. Daian, P., et al.: Flash Boys 2.0: frontrunning, transaction reordering, and consensus instability in decentralized exchanges. arXiv preprint arXiv:1904.05234 (2019)

11. Di Angelo, M., Salzer, G.: Mayflies, breeders, and busy bees in Ethereum: smart contracts over time. In: Third ACM Workshop on Blockchains, Cryptocurrencies and Contracts (BCC 2019). ACM Press (2019)

12. Fröwis, M., Fuchs, A., Böhme, R.: Detecting token systems on Ethereum. In: Goldberg, I., Moore, T. (eds.) Financial Cryptography and Data Security (2019)

13. Harrigan, M., Shi, L., Illum, J.: Airdrops and privacy: a case study in cross-blockchain analysis. In: 2018 IEEE International Conference on Data Mining Workshops (ICDMW), pp. 63–70. IEEE (2018)

14. Hill, S., Provost, F., Volinsky, C.: Network-based marketing: identifying likely adopters via consumer networks. Stat. Sci. **21**(2), 256–276 (2006)

15. Howell, S.T., Niessner, M., Yermack, D.: Initial coin offerings: financing growth with cryptocurrency token sales. Technical report, National Bureau of Economic Research (2018)

16. Katz, M.L., Shapiro, C.: Systems competition and network effects. J. Econ. Perspect. **8**(2), 93–115 (1994)

17. Merkle, R.C.: A digital signature based on a conventional encryption function. In: Pomerance, C. (ed.) CRYPTO 1987. LNCS, vol. 293, pp. 369–378. Springer, Heidelberg (1988). https://doi.org/10.1007/3-540-48184-2_32

18. Nagele, J., Schett, M.A.: Blockchain Superoptimizer (2018). http://www.maria-a-schett.net/talks/2019_04-MaS_imdea.pdf. Accessed 24 June 2019

19. nharrison: Airdrop Update 2 (2017). https://steemit.com/ethereum/@nharrison/what-you-need-to-know-about-the-omisego-airdrop. Accessed 25 June 2019

20. Victor, F., Lüders, B.K.: Measuring Ethereum-based ERC20 token networks. In: Goldberg, I., Moore, T. (eds.) Financial Cryptography and Data Security (2019)

21. Vladimirov, M., Khovratovich, D.: ERC20 API: An Attack Vector on Approve/Transfer From Methods. https://docs.google.com/document/d/1YLPtQxZu1UAv O9cZ1O2RPXBbT0mooh4DYKjA_jp-RLM. Accessed 25 June 2019

Blockchain Driven Platform for Energy Distribution in a Microgrid

Arjun Choudhry(✉)🆔, Ikechukwu Dimobi, and Zachary M. Isaac Gould

Virginia Polytechnic Institute and State University, Blacksburg, USA
{aj07lfc,ikdimobi,gould}@vt.edu

Abstract. Ever since the Digital Revolution, the World population has been growing at an ever-increasing rate, with the last billion being amassed in only 9 years. The increase in population only exacerbates the current challenge of meeting peak-energy demands while accommodating an increasing amount of distributed energy resources such as solar panels. Utilities must learn how to encourage energy-efficient practices to reduce demand, as well as how to effectively coordinate interaction between consumers, renewable-energy sources, and the grid. With earnings currently communicated on monthly billing cycles, existing solutions such as net-metering, feed-in-tariffs and demand response are neither transparent nor resolute enough. They are also not flexible enough to accommodate community shared energy resources. This paper aims to bridge this gap by developing a secure energy-distribution model for multi-family apartment buildings and neighborhoods that share a solar panel array. EOSIO smart contracts will be used to transparently and dynamically distribute energy using tokens. At each time-step, the residents are allocated tokens worth a fair portion of total energy production minus their unit's consumption. Excess energy is sold to neighbors at just below the retail rate, instead of back to the grid at much lower cogeneration rates, thus constantly incentivizing efficiency and energy sharing. This will pave the way towards community scale solar and even communal food and water resources where residents can equitably share in both costs and profits. The model presented in the paper was validated through the award winning entries in the 2019 EOSIO Hackathon as well as the Virginia Tech's grand prize winning entry in the 2019 Solar Decathlon Design Challenge.

Keywords: Microgrid · Blockchain · Resource distribution

1 Introduction

The dawn of the digital revolution coincided with a rapid rise in the human population, with it climbing towards an ever-rising crescendo. Unsurprisingly, this has aggravated the demand on the existing finite resources and pushed them to the brink of depletion. However, this adversity has also paved way for novel ideas and cleaner methods to guide the human race on the path of self-sustainability. This is especially true in the case of energy resources, which is undergoing a platonic shift towards renewable energy sources such as solar, wind and biogas.

C. Pérez-Solà et al. (Eds.): DPM 2019/CBT 2019, LNCS 11737, pp. 271–288, 2019.
https://doi.org/10.1007/978-3-030-31500-9_18

Solar energy sources, specifically photovoltaic solar technology, has seen a widespread adoption in the residential commercial and industrial building domains, with them also being the prime energy sources in many of these buildings. However, most of these buildings were designed to operate in isolation to their neighborhood with regards to energy distribution. They were intended to facilitate a solitary self-sustainable space, such as an office, industry or a single family home with net-zero consumption. In other words, the buildings were not designed to follow the energy distribution protocols that form the basis of a practical residential community. Moreover, the advent of smart-grids and the ever-rising adoption of energy generation resources like solar panels has lead to calls for better resource sharing and peak-load balancing paradigms. This paper presents a novel model that promotes interaction amongst not only each individual units in a building but also amongst several buildings in a community. The bedrock of the model lies in the dynamic distribution of the total solar energy produced by the common solar panels, amongst different entities, as well as the token incentivization engine, which rewards each unit owner with tokens corresponding to the efficiency of their energy consumption pattern.

The developed model draws inspiration from how trees interact with their surroundings in order to continually sustain their existence and achieve growth. Akin to the energy flows in nature, the units in the community are also designed to encourage and promote interaction on a granular level. The interaction mostly takes the form of energy distribution amongst different utilities at each timestep. Each unit is assigned a portion of the solar energy produced from the solar panels, by factoring in the house characteristics such as size and the walls exposed to sunlight. The assigned energy is then offset by the energy consumed by the unit to achieve a net surplus/deficit score for each unit at each timestep. The surplus/deficit energy further serves to categorize the unit as an energy producer or energy consumer. To quantify the participation of units at each time-step, community stable tokens (ECOT) are assigned to each unit corresponding to the energy transactions made at that time-step. These tokens are analogous to fiat currencies and are designed to access various community amenities and pay their monthly rent. Although the unit owner receives a fixed proportion of the total energy produced by the panels, they lose the accumulated tokens based on their consumption patterns, hence incentivizing energy efficient behavior. The rules governing the token exchanges are written in a smart contract and then deployed on the EOSIO blockchain platform. Hence, the blockchain platform, the EOSIO contract and the distribution algorithm act as the primary actors that ensure the sanity of the network, thus ensuring a data-driven, decentralized and automated network.

Finally, the platform and the engine are presented to the residents in the form of a black box, with a web application created to abstract the underlying layers. This application empowers both the individual residents as well as the community administrators by allowing them to track their day to day operations. It also provides visual representation of the tokens gained/lost by each unit at each timestamp as well as the total tokens gained/lost until that timestamp.

While the components of the black-box ensure transparency due to the lack of a middle-man in the whole network, the web application further enhances the transparency by providing real time insights to the user about their energy consumption patterns as well as their net token balance. It also provides a way for users to pay their rent manually via the portal or subscribe to the monthly payment scheme governed by the smart contracts.

To validate the functioning and effectiveness of the developed model, Virginia Tech's grand prize entry in the 2019 Solar Decathlon Design Challenge, TreeHAUS [6] is used as a case study. The model mentioned in this paper forms an integral component of the TreeHAUS by managing flow of energy and tokens in the community. For demonstration purposes, the community is envisaged as a network of 12 units, with an equal distribution of 1, 2, 3 and 4 bedroom unit types.

2 Preliminaries

2.1 Microgrid/Community

A *microgrid* is defined as a localized network of energy sources that have the capability to operate either synchronously with a centralized grid or disconnected from the grid (island mode). In island mode, the microgrid is self-sustainable without the need of external energy supply. The microgrid may itself comprise of networks varying in complexity, size and the manner of governance and allows investment in novel technologies to enhance the energy distribution system. The benefits of improving the energy distribution methodologies trickles into reduced peak load demands and congestion handling.

2.2 Prosumers

The model defined in the paper leverages the underlying blockchain platform and dynamically distributes the tokens by categorizing the units as either *Producers* or *Consumers* at each time-step. The metric that governs the categorization is the delta energy consumption, i.e. the difference between the allocated solar production and the total energy consumption at each time-step. A unit having a net-positive delta energy consumption is classified as a *Producer* and receives incentives, in the form of tokens, at that time-step. On the other hand, a *Consumer* receives energy (either from the grid or a neighboring house), and thus loses a proportionate amount of tokens at that time-step.

2.3 EOSIO

EOSIO [3] is a state-based blockchain protocol that deploys a Delegated Point-of-Stake (dPOS) consensus algorithm. Blocks are produced every 0.5 s, with it also boasting faster smart contract computation due to its use of Web Assembly (WASM) virtual machine. To meet the Byzantine Fault Tolerance criterion, producers are allowed to sign all the blocks, such that the two blocks differ in timestamp or block height. Once 15 producers have signed a block the block is deemed irreversible [3]. It is one of the fastest blockchain platforms and is designed to accommodate both vertical and horizontal scaling.

3 Related Work

As an emerging topic, the use of blockchain technology to solve power grid problems has been explored in both real world pilot projects and research endeavours. A very notable project is the Brooklyn Microgrid project [11]. LO3 energy operated a decentralized microgrid energy market without the need for central intermediaries, however the current unsupportive regulatory atmosphere was highlighted as a challenge. Other similar projects include PowerLedger [1] and WePower [2], wherein these platforms allow users to trade renewable energy and energy credits between each other. ETHome, an ethereum blockchain based energy community controller was also prototyped using off-the-shelf devices as detailed in [14].

Research efforts in this field include [16] where a P2P trading mechanism was designed to reduce exchange of energy between microgrids and the municipal grid. In [12], a decentralized game theoretic approach, enhanced by the use of blockchain for microgrid demand side management is proposed. Prospect theory, and specifically the framing effect, is employed in [9] to simulate prosumer behavior in relation to P2P trading and utility-determined dynamic retail pricing. The proposed approach differs from these precedents by de-emphasizing consumer behavior in favor of analytical ranking of prosumer surplus or deficit and automated smart-contract execution of energy trading. This incentivizes more efficient consumption patterns and reduces the cognitive load of prosumers managing their own energy transactions on fine timescales.

The authors in [15] develop a new approach to consensus reaching in the blockchain for peer to peer energy local energy markets. Electric vehicles play a huge part in the future of microgrids and [10] proposes a consortium blockchain solution for electricity trading among plug-in hybrid electric vehicles. Furthermore, hyperledger fabric is used in [13] for a laboratory demonstration of P2P energy trading between prosumers with solar panels. Other researchers have taken steps to simulate the performance of blockchain driven microgrids using an agent-based approach [7]. The results of their simulation suggest that there may be a bottleneck in validation time per transaction as the number of validators on the Ethermint blockchain network increases. There are a couple of critical differences in the current work, especially with respect to the behavior model and blockchain platforms of choice. This work avoids auction-based approaches like the double auction model utilized in [7], instead ranking prosumers by net surplus or deficit and automatically distributing incentives to participating households using a smart contract. This reduces the need for more sophisticated forecasting or bidding agents in the users' end.

On the other hand, the continuous advancements in the still nascent quest of expanding solar electricity generation in communities has majorly seen the adoption of two different types of ownership structure and energy distribution models. The ownership structure deployed in multifamily homes can be broadly divided into the following two categories:

- Direct Ownership: Under this model, the building owner or the building occupants pay for the setup and thus own the entire solar energy produced.
- Third Party Ownership: This model frees up the unit owners from the liability of building, administrating and maintaining the solar plant. Instead, a for-profit or non-profit organization maintains the solar plant, with unit owners leasing a part of the plant, thus tapping into the solar power generated from that portion of the solar panel. Majority of the real world projects deploy this model due to the reduced work on the unit owner's end.

On the other hand, the solar distribution models for a multi-family building can be divided into the following two types:

- Common Utility Sharing: Under this model, the solar energy is used to power the common resources in a community, such as hallways, parking spaces and elevators. This ensures that the gains from the solar system are equally shared amongst the units.
- Individual Participation: This model allows only the interested units to participate in the solar program. Hence, the solar panel is directly connected to the electric meters of the interested parties, which helps in offsetting a portion of their monthly electric bills.

Most of the buildings/communities today deploy a variation of the above mentioned distribution models. Even though these models provide autonomy to the unit owners on choosing the proportion of the solar energy to be distributed among each unit, they fail to leverage the prevalent ecosystem provided by the community to provide dynamic distribution of the produced solar energy. In other words, the existing models allocate a predetermined portion of the total solar energy produced to each unit at each time-step. This energy is then used to offset their own energy demands. In the case of a unit's solar production exceeding it's consumption at a certain time-step, the excess energy is either stored in a battery for use at a later time-step or sold back to the grid.

The model proposed in this paper takes a different approach. It is driven by the fact that, using the power of blockchain smart contracts, excess energy can be sold to neighbors, behind the meter, at just below retail rate ($/kWh), instead of at much lower rate to the grid. This results in savings for net-consumers and increased revenue for net-producers. Hence, even though each unit owner is allocated a fixed portion of the total solar panel area, the portion of the solar energy actually distributed to the unit owners doesn't conform to that proportion at all times. This dynamic distribution allows for a high consumption unit to get a proportion of solar energy larger than its present allocation if the other units produce more than their own consumption at that time-step, with monetary incentives flowing in the opposite direction. The rate structure described above is important because on a community level and depending on the jurisdiction, the revenue gained for selling energy to the grid can be very little compared to the rate paid for buying energy. Particularly in Blacksburg, the area which TreeHaus was designed for, the community would be paid at co-generation/wholesale rate for energy fed back to the grid [8] which is only a fraction of the average retail rate. Therefore, the residents benefit through transparently earning money for good consumption habits, whereas the overall community benefits as improved resource efficiency leads to reduced costs.

4   Proposed Model

The overall architecture proposed in the paper can be conceptualized in 4 separate layers, with each layer built on top of the previous layer. Moreover, the top layer serves to abstract the bottom layer's functionality and provides infrastructure for the smooth running of the layer that it encapsulates. Figure 1 shows a bird's eye view of the architecture and the technologies enveloped in each layer.

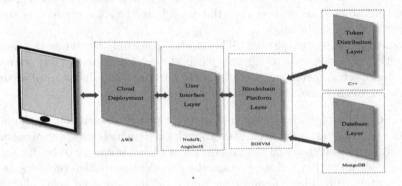

Fig. 1. Architectural overview of the proposed model and the interaction of the four layers with each other

The modular structure ensures that each of the layers is tasked with a specific function, which can be summarized as follows:

- **Database Layer:** This layer tracks the total energy consumed by each unit and provides an interface to other layers to access the stored data.
- **Token Distribution Layer:** This layer sources the energy supply/demand information from the *Database Layer* and dynamically allocates the solar energy produced at each time-step.
- **Blockchain Platform Layer:** This layer provides governance to the whole system through the use of *Smart Contracts*, while also acting as an interface between all the other three layers. Furthermore, this layer is also responsible for distributing the incentives to the producers at each time-step.

- **User Interface Layer:** This layer overlays the preceding layers and provides an interactive dashboard for the users to view their token balance and energy consumption trends, while also allowing them to pay their rent using their available tokens.

Figure 2 details the integration of the proposed model with the external peripherals. The *Blockchain Platform* forms the central binding layer and acts as an interface between the other layers, utility appliances and the central grid. It can be envisaged as a virtual net that administers governance transparently through the running of smart contracts. This transparency is further highlighted by empowering the user through visual representation of their daily energy usage trends in the deployed web application. At each time-step, the desired energy required by each unit is estimated using their previous consumption patterns sourced from the *Database Layer*. This estimated consumption value is then tallied across with their allocated portion of the total solar energy production, in the *Token Distribution Layer*, to classify them as either *Producers* or *Consumers*. Furthermore, the *Producers* and *Consumers* are ranked separately based on their delta energy percentage values, and tuples containing energy distribution information are generated such that the unit with a higher energy efficiency score receives greater incentives.

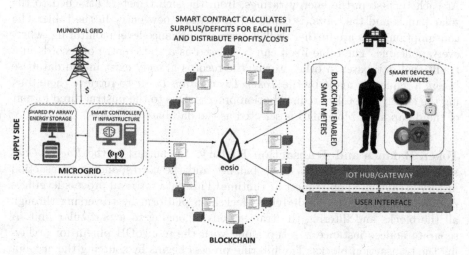

Fig. 2. Figure explaining the model integration with the external peripherals, the microgrid and the user-interface

Corresponding to the flow of deficit energy between the *Producers* and *Consumers*, there is also a transfer of incentives, in the form of tokens, in the opposite direction. The transfer of tokens is governed by Smart Contracts running on the EOSIO blockchain platform, thus absolving the system of any middleman, while ensuring the sanity of the whole network. Excess energy left, after meeting the demands of the *Consumers*, is sold to the central grid for further incentives, while the current energy consumption readings are stored in the Database for use during the next time-step. Meanwhile, the web application provides each unit owner with a microscopic as well as a macroscopic view of their energy consumption trends and token balance history, while allowing them to access other community resources using their accumulated token balance.

4.1 Database Layer

The database layer is the single reference point for the proposed model to gain access to each of the unit's energy consumption information, which guides the energy redistribution process. The database essentially contains only a solitary table that tracks the energy attributes of each house at each time-step. However, for ease of documentation, the process of estimating each individual unit's token balance and their transaction history is also covered in this section. Furthermore, the gathered data itself encompasses separate use-cases and follows completely different workflows with respect to their preparation and ultimate consumption.

Energy Consumption Table. This table is tasked with containing a holistic overview of the energy attributes for each unit at each time-step. These attributes enclose the solar production statistics as well as the energy consumption figures. At each time-step, the energy ratings, from the smart meters attached to the solar panels and the units, is captured to create a new entry in the table. The consumption stats are further drilled down to appliance-level monitoring, wherever applicable. The table itself can be viewed as a static entry of records in a traditional database. In other words, the records already exist in the database before a request is made by the *Token Distribution Layer* to retrieve them during the energy and token redistribution process. Due to the unstructured format of the datapoints, MongoDB is selected as the database of choice.

Token Balance and Transaction History. In contrast to the *Energy Consumption Table*, each unit's token balance and the corresponding transaction history is calculated on request at runtime. This data retrieval process leverages the immutability of the underlying blockchain platform by traversing through all the blocks and filtering the transactions belonging to a particular unit. A separate nodeos instance is setup along with the mongoDB plugin for archiving the transaction blocks. The filtering process begins by sourcing the account name of the unit from the public hash key associated with the user and further querying the instance to retrieve the transactions. The query filters through the account name of the sender and recipient in each transaction to match the account name sourced above. The filtered records are then traversed to calculate the token balance of the particular unit, this traversal process is expounded in Sect. 4.2. Finally, the token balance as well as the complete transaction history is displayed to the user. This process ensures that only personnel with a valid key can get access to the corresponding token balance information. The information retrieved acts as a reference for the web application to display the token balance for each unit at each time-step, thus increasing the transparency of the overall system.

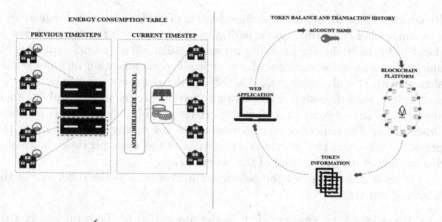

Fig. 3. Figure explaining the use-case and the interaction lifecycle for the Energy Consumption Data (left) and the user's token balance information (right)

Figure 3 depicts a representation of the data interaction amongst the different layers. The *Energy Consumption Table* is updated at each time-step using the reading from smart meters attached to each unit and appliance. The last recorded energy rating for each house is further sent as input to the *Token Distribution Layer* for redistribution of the tokens and solar energy produced. On the other hand, the token balance and transaction history is only formulated when the user requests for it through the web application. Hence, the lifecycle for retrieving the data is initiated at the *User Interface Layer* and traverses through the *Blockchain Platform Layer* before the retrieved data is displayed to the user.

4.2 Token Distribution Layer

This layer is sandwiched between the *Blockchain Platform layer* and the *Database layer* and acts as the brain of the whole model. It gathers the requisite energy readings for each unit, from the database, and performs priority matching of units having a surplus amount of net energy (*Producers*) and units in demand of extra energy (*Consumers*). Upon creation of tuples, corresponding to the possible energy transactions, the produced solar energy is distributed in accordance with these generated tuples, with incentives flowing in the opposite direction.

The distribution algorithm is a 3-step process. The first process involves classifying the units as either *Producers* or *Consumers* based on the allocated solar energy produced at that time-step and the net energy consumption of the house in the previous time-steps. The next step involves ranking the *Producers* and *Consumers*, with the goal of better incentivizing the units with better energy efficiency profiles. The final step presides over the formation of unit tuples, where each tuple symbolizes the combination of units participating in energy transactions. These tuples are then passed over to the EOSIO blockchain platform for actual distribution. The 3 step process is further elaborated below:

Prosumer Classification. The classification of units as either a *Producer* or a *Consumer* filters the units participating in the resource distribution process at that time-step, while also providing an early indication of the different possible unit combinations for the flow of resources. The classification of units is also aligned with the ultimate goal of satisfying the delta energy requirements of each unit at each time-step, while trying to incentivize the units supplying their allocated solar energy at that time-step. Hence, classification of units as either *Producers* or *Consumers* only accounts for the net energy consumption at the previous time-step. This avoids repetitiveness of the same previous datapoints due to the iterative procedure of the overall model.

The metric that decides the polarity of a *Prosumer* is the difference of the following two terms:

- `Allocated Solar Energy`: This is the proportion of the total energy that is produced by the photovoltaic solar array panel at each time-step. The proportion of the total solar energy received by each house is a fixed number that is determined by factors such as floor size and external walls and is entered into the Smart Contract.
- `Net Energy Consumption`: This is a measure of the unit's net energy consumption at the previous time-step. This value is retrieved from the *Database Layer* at each time-step.

A unit that achieves a positive difference of the above values is categorized as a *Producer*, whereas a unit with a negative difference score is classified as a *Consumer*. Only a house labelled as a *Producer* or a *Consumer* participates in the resource distribution process at that time-step.

Prosumer Ranking. The next step involves sorting the list of *Producers* and *Consumers* based on the surplus energy that can be supplied and the total energy demanded respectively. The basis for sorting the lists are as follows:

- `Producers`: The sorting is based on the amount of the surplus energy available for distribution as a percentage of the total solar energy allocated to the unit. The *Producers* are sorted in descending order of the defined metric.
- `Consumers`: The sorting is based on the amount of excess energy demanded as a percentage of the total solar energy allocated to the unit. The *Consumers* are sorted in ascending order of the defined metric.

The idea behind sorting the *Prosumer* list stems from the motivation that units having better energy profiles should be given preference during the formation of tuples. In other words, *Producers* that save a higher percentage of the total solar energy allocated and *Consumers* that demand a lower percentage of the total solar energy allocated are prioritized in the tuple formation process.

A higher priority during the tuple formation process leads to a unit being assigned to a higher paying slab, which ultimately leads to better incentivization. This also opens up the domain of behavioral economics, where the unit owners try to improve their energy consumption patterns so as to attain a higher priority in the tuple formation step. A community-wide infestation of such an approach leads to huge energy savings as well as reduced peak-loads demands.

Tuple Formation. The final step involves taking the sorted *Producer* and *Consumer* lists and forming tuples based on that list. The tuples carry information pertaining to the transaction of the solar energy as well as the tokens. The information itself consists of the *Producers* and *Consumers* involved in the transaction as well as the price tier associated with the involved entities. The central grid also plays an integral role in the overall process, with it acting as an energy source in some transactions as well as an energy sink in other transactions.

Table 1. Representation of the different Price Tiers and their description

Price tier	Description	Price/Unit
P1	Buying from the community	$0.12/KwH
P2	Buying from the central grid	$0.14/KwH
S1	Selling to the community	$0.12/KwH
S2	Selling to the central grid	$0.06/KwH

Table 1 represents the four price slabs as well as the price associated with each slab. The slab association is dependent on whether the *Producer/Consumer* is selling/buying the surplus/demanded energy to/from the community or the grid. *Producers* are assigned either the S1 or S2 tag depending on whether their surplus energy is sold to the community or grid respectively. On the other hand, *Consumers* are assigned either the P1 or P2 tag depending on whether they received their deficit energy from the community or grid respectively. As is evidenced from the pricing charts, a *Consumer* pays a higher price to buy the same amount of energy from the grid as compared to buying from the community, whereas a *Producer* receives a higher incentive for selling to another unit in the community as compared to selling to the grid. Therefore, energy transactions within the community is much more financially beneficial for both the *Consumers* and *Producers* than transactions involving the centralized grid.

Figure 4 represents the process involved in the formation of a tuple. The *Producers* are sorted according to the surplus energy as a percentage of the total solar energy received, whereas the *Consumers* are sorted according to the energy demanded as a percentage of the total solar energy. The overall process inherently tries to maximize the financial benefit for the *Producers* and *Consumers* alike by prioritizing energy transactions within the community. The process involves continuously picking the units from the sorted *Consumer* list in serial order and fulfilling their energy requirements from the sorted *Producer* list until either of the list becomes empty. The units leftover on the remaining list then either sell the surplus energy to the grid or receive the deficit energy from the grid.

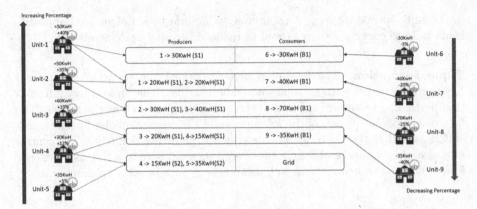

Fig. 4. Tuple formation process for a set consisting of five Producers and four Consumers, sorted according to the efficiency of their energy consumption profile at a specific time-step

4.3   Blockchain Platform Layer

This layer forms the backbone of the entire system by providing infrastructure and governance to the entire model. The overall network consists of three types of participants, the central grid, the residential units and the community administrative node. Each unit is represented by a unique node on the EOSIO platform and can act as an energy consumer and producer at different time-steps. The dPOS consensus algorithm of the EOSIO platform empowers a system where the community elects the block producers, from the pool of participating units, on a monthly rotation cycle, based on the energy efficiency profiles of the units. This election mechanism not only imparts a democratic system to the overall model, but further motivates units to improve their energy consumption profiles to receive additional incentives. The community administrative node is associated with shared resources such as fitness centers, thus allowing the unit owners to record payments to access the facilities. Hence, unlike the other two participants, the community administrative node doesn't partake in the energy distribution process. The photovoltaic panels and the subsequent solar energy produced serves as the asset in the system and is represented on the chain through the ECOT tokens. Each ECOT token has a fixed predetermined value of \$0.01, thus absolving the tokens of price depreciation over time and imbibing the users with trust to keep their tokens in the network for longer time durations.

On the other hand, the Smart Contract is prewired with the requisite characteristics of each unit, so as to enable the platform to estimate the proportion of the total solar energy that will be allocated to each unit. The major operations provided by the contract are exchange of tokens amongst houses in conjunction with the units of energy exchanged (*function: exchangeTokens([energyTuples])*), recording payments for gaining access to the community utilities (*function: gainAccess(for:, to:)*) and retrieving all historical transactions for a certain unit by leveraging the mongoDB plugin (explained in the Sect. 4.1). The *exchange-Tokens([energyTuples])* function takes the tuples as inputs and iterates over the tuples to transfer tokens from the *tuple[consumer]* to *tuple[producer]*. Meanwhile, the *gainAccess(for:, to:)* function requires two inputs, the facility *for* which the access has been requested and the unit *to* which the access will be granted on execution of a successful transaction. Lastly, the decision to run the resource intensive mongoDB plugin on a separate node was taken to ensure the smooth operation of the main chain as the plugin is designed to shut on occurrence of a failed insert request. Since, the overall system is governed through the use of the contract, the subscription to the contract is mandated at the time of leasing the apartment. Figure 5 presents a visual representation of the manner in which the EOSIO platforms binds the whole process of distributing resources amongst units in a community. At each time-step, each unit is allocated a portion of the total solar energy generated in accordance to the values stored in the *Smart Contract*. These values then form the basis for the *Distribution Layer* to create tuples, which are then further transmitted back to the *Blockchain Layer*. The returned tuples are then aggregated with the initial allocated solar energy to arrive at the final distribution numbers for each unit at each time-step. These values are then transmitted to the actuators connected to the solar panels to activate the distribution of the solar energy produced at that time-step. Meanwhile, the *Smart Contract* leverages the EOSIO platform to redistribute tokens amongst the *Producers* and *Consumers*.

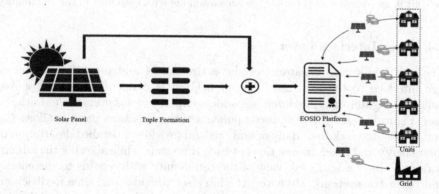

Fig. 5. Figure explaining the integration of the Smart Contracts, deployed on the EOSIO platform, in the proposed model as well as their role in governing the redistribution of tokens and the produced solar energy

The choice of the blockchain platform was driven by the need for low energy footprint as well as fast transactions, with EOSIO scoring high on both fronts. The overall value addition of the blockchain platform is provided by the high transparency lended to the economic model as well as the democratization of the token transfer model, thus drifting away from a centralized structure. The fact that EOSIO also allows archiving of transactions through plugins like mongoDB lends to further future enhancements, through intelligent energy consumption analysis and smart energy allocation.

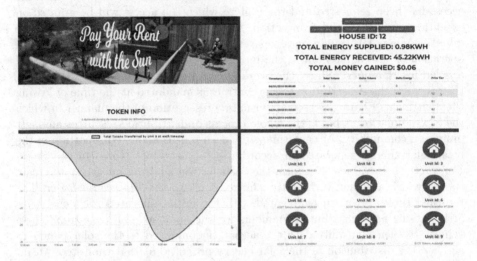

Fig. 6. Snapshots of the web application (top to bottom). Home page of the Residents section, a microscopic view of the energy distribution history for Unit 12 and links to pay rent/deposit further ECOT tokens, visual representation of the tokens transacted by Unit 9, an overview of the ECOT tokens available with each unit in the community

4.4   User Interface Layer

The web application encapsulates the entire model and presents an interactive interface to the end user. The application is developed to mimic a real builders website, thus providing a seamless experience for even the uninitiated user. Figure 6 shows the various components of the *Residents* page. It allows the unit owners to track their daily operations and provides a detailed description of their energy and token transactions at each time-step, while allowing the administrator to have a bird's eye view of the community with regards to the energy and token transactions. Moreover, the interface provides real time feedback on the amount of tokens transacted by the unit until that time-step, while also enabling users to access shared utilities and pay their rent using their balance tokens.

5   Results

The experimental setup involves simulating a real community comprising of units having different room characteristics and electric consumption. Figure 7 shows the different units that make up the community, with there being an equal number of 1, 2, 3 and 4 bedroom (BR) units, each having low, medium and high electric consumption patterns. The data is sourced from the US Department of Energy's Office of Energy Efficiency and Renewable Energy as part of their OpenEI open data catalog [5]. Baseline, high, and low energy profile simulation benchmarks containing hourly load profiles for the TMY3-724113 weather station in Blacksburg, VA were normalized by floor area and corrected for gas appliances to better match the electrical consumption of units at the planned site. Hourly PV production values for a shared 50 kW array were derived from the National Renewable Energy Lab's PVWatts tool [4]. The simulation is performed over the course of an year, thus incorporating an exhaustive scope of all the weather cycles and varying energy production capacities of the solar panel.

Unit-1	Unit-2	Unit-3	Unit-4
1BR,Low Consumption	2BR,Low Consumption	3BR,Low Consumption	4BR,Low Consumption
Unit-5	Unit-6	Unit-7	Unit-8
1BR, Medium Consumption	2BR, Medium Consumption	3BR, Medium Consumption	4BR,Medium Consumption
Unit-9	Unit-10	Unit-11	Unit-12
1BR,High Consumption	2BR,High Consumption	3BR,High Consumption	4BR,High Consumption

Fig. 7. Energy consumption profile for each unit in the TreeHAUS community

The value addition of the proposed model is observed from the view point of the grid stability as well as the financial gains for the unit owners. In the absence of the proposed model, the centralized grid would act as the middleman. This correlates to the grid buying the excess energy from the units at a reduced price (S2, \$0.06/KwH) and then selling it at a higher price (P2, \$0.14/KwH), thus leading to an eventual cashflow out of the community. On the other hand, the proposed model prioritizes the redistribution of energy within the community. The *Prosumers* exchange energy at just below the retail price, which leads to increased profits for both the buyer and seller. Figure 8 shows the redistribution statistics of the excess solar energy produced by the units. As can be seen, a major chunk of the excess solar energy produced by each unit is now categorized in the subsequent S1 and P1 price tier. The net savings incorporates the following three metrics:

- **P1 Energy:** The amount of energy that was bought from the community (at \$0.12/KwH) instead of the grid (at \$0.14/KwH), thus leading to savings of \$0.02/KwH.
- **S1 Energy:** The amount of surplus energy produced that was sold to other units in the community at \$0.12/KwH.
- **S2 Energy:** The amount of surplus energy produced that was sold to the grid at the reduced price of \$0.06/KwH.

The annual savings for each unit is calculated as follows:

$$NetSavings = (P1_{Energy} * 0.02) + (S2_{Energy} * 0.06) + (S1_{Energy} * .12)$$

Price Tier	Units											
	Unit-1	Unit-2	Unit-3	Unit-4	Unit-5	Unit-6	Unit-7	Unit-8	Unit-9	Unit-10	Unit-11	Unit-12
S1 (KwH)	1686.236	3585.609	5196.154	5510.4	233.1231	235.0531	820.4709	930.5318	0	0	0	0
S2 (KwH)	381.6829	73.04922	161.0271	428.9488	414.9029	409.8829	869.6985	1052.83	0.835075	1.532191	2.243508	0
P1 (KwH)	4.000915	10.72421	14.30761	16.78507	145.985	642.3816	803.9528	884.8265	784.6286	4433.959	5618.769	5533.396
Net Savings($)	225.2	434.78	633.46	687.3	55.77	65.64	166.7	192.52	15.74	88.72	112.5	110.67

Fig. 8. The total annual energy distributed within the community (S1), sold to the grid (S2) or bought from the community (P1) for each of the 12 units

Figure 8 shows the annual net savings of each unit in the community. As can be seen, the low consumption units (Units 1–4) save a higher proportion of the solar energy produced and hence are prioiritized for S1 trade during the tuple formation process, thus leading to higher profits for these units. The medium consumption units (Units 5–8) have a fair proportion of the S1 and S2 trades contributing to their profit margins, whereas the higher consumption units (Units 9–12) benefit through buying energy from the community at a reduced price. Figure 9 also shows the S1 and P1 trends for all the units for the month of August. The higher consumption units dominate the P1 charts, receiving energy from the low and medium consumption units, as evidenced by the S1 charts. This eventually leads to everyone tapping in on the financial benefits offered by the proposed model. The overall community also benefits, as evidenced by Fig. 10. It shows a birds eye view of the total solar energy distributed within the community as well as the excess energy supplied back to the centralized grid. This leads to reduced energy demands by the overall community, from the grid, contributing to better load-management and congestion handling.

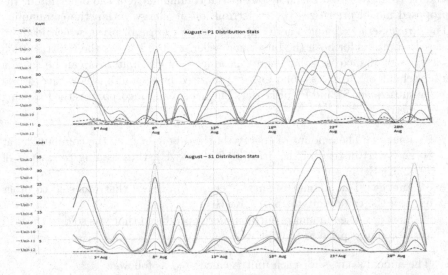

Fig. 9. Daily timeseries representation of the total energy bought by each unit from within the community (top) and the total energy sold to other units in the community

Fig. 10. Overview of the benefits to the community from the viewpoint of energy distribution and financial savings

6 Conclusion

The model proposed in the paper introduces a novel approach towards distribution of a common resource pool amongst different units in a community. Although the value addition of the model proposed is explored through the distribution of solar energy, the model can be easily extended to incorporate other shared resource pools such as biogas production. Moreover, the modularity of the model can be leveraged to easily scale up by integrating with IoT devices to achieve appliance level control. By gamifying the community through energy comparisons and dynamic price-tiers, there is a socio-behavioral opportunity leveraged by the model that can further improve the community's energy efficiency. Overall, the residents benefit through transparently earning money for good consumption habits, whereas the overall community benefits as improved resource efficiency leads to reduced costs and more robust and renewably-powered infrastructure.

References

1. Power Ledger Whitepaper (2018). https://www.powerledger.io/
2. WePower White Paper (2019). https://wepower.network/
3. EOS.IO White Paper (2019). https://github.com/EOSIO/Documentation/blob-/master/TechnicalWhitePaper.md
4. National renewable energy lab's PVWatts tool (2019). https://pvwatts.nrel.gov
5. OpenEI Open Data Catalog (2019). https://openei.org/doe-opendata/dataset/commercial-and-residential-hourly-load-profiles-for-all-tmy3-locations-in-the-united-states
6. TreeHAUS, International Solar Decathlon 2019 (2019). https://www.solardecathlon.gov/2019/assets/pdfs/dc19-virginiatech-projectsummary.pdf
7. Brousmichc, K.L., Anoaica, A., Dib, O., Abdellatif, T., Deleuze, G.: Blockchain energy market place evaluation: an agent-based approach. In: 2018 IEEE 9th Annual Information Technology, Electronics and Mobile Communication Conference (IEMCON), pp. 321–327. IEEE (2018)
8. David, L.: Bennet: personal communication, March 2019
9. El Rahi, G., Etesami, S.R., Saad, W., Mandayam, N.B., Poor, H.V.: Managing price uncertainty in prosumer-centric energy trading: a prospect-theoretic stackelberg game approach. IEEE Trans. Smart Grid **10**(1), 702–713 (2017)

10. Kang, J., Yu, R., Huang, X., Maharjan, S., Zhang, Y., Hossain, E.: Enabling localized peer-to-peer electricity trading among plug-in hybrid electric vehicles using consortium blockchains. IEEE Trans. Ind. Inform. **13**(6), 3154–3164 (2017). https://doi.org/10.1109/TII.2017.2709784. http://ieeexplore.ieee.org/document/7935397/

11. Mengelkamp, E., Gärttner, J., Rock, K., Kessler, S., Orsini, L., Weinhardt, C.: Designing microgrid energy markets: a case study: the Brooklyn Microgrid. Appl. Energy **210**, 870–880 (2018). https://doi.org/10.1016/j.apenergy.2017.06.054. https://linkinghub.elsevier.com/retrieve/pii/S030626191730805X

12. Noor, S., Yang, W., Guo, M., van Dam, K.H., Wang, X.: Energy demand side management within micro-grid networks enhanced by blockchain. Appl. Energy **228**, 1385–1398 (2018). https://doi.org/10.1016/j.apenergy.2018.07.012. https://linkinghub.elsevier.com/retrieve/pii/S0306261918310390

13. Pipattanasomporn, M., Kuzlu, M., Rahman, S.: A Blockchain-based platform for exchange of solar energy: laboratory-scale implementation, p. 8. IEEE (2018)

14. Schlund, J., Ammon, L., German, R.: ETHome: open-source blockchain based energy community controller. In: Proceedings of the Ninth International Conference on Future Energy Systems - e-Energy 2018, Karlsruhe, Germany, pp. 319–323. ACM Press (2018). https://doi.org/10.1145/3208903.3208929. http://dl.acm.org/citation.cfm?doid=3208903.3208929

15. Siano, P., De Marco, G., Rolan, A., Loia, V.: A survey and evaluation of the potentials of distributed ledger technology for peer-to-peer transactive energy exchanges in local energy markets. IEEE Syst. J. 1–13 (2019). https://doi.org/10.1109/JSYST.2019.2903172. https://ieeexplore.ieee.org/document/8671697/

16. Zhang, C., Wu, J., Zhou, Y., Cheng, M., Long, C.: Peer-to-Peer energy trading in a Microgrid. Appl. Energy **220**, 1–12 (2018). https://doi.org/10.1016/j.apenergy.2018.03.010. https://linkinghub.elsevier.com/retrieve/pii/S0306261918303398

Practical Mutation Testing for Smart Contracts

Joran J. Honig[1,3]([✉]) [iD], Maarten H. Everts[1,2] [iD], and Marieke Huisman[1] [iD]

[1] University of Twente, Enschede, The Netherlands
{maarten.everts,m.huisman}@utwente.nl
[2] TNO, The Hague, The Netherlands
[3] ConsenSys, New York, USA
joran.honig@consensys.net

Abstract. Solidity smart contracts operate in a hostile environment, which introduces the need for the adequate application of testing techniques to ensure mitigation of the risk of a security incident. Mutation testing is one such technique. It allows for the evaluation of the efficiency of a test suite in detecting faults in a program, allowing developers to both assess and improve the quality of their test suites. In this paper, we propose a mutation testing framework and implement a prototype implementation called Vertigo that targets Solidity contracts for the Ethereum blockchain. We also show that mutation testing can be used to assess the test suites of real-world projects.

Keywords: Mutation testing · Smart contract · Solidity

1 Introduction

Recent developments in distributed computing have resulted in platforms that support the execution of so-called "smart contracts". In this paper, we will look specifically at smart contracts written in the Solidity programming language [12], for the Ethereum blockchain. These smart contracts are programs that are executed by the nodes in the Ethereum network and can deal with votes, money transfers and even digital collectibles, such as Cryptokitties [3]. Because smart contracts exist on the Ethereum blockchain, they are openly readable and executable by any participant in the network. Furthermore, once a contract is uploaded it cannot be changed anymore, it is immutable. These properties force smart-contract developers to be very conscientious about the correctness and security of their code.

To reason about the correctness and security of smart contracts, developers employ different testing techniques. One of these techniques is unit testing, a commonly accepted practice in traditional software development. It has clear benefits and allows teams around the world to develop and release their software with confidence.

© Springer Nature Switzerland AG 2019
C. Pérez-Solà et al. (Eds.): DPM 2019/CBT 2019, LNCS 11737, pp. 289–303, 2019.
https://doi.org/10.1007/978-3-030-31500-9_19

However, it is often very difficult to provide adequate guarantees over the correctness and security of a system using unit tests. As a result, critical vulnerabilities have frequently been discovered in live contracts. Notable security incidents include the hack of "The DAO" [14], where an attacker was able to exploit a re-entrancy vulnerability, allowing the attacker to withdraw 60 million USD worth of Ether, the base currency of Ethereum. Another example is the Parity wallet bug [8], where a user accidentally self-destructed code on which many other contracts were dependent, freezing assets worth 280 million USD at the time.

One factor that makes the application of unit testing difficult to do well is accurately described by Dijkstra, who once wrote, "Program testing can be used to show the presence of bugs, but never to show their absence!" [19]. This paints a grim picture, as it indicates that even if there are a multitude of tests, then there is still no guarantee that there are no bugs in the program. However, it also emphasises what unit tests can be good at, namely showing the presence of bugs. One needs to be able to gauge the effectiveness of a test suite at detecting bugs, to effectively apply unit testing to smart contracts. A good metric can give insight into both the guarantees given by a test suite and possible areas of improvement. One metric that is frequently used is code coverage, which gives the developer a concrete idea of which parts of the code are covered. However, code coverage is flawed and does not give a direct indication of the test suite's effectiveness for detecting bugs. An example of a flawed test suite, which scores well with this metric is one without assertion statements, which could cover all the lines of code, but would not give any guarantees whatsoever.

Mutation testing [23] is a technique which aims to mitigate this problem. Instead of using the amount of code covered as a heuristic for the quality of the test suite, it measures the suite's effectiveness at detecting faults in the source code directly. It does this by procedurally introducing small mutations in the source code, for example, a change from ">" to "<=", and executing the test suite for each such mutation. After having tested all the mutations, the ratio between the detected mutations (also called "killed" mutations), and the undetected mutations (also called "surviving mutations") can be used as a metric to evaluate the quality of the test suite.

This metric can provide valuable input in multiple stages of the software development lifecycle. A threshold value for the mutation score can be used to set a quality gate before a software release. Additionally, the information on mutants that have survived (i.e. have not been killed) can be used to increase the quality of the test suite, as these mutants identify specific edge cases which are not yet covered by the current test suite. An overview of the surviving mutations is also useful for a security review, as these mutants indicate locations where the program behaviour might not match the expectation of the developer.

This paper introduces Vertigo, a mutation testing tool written for Truffle and Solidity. We will describe the design choices, limitations, and further work for this implementation. Additionally we provide and discuss the application of mutation testing in the smart contract development lifecycle.

The two main contributions of this paper are:

- We provide a detailed design and implementation of a mutation testing framework agnostic of its test environment. The implementation is designed to be extensible to facilitate the implementation of future research.
- We propose new mutation operators that can be applied to Solidity.

2   Background

This section provides an overview of the current state-of-the-art for mutation testing and Ethereum smart contracts.

2.1   Mutation Testing

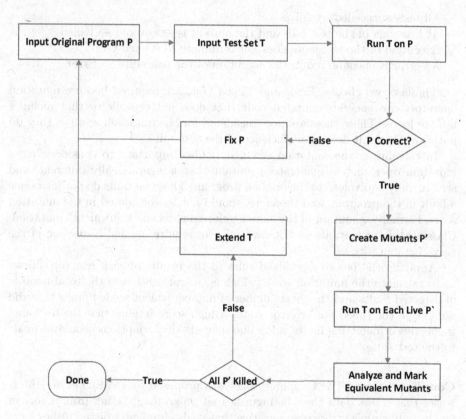

Fig. 1. Mutation analysis process adopted from [26]

Jia [23] provides an apt description of mutation testing: "Mutation testing is a fault-based testing technique which provides a testing criterion called the "mutation adequacy score". The mutation adequacy score can be used to measure the effectiveness of a test set in terms of its ability to detect faults".

The Process. The standard mutation testing process is described in Fig. 1. Initially, a set of mutants p' is generated based on a set of mutation operators. Mutation operators are transformation rules that apply a syntactic operation to the original program p to generate the mutated program p'. As mentioned in the previous section, an example of such a mutation rule is to substitute ">" by "$<=$".

The original program is then checked for correctness. The correctness of p under the test suite is required, as it is not possible to distinguish a program p from a mutated program p' using a test suite that classifies p as faulty.

If the original program is correct, then the test suite is applied to the different mutants. Executing the test suite on one particular mutant can lead to different results: Alive, Killed, Timed Out and Error. The following are rules used to determine the state of a mutation:

- All tests succeeded → Alive
- At least one of the test fails and the mutant is discovered → Killed
- Execution of the test suite does not terminate → Timed Out
- An error is encountered in the execution of the test suite → Error

The last two classes, Error and Timed Out, are required because mutation operators can generate mutated code that does not compile or that includes infinite loops. These cases are not considered for the mutation score as they do not show the fault detection efficiency of the test suite.

In calculating the mutation score, it is also important to consider that a mutation operator can generate a mutant that is syntactically different, and semantically equivalent to the original program. These mutants do not introduce a fault in the program, and therefore, should not be considered in the mutation score. Formally, a mutant of this kind is referred to as an "Equivalent" mutation. Classification of mutants as "Equivalent" can require manual inspection of the surviving mutants.

Applying the before mentioned rules to the results of each test run, allows us to calculate the mutation score, which is the ratio between the total number of detected faults and the total number of non-equivalent seeded faults that did not end in the timed out or error state. A high score implies that the test suite is effective at detecting faults, a low one suggests that adaptions should be made to the test suite.

Coupling Effect. The Coupling Effect was proposed by DeMillo et al. [16]. It states that "Test data that distinguishes all programs differing from a correct one by only simple errors is so sensitive that it also implicitly distinguishes more complex errors". This effect allows us to limit our mutation operators to only consider simple syntactic changes, as the tests killing simple errors would also kill more complex errors. Extending the analysis to include mutation operators that introduce complex changes will only increase the number of mutants to test and should not affect the mutation score. Thus, one can assume that test cases unable to detect simple mutations, also are more likely to permit more complex errors to remain undiscovered. This insight adds value to mutation testing as a technique for bug finding, rather than just for code and test quality.

2.2 Ethereum Smart Contracts

Ethereum. Similar to the well-known blockchain protocol Bitcoin, Ethereum is a blockchain that can keep track of transactions and balances.

A key aspect of Ethereum is its ability to have contract accounts, in addition to regular user accounts. These contract accounts have code associated with them, and every time a transaction is sent to such an account, the nodes in the Ethereum network evaluate this code. These contracts allow the implementation of co-owned wallets or the governance of an organisation. The interactions with these contracts are, because of the decentralised and open nature of the Ethereum network, less prone to problems like corruption and secrecy.

Smart contracts have a few quirks and properties which make them both useful and challenging to secure. The first property is that all smart contracts are uploaded to a public blockchain, which means that the code, is immediately available to all adversaries, even if it is only bytecode.

The second aspect is that Ethereum smart contracts that are in use can hold tremendous amounts of value, and even small bugs allow attackers to freeze[1] or steal the currency stored in the contract.

This has led to the implementation of several security analysis tools, which automatically check for the presence of known weaknesses and vulnerability patterns. These techniques aim to improve the security of the software by looking for patterns that are known to be vulnerable.

Truffle. Truffle is a framework that aims to provide a development environment for Solidity developers. For the purpose of this paper, its most relevant aspect is that it allows developers to write and execute unit tests. These tests can be written in Javascript using the Mocha framework, or in pure Solidity. Truffle then uses another tool called Ganache, a Javascript Ethereum node which supports additional JSON RPC calls for the purposes of testing, to execute these unit tests.

3 Design

To evaluate the feasibility of mutation testing in the context of Ethereum-based smart contracts, we created Vertigo, a mutation testing framework for smart contracts. Vertigo is available on Github[2]. Vertigo will be made available as open source.

This section describes the design and implementation of Vertigo. Figure 2 shows the structure and interaction of the base components in Vertigo. The three main sections shown in Fig. 2 are discussed in order.

---

[1] With *freeze*, we mean the act of blocking users from accessing the currency stored in the contract.

[2] https://github.com/JoranHonig/vertigo.

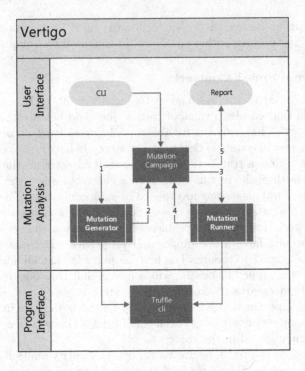

Fig. 2. Structure diagram for components in Vertigo

3.1 User Interface

Through the command line interface (CLI) users are able to initialise mutation testing campaigns. It is possible to configure the Ethereum networks that will be used for the test runs, through the use of command line options. Using a command line option, Vertigo can also be instructed to apply the mutation sampling technique, which will be elaborated in Sect. 3.2.

By default, Vertigo will report all mutants that have survived a mutation campaign. Optionally, a full report with the other mutations, can be written to a file.

3.2 Mutation Analysis

The mutation analysis compartment is where most of the work happens, it is comprised of three interconnected components. These components implement everything from generating mutants to directing the testing process. The edges visible in Fig. 2 are numbered according to the data flow through the mutation analysis components.

Table 1. Increments mirror substitution matrix

Original	Result
+=	=+
-=	=-

Table 2. Mutation operators

Mutation operator	Description
Condition Boundary	Replaces an conditional operation with its inclusive or exclusive counterpart
Condition Negation	Replaces a conditional operation with its inverse
Math Inversion	Replaces a math operator with its inverse
Increments Inversion	Replaces an increments statement with its inverse
(a) Mutation operators inherited from PIT	

Mutation operator	Description
Increments Mirror	Replaces an increments statement with it's mirror
Modifier Removal	Removes a modifier application
(b) Vertigo mutation operators	

Mutation Campaign. The mutation campaign is the central component in Vertigo that directs and manages the mutation testing of a project. What follows is an overview of the steps taken by the mutation campaign while performing mutation testing.

The very first step is the application of the test suite on the original program. If the given test suite fails on the original contracts, then further analysis is halted, as it will be impossible to determine if the test suite distinguishes a mutated program from the non-mutated program.

During the next step in the mutation testing process, the mutation campaign will request the Mutation Generator component to generate a set of mutations for the project.

The Mutation Campaign implements functionality that allows it to apply custom filters on the set of generated mutations. In Vertigo we implement one initial filter that selects a random sample from the available mutations for further analysis. This is an implementation of the optimisation technique "Mutant Sampling" [23], which is based on the assumption that the mutation score of a random sample of the mutants will reflect that of the entire set of mutations. Wong [29] showed that a sample of 10% of the entire mutation set is only 16% less effective than an analysis of all the mutants, with respect to the mutation score.

To facilitate the implementation of mutant sampling, we have implemented an extensible interface for filters. This same interface can also be used to implement other filtering techniques, such as mutant clustering [22] or equivalent mutation detection [23].

The generated and filtered mutations are passed to the Mutation Runner, which applies the test suite on the mutated contracts storing the result of each test run.

Finally, the mutation campaign will return the results to the user interface, where these results are formatted and displayed to the user.

Mutation Generator. The Mutation Generator component is responsible for the generation of mutants, which are generated through the application of so-called mutation operators. These mutation operators [23] are transformation rules that can be used to generate mutant programs p' from a program p. An example mutation operation could specify the replacement of "==" with "!=". The mutation operators implemented in Vertigo have been both specifically designed for the context of smart contracts (Table 2b) and based on existing work (Table 2a).

PIT [9], a state of the art mutation testing tool that works on Java, has a well-documented set of mutation operators. While it implements a wide range of mutation operators, not all of them are enabled by default. They try to limit the developer's exposure to equivalent mutations, by only enabling the mutation operators which generate lower amounts of equivalent mutations. Vertigo leverages a few selected mutation operators from the PIT default mutation operators; these are visible in Table 2a.

Additionally, we propose and implement the addition of a set of new mutation operators, relevant to the Solidity programming language and smart contracts. They are designed to simulate bugs and security issues applicable to the Solidity programming language. These mutation operators can be found in Table 2b and will be discussed in the following paragraphs.

Increments Mirror. A recent development in Smart Contract security has been the standardisation and formulation of a weakness registry for smart contracts and Solidity [11]. One issue illustrated in this registry [13] describes a typographical error introduced when "=+" is written instead of "+=". In this case the statement is interpreted as an assignment with a unary operator following it, whereas the developer intends to write an incrementation operation. This Mutation Operation tries to introduce the same faults in the target program, substituting incrementations according to Table 1

Modifier Removal. Solidity allows the use of modifiers. These are functions that can be used to wrap the functionality of other functions, similar to decorators in python. One pattern used in smart contract development is that of ownership. A contract that implements this functionality has a variable that stores the address of the account that is currently the owner of the contract, and functionality to transfer ownership. This contract will also implement a modifier which is called onlyOwner; this modifier will change every function that invokes it to revert if the caller is not the owner of the contract. The ownership paradigm, allows smart contract developers to protect administrative functions in a smart contract.

Forgetting to add a modifier is a simple mistake that a developer could easily make. Because of this, and the serious effect it can have on the security of a contract we include a mutation operator which simulates the omission of modifiers.

Mutation Runner. The Mutation Runner component implements the logic that allows Vertigo to apply mutations to the test project and then apply these tests on the project. There are two main aspects of this component that are of interest: Its ability to perform multiple test executions in parallel, and the mutation application and autonomous configuration.

The mutation application process is described by the following steps:

1. Vertigo creates a temporary directory and copies the files from the initial project to it.
2. The mutation is applied to the source code
3. The Truffle configuration is adapted allowing Vertigo to interpret test results
4. The Truffle test command is invoked directed at the mutated project

Because Vertigo uses temporary directories, the original project directory will never need to be written to. The use of temporary directories allows parallel application of multiple mutations; additionally, there is no risk of a mutation persisting if Vertigo fails unexpectedly.

Parallelisation. The user is able to indicate the networks that will be used for the mutation testing process. These networks provide a virtual environment used for the execution of the tests. Parallel testing is enabled if Vertigo is provided with multiple networks, in which case each instance of a runner is dynamically assigned a free network for each test run.

3.3 Program Interface

Vertigo already implements an interface with Truffle and its CLI right now, but this is extensible and allows integration of different platforms with a similar interface.

The first element exposed by the Truffle interface is the analysis of source code and the generation of an abstract syntax tree (AST). The information in the AST is used by the Mutation Generator to find possible mutations.

The second element is the application of the test suite on a project, which allows the Mutation Runner component to check if the test suite distinguishes the original program from a mutated one.

Ganache. Truffle uses Ganache to model an Ethereum blockchain. Ganache extends the traditional Ethereum interface by adding some specific JSON RPC calls that allow Truffle to execute each test in a clean environment. During the development of the Vertigo framework, we found that Truffle, through its support for the Mocha framework, also supports the writing of tests that do not use this clean environment. This results in a situation where although a program

Table 3. Mutation test results

	Lived	Error	Timeout	Killed	Equivalent	Total	Mutation score	Code coverage	Duration
Aragon OS	0	27	1	306	0	334	100%	99%	129 min
Openzeppelin-solidity	30	55	3	303	5	391	92%	100%	260 min

might be correct under a test set T, it might not be under consecutive runs of the test set T. As a solution, we propose that developers refrain from using this functionality. Alternatively, if the only side effect caused by the tests is an alternative distribution of resources over the test accounts initialised by the Truffle framework, then the user can decide to initialise Ganache with the option -e to increase the default balance of the test accounts. This change will increase the number of times that T can successfully be applied.

4   Evaluation

In this section we apply mutation analysis on two popular smart contract projects. Moreover, we provide a discussion on the analysis results provided by Vertigo.

4.1   Experiments

We analysed two popular smart contract projects, both of which boast impressive code coverage results (see Table 3). Since both projects are very well maintained and extensively tested, any analysis results can provide insight into the applicability of mutation testing to real-world smart contract projects.

The first project we analyse is AragonOS, which is described as "a smart contract framework for building decentralised organisations, dapps, and protocols" [1]. The second is openzeppelin-solidity [7], a library of contracts for secure contract development.

The results of these test runs can be found in Table 3. The mutation score is calculated using the following formula:

$$score = \frac{(killed)}{(lived - equivalent) + killed} \times 100\% \tag{1}$$

The experiments were executed on a Ubuntu 18.04 machine with the following available resources: Threadripper 1950X, 32 GB RAM and a 1 TB NVMe SSD. The mutation testing was performed using 16 parallel testing networks.

4.2   Discussion

The results show that AragonOS outperforms openzeppelin-solidity on mutation score even though openzeppelin-solidity would seem to perform better when code coverage is used as a metric. If the developers of either project decide that they want to improve the quality of their test suite, then code coverage will not be able to provide much insight, as both projects have a coverage of 99 or 100%.

However, Table 3 shows that openzeppelin does not have a perfect mutation score; thus, the application of mutation testing would allow the developers to improve their test suite. Besides the mutation score, the developers can also use the surviving mutations to identify the portion of code that requires improved tests.

Additionally, the 30 surviving mutations for openzeppelin-solidity demonstrated yet uncovered edge cases in the behaviour of the program. An example of such an edge case is a function which uses a Solidity modifier to implement access control. The survival of a mutant that removes such a modifier, indicates that the test suite might not check whether an access control policy is enforced for this function. Without an automated technique like mutation testing, discovering these edge cases would have been strenuous and time-consuming.

While the results show that mutation testing can be used to improve openzeppelin-solidity, the analysis was unable to find a mutant which would not be caught by the test suite of Aragon OS. This result indicates that the developers of this project were able to implement a high-quality test suite without the help of mutation testing.

During the inspection of the analysis subjects, we realised that the smart contract projects sparingly made use of plain arithmetic operations. Rather, the projects employ the use of safe math libraries. A safe math library is a library that implements simple arithmetic operations, like addition and subtraction. These libraries extend normal arithmetic behaviour with safety checks. For example, the add functions will check if the result of an addition is overflowed and raise an exception that is the case. As a result, most of the business logic will use functions like add() and subtract(), instead of the binary operators $+$ and $-$. This can have an impact on some of the mutation operators which target mathematical operators, as the usage of plain arithmetic operators will be less frequent. Introduction of mutation operators that extend the behaviour of the arithmetic mutation operators to the safe math functions is a topic for future work (see Sect. 6.1).

Finally, the results in Table 3 show that a remarkably low number of equivalent mutants have been generated. This shows the effectiveness of the mutation operators taken from PIT [9] (visible in Table 2a). Additionally, there were no equivalent mutants for the mutation operators in Table 2b. The low amount of equivalent mutants being generated by the mutation operators in Vertigo is beneficial for the application of mutation testing, as the manual process of determining equivalency for smart contracts can be time-consuming.

5   Related Work

Securing smart contracts is a difficult task, for which multiple approaches have been proposed and implemented. These tools apply techniques such as symbolic execution (Mythril [6], Oyente [24], Maian [25], Manticore [5]), or perform data and control flow analysis (Securify [27], Slither [10], Vandal [15]). A common aspect of these tools is that they leverage common patterns that describe vulnerabilities, to report on the existence of vulnerabilities or bugs. This makes them complementary to the application of unit testing and mutation testing, as they target the identification of common bugs, whereas unit testing allows for the testing of business logic.

Besides fully automated approaches, there are also verification approaches where a user is asked to define invariants and properties which are then validated (K-framework [21] and Verisol [28]). Such approaches are complementary to mutation testing, as mutation testing can also be applied to the evaluation of specifications [17].

There have also been some developments in the application of mutation testing to smart contracts. Eth-mutants [4] is a project that implements a proof of concept mutation analyzer for Solidity. However, key optimisations like parallelisation are not present in this project, such optimisations are necessary to make mutation testing practical for real-world projects. Additionally, Vertigo is designed to be extensible with various proposed optimisations [23] in mind.

Another project is Universalmutator [20], which has recently been extended with mutation rules for Solidity. Universalmutator is a project aimed at rapid development of mutations for different programming languages, using regular expression based substitution rules. This approach is focused on the generation of mutants, whereas Vertigo aims to implement the entire mutation testing process.

6 Conclusion

In this paper, we studied the design of a mutation testing framework and the application of mutation testing to the domain of smart contracts. We showed that mutation testing allows developers to both gauge the effectiveness of their test suite and improve the test suite in order to increase this effectiveness. Additionally we provide analysis results for two major smart contract projects, openzeppelin-solidity [7] and AragonOS [1]. For this paper, we created the tool Vertigo, an implementation of the proposed designs.

6.1 Future Work

The implementation of Vertigo relies upon functionality exposed by Truffle through its command line interface. This interface does not yet support running individual tests and code coverage reporting. These features can provide essential information for optimisations that allow Vertigo to "do less" [23].

Specifically, the addition of these features will allow for the implementation of the following two optimisations, which are also available in PIT [9]:

- Instead of running the entire test suite for each mutant, Vertigo should determine the specific tests that cover the line on which the mutant introduces a syntactic change. The results of tests that do not cover this line should not depend on the change, and therefore these tests provide little value when executed on the mutant.

– Instead of running the tests in the test suite in a random or pre-set order, Vertigo should optimise the order, so fast tests are executed earlier. Once a test fails, the execution of the other test can be omitted since they do not provide additional value over the previously executed test.

For this paper, we selected a limited set of mutation operators to provide a prototype that is applicable to real-world projects. A topic for further research is to evaluate existing mutation operators, and design new mutation operators specifically for the analysis of smart contracts. An evaluation of mutants with respect to their representativeness of the mutation score allows for the application of mutant selection as an optimisation technique.

Vertigo now supports mutation of Solidity smart contracts, and leverages the Truffle framework to execute the tests; thus, Vertigo is limited to the mutation testing of Solidity smart contracts for ethereum. A possible extension to Vertigo is the design of mutation operators for other smart contract languages and the implementation of interfacing logic for other test frameworks. Such an extension allows for a more uniform application of mutation testing over different blockchain platforms.

Furthermore, examining security critical bug classes like those described in the SWC registry [11] to design mutation operators that try to introduce vulnerabilities is promising. Daran and Thévenod-Fosse [18] have examined to what extent seeded faults represent actual faults. A possible topic for further research is to measure the extent to which mutation operators, like *Modifier Removal* in Table 2b, represent actual security critical faults.

Finally, in Sect. 4 we discussed the impact of so-called safe math libraries on the mutation analysis results. Specifically, we saw that the use of plain arithmetic operators was replaced by the use of functions that extend the basic behaviour of these arithmetic operators in the tested projects. Vertigo can be extended to more extensively cover these projects by implementing one of the following approaches:

– the implementation and design of mutation operators that reflect the behaviour of the mathematical operations for commonly used safe math functions
– An extension of Vertigo to allow for project specific mutation operators, allowing the developer to express the mathematical mutation operators for their specific project

Additionally, a new mutation operator can be designed for projects using safe math libraries, that introduces faults that mimic developers forgetting to use safe math libraries; a problem that has resulted in the well known "batchOverflow" bug [2].

References

1. aragonOS. https://hack.aragon.org/docs/aragonos-intro.html
2. Batch overflow vulnerability - CVE-2018-10299. https://cve.mitre.org/cgi-bin/cvename.cgi?name=CVE-2018-10299
3. CryptoKitties. https://www.cryptokitties.co/
4. eth-mutants: a mutation testing tool for smart contracts. https://github.com/federicobond/eth-mutants
5. Manticore. https://github.com/trailofbits/manticore
6. Mythril. https://github.com/consensys/mythril
7. openzeppelin-solidity. https://github.com/OpenZeppelin/openzeppelin-solidity
8. Parity Bug Security Alert. https://www.parity.io/security-alert-2/
9. PIT Mutation Testing. http://pitest.org/
10. Slither: Static Analyzer for Solidity. https://github.com/crytic/slither
11. Smart Contract Weakness Classification and Test Cases. https://smartcontractsecurity.github.io/SWC-registry/
12. Solidity. https://github.com/ethereum/solidity
13. SWC-129. https://smartcontractsecurity.github.io/SWC-registry/docs/SWC-129
14. The DAO Attacked: Code Issue Leads to $60 Million Ether Theft - CoinDesk. https://www.coindesk.com/dao-attacked-code-issue-leads-60-million-ether-theft
15. Brent, L., et al.: Vandal: a scalable security analysis framework for smart contracts. CoRR (2018)
16. Budd, T.A., DeMillo, R.A., Lipton, R.J., Sayward, F.G.: The design of a prototype mutation system for program testing. In: Proceedings of the AFIPS National Computer Conference, vol. 74, pp. 623–627 (1978)
17. Budd, T.A., Gopal, A.S.: Program testing by specification mutation. Comput. Lang. **10**(1), 63–73 (1985). https://doi.org/10.1016/0096-0551(85)90011-6
18. Daran, M., Thévenod-Fosse, P.: Software error analysis. In: Proceedings of the 1996 International Symposium on Software Testing and Analysis - ISSTA 1996, vol. 21, pp. 158–171. ACM Press (1996). https://doi.org/10.1145/229000.226313
19. Dijkstra, E.W.: Ewd 249 Notes on Structured Programming, 2nd edn. Department of Mathematics, Technische Hogeschool Eindhoven (1970)
20. Groce, A., Holmes, J., Marinov, D., Shi, A., Zhang, L.: An extensible, regular-expression-based tool for multi-language mutant generation. In: Proceedings of the 40th International Conference on Software Engineering Companion Proceeedings - ICSE 2018, pp. 25–28. ACM Press (2018). https://doi.org/10.1145/3183440.3183485
21. Hildenbrandt, E., et al.: KEVM: a complete semantics of the Ethereum virtual machine. In: 2018 IEEE 31st Computer Security Foundations Symposium, pp. 204–217. IEEE (2018). https://doi.org/10.1109/CSF.2018.00022
22. Hussain, S.: Mutation clustering. Master's thesis, King's College London, UK (2008)
23. Jia, Y., Harman, M.: An analysis and survey of the development of mutation testing. IEEE Trans. Softw. Eng. **37**(5), 649–678 (2011). https://doi.org/10.1109/TSE.2010.62
24. Luu, L., Chu, D.H., Olickel, H., Saxena, P., Hobor, A.: Making smart contracts smarter. In: Proceedings of the 2016 ACM SIGSAC Conference on Computer and Communications Security - CCS 2016, pp. 254–269. ACM Press, New York (2016). https://doi.org/10.1145/2976749.2978309

25. Nikolic, I., Kolluri, A., Sergey, I., Saxena, P., Hobor, A.: Finding the greedy, prodigal, and suicidal contracts at scale. In: Proceedings of the 34th Annual Computer Security Applications Conference. ACSAC 2018, pp. 653–663 (2018). https://doi.org/10.1145/3274694.3274743

26. Offutt, A.J., Untch, R.H.: Mutation 2000: uniting the orthogonal. In: Wong, W.E. (ed.) Mutation Testing for the New Century, pp. 34–44. Springer, Boston (2001). https://doi.org/10.1007/978-1-4757-5939-6_7

27. Tsankov, P., Dan, A., Cohen, D.D., Gervais, A., Buenzli, F., Vechev, M.: Securify: practical security analysis of smart contracts. In: Proceedings of the 2018 ACM SIGSAC Conference on Computer and Communications Security, CCS 2018 (2018). https://doi.org/10.1145/3243734.3243780

28. Wang, Y., et al.: Formal specification and verification of smart contracts for Azure blockchain. CoRR (2018)

29. Wong, W.E.: On mutation and data flow. Ph.D. thesis (1993)

CBT Workshop: Payment Systems, Privacy and Mining

Online Payment Network Design

Georgia Avarikioti[✉], Kenan Besic, Yuyi Wang, and Roger Wattenhofer

ETH Zürich, Zürich, Switzerland
{zetavar,besick,yuwang,wattenhofer}@ethz.ch

Abstract. Payment channels allow transactions between participants of the blockchain to be executed securely off-chain, and thus provide a promising solution for the scalability problem of popular blockchains. We study the online network design problem for payment channels, assuming a central coordinator. We focus on a single channel, where the coordinator desires to maximize the number of accepted transactions under given capital constraints. Despite the simplicity of the problem, we present a flurry of impossibility results, both for deterministic and randomized algorithms against adaptive as well as oblivious adversaries.

Keywords: Blockchain · Payment channels · Layer 2 · Network design · Online algorithms · Competitive ratio

1 Introduction

Recently, blockchain systems and cryptocurrencies such as Bitcoin [22] and Ethereum [6] have gained in popularity – in research, in economy and even in the general public. With the increased number of transactions, scalability has become a serious problem [8,10]. The maximum transaction throughput on the bitcoin network is approximately ten transactions per second. Other networks can process up to hundreds of transactions per second. In contrast, current digital payment systems (e.g. credit cards, WeChat Pay, etc.) handle tens of thousands of transactions per second [28].

Several solutions have been proposed to address the limitation on the transaction throughput on blockchain systems. Sharding [17,19] is a so-called first-layer (on-chain) solution. However, the most promising approach are payment channels [10,24,27]. Payment channels allow transactions between two parties of a blockchain system to be executed off-chain. Furthermore, the existence of multiple two-party channels leads to the creation of payment networks, where transactions between a sender and a receiver can be executed through a path of channels in the network, even if the two parties have no direct channel with each other. Payment networks [9–13,20,24] operate *on top* of the blockchain, introducing a second layer, and are thus known as *Layer 2* solutions.

Even though payment networks are efficient and enable high transaction throughput on blockchain systems, they also demand from users to lock a lot of capital a priori. In addition, complex routing algorithms are needed to discover

C. Pérez-Solà et al. (Eds.): DPM 2019/CBT 2019, LNCS 11737, pp. 307–320, 2019.
https://doi.org/10.1007/978-3-030-31500-9_20

routes with enough capital capacity from sender to receiver. To address these issues, multiple proposals emerged that suggest the use of a central operator, in theory [1,3] and in practice [16,23]. Concretely, [3] studied the complexity of a Payment Service Provider (PSP) from an algorithmic perspective. Similarly, [1] examined the optimal graph structure and fee assignment to maximize a PSP's profit.

Similarly to [1,3], we study the problem from the perspective of a PSP. The PSP creates the payment network and opens the channels between interacting parties in which the PSP locks the required capital, acting as a creditor. Since opening a channel is associated with cost (registration with the blockchain), the PSP charges fees to the customers using its channels. Additionally, the PSP decides which channels to open and how the capital will be distributed on the channels. The PSP's objective is to maximise its profit under capital constraints.

However, both [1] and [3] assumed to know all future transactions. While such assumptions may be valid in many situations, our paper takes a different route. We want to know how a PSP should set up channels *without any assumption of future payments*. We wonder to what extent a PSP can still do a good job, and in which cases planning ahead is hopeless. In other words, we study the so-called *online* version of setting up payments channels with a single PSP.

Our Contributions

We study the single channel case, in which the PSP is called to decide which transactions to accept to maximize its profit on a single channel given capital constraints. First, we show there is no randomized online algorithm against any adaptive online adversary. Then, we consider algorithms against oblivious adversaries; we show that there is neither a competitive deterministic nor a competitive randomized algorithm. We derive our results for the randomized case from analysing deterministic algorithms with advice. Next, we consider resource augmentation [15], an approach that relaxes the capital constraints to achieve better competitiveness. However, we prove there is no competitive deterministic algorithm, even with double the capital. Furthermore, we approach the problem as a minimization problem, where we want to minimize the number of rejected transactions. Similarly to the maximization problem, we prove there is no competitive randomized algorithm against oblivious adversaries.

2   The Model

In this section, we begin with a brief introduction on the payment channel operation and the current payment network design rationale. Then, we present the graph theoretic model for the payment network design problem, as addressed in this paper. Later, we define the problem variants and lastly, we present the different adversarial models we consider in this work.

2.1 Payment Channels and Networks

A payment channel is a construction that is established between two participants of a blockchain system and allows them to interact off-chain while maintaining the security guarantees of the blockchain. Thus, a payment network
enables the exchange of capital without committing every transaction on the
blockchain, and therefore addresses successfully the scalability problem of the
underline blockchain.

To establish a payment channel, the participants publish a funding transaction, which is in essence an on-chain joint account where the participants of the
channel lock their capital. As long as the funding transaction is securely included
in the blockchain, the channel parties can execute transactions safely off-chain
by updating the state of the channel, i.e., the distribution of capital between the
participants. The new state of the channel is described in an update transaction
which is signed by all the channel parties.

Multiple channels form a *payment channel network*. To enable the transaction
execution between two nodes that are not directly connected in the network,
routing the transaction along a path of directly connected nodes is allowed. In
such a case, we demand atomic execution of the transaction; either all payments
in the path will be executed or none. To guarantee the atomic execution of
transactions, Hashed Timelocked Contracts (HTLCs) [10,24] can be used.

Routing in a payment network alleviates the necessity for direct links and
hence the need for multiple channels (transactions). However, the update transactions in a payment channel are executed in private thus it is impossible to
know the current distribution of the capital on a channel. This in turn leads to
complex routing schemes, because often the chosen route is depleted.

In our setting, we assume that a central authority, a Payment Service
Provider (PSP), can open channels between two nodes and therefore knows how
much capital is on each edge and how it is distributed between the nodes. The
nodes are the PSP's customers. The PSP, as the creator of the network, bares
the costs of opening channels. Further, we assume the PSP acts as a creditor,
and hence locks the necessary capital in the network and then periodically gets
paid by the customers (either in crypto or fiat money). In our model, we assume,
without loss of generality, that the cost for opening (or updating) a channel is
1, which means that the total cost is the total number of funding transactions
and channel updates in the network. Therefore, the fee that the PSP can charge
a customer cannot be greater than 1 per transaction. Otherwise, the customer
would not execute the transaction through the payment network, but would
instead directly use the blockchain.

2.2 Graph Model

We adapt the model of Avarikioti et al. [3]. We define the network as an undi-
rected graph $G = (V, E)$ with a set of edges E and a set of nodes V. A node
$v \in V$ is a participant in the network whereas an edge $e = (u, v) \in E$ is a channel
between two nodes $u, v \in V$. For each channel $(u, v) \in E$, we denote by C_l and

C_r the capital available to node u and node v, respectively. Therefore, every time a transaction is executed through the channel (u, v), the capitals C_l and C_r are updated according to the distribution of the capital in the latest update transaction of the channel. Note, that the total capital of a channel does not change, i.e., the sum of the capital C_l and C_r remains the same. Further, the capital moves on the channel like the balls in a row of an abacus: if u wants to send capital c to v on channel (u, v) then the new distribution of the capital on the edge (u, v), after the off-chain execution of the transaction, will be $C_l - c$ and $C_r + c$, respectively.

We denote by $t = (s, r, v)$ a transaction t that moves capital of value v from sender $s \in V$ to receiver $r \in V$. Furthermore, in the simple case where we examine only a single channel, i.e., the graph has a single edge, we denote by $\langle C_l; C_r \rangle$ the capital distribution on the single edge.

2.3   Problem Variants

In this section, we present the problem variants, as originally introduced in [3].

Problem 1 (General Payment Network Design)
INPUT: *Capital C, profit P, the sequence of n transactions $t_i = (s_i, r_i, v_i)$ with $1 \leq i \leq n$, each containing the sender node s_i, the receiver node r_i and the value v_i of the transaction t_i.*
OUTPUT: *Strategy $S = \{0,1\}^n$, a binary vector where the ith position is 1 if we choose to execute the ith transaction of the input and 0 else. The graph $G(V, E, C_l, C_r)$ is the network we created to execute the chosen transactions, where V is the set of senders and receivers that participate in any transaction, E is the set of channels we open and C_l, C_r the capital on each side of each edge. Each transaction can be routed arbitrarily in G, denoted by $S_e = \{-1, 0, 1\}^n$, for all $e \in E$, i.e., $S_e(i) = 1$ (or -1) if transaction i is routed through edge e from left to right (from right to left, respectively) and $S_e(i) = 0$ if transaction i is not routed through edge e.*

Our goal is to return (if it exists) a strategy S, a graph G and a routing S_e subject to the following constraints:

1 $|S| - |E| \geq P$
2 $\forall e \in E, \forall j \in \{1, 2, \ldots, n\}, -C_l(e) \leq \sum_{i=1}^{j} S_e(i) v_i \leq C_r(e)$
3 $\sum_{e \in E} C_l(e) + C_r(e) + |E| \leq C$

where $|E|$ denotes the cardinality of set E, i.e., the number of opened channels, and $|S|$ denotes the L_1 norm of vector S, i.e., how m any transactions are executed through the payment network.

Problem 2 (Single Channel).
Given a sequence of n transactions $t_i = (s, r, v_i)$, where s and r are the nodes of the single edge e, a capital assignment $C_r(e), C_l(e)$, and a profit P, decide whether there is a strategy S such that $|S| \geq P$ and $\forall j \in [n], -C_l(e) \leq \sum_{i=1}^{j} S(i) v_i \leq C_r(e)$.

Problem 1 is general and, as shown in [3], difficult to tackle even in the offline setting where the future transactions are known apriori. For this purpose, Problem 2 was defined, as a subcase of the general problem that is of interest in our setting. We study this problem in the online setting, where we assume no prior knowledge about the future transactions.

2.4   Adversary Models

In literature [4], there are three types of adversaries concerning online problems: the oblivious adversary, the adaptive online adversary and the adaptive offline adversary. These adversarial models are equally powerful regarding deterministic algorithms but may yield drastically different results when considering randomized algorithms. We define the adversarial models below with respect to our setting.

Definition 3 (Oblivious adversary). *An oblivious adversary provides a sequence of transactions before an online algorithm starts its computations.*

Definition 4 (Adaptive online adversary). *The adaptive online adversary provides the next transaction based on the decision an online algorithm makes (accept or reject the previous transaction) but serves it immediately.*

Definition 5 (Adaptive offline adversary). *The adaptive offline adversary provides the next transaction based on the decision an online algorithm makes (accept or reject the previous transaction) and serves the output at the end such that it acts optimally.*

Note that an adaptive offline adversary (Definition 5) knows the randomness of the online algorithm, in contrast to the adaptive online adversary (Definition 4).

3   Single Channel

In this section, we consider the online version of Problem 2. In other words, we study whether there is a profitable strategy for accepting and rejecting transactions (online) given capital constraints.

To this end, we define the competitive ratio of an algorithm, denoted by c, as the value that upper bounds the ratio between the profit of the optimal offline solution, $OPT(I_x)$, and the profit of the online algorithm, $ALG(I_x)$, for any input sequence I_x, i.e.,

$$\forall x, \ c \geq \frac{OPT(I_x)}{ALG(I_x)}$$

The profit, in both cases, represents the number of accepted transactions.

The rest of the section is structured as follows: First, we discuss the existence of competitive algorithms against adaptive adversaries. Later, we mainly focus on algorithms, deterministic or randomized, against oblivious adversaries. In particular, we show lower bounds on the advice bits for competitive deterministic algorithms with advice. Then, we use the relationship between advice and randomization to discuss the competitiveness of randomized online algorithms. Last, we consider resource augmentation, i.e. we assume an online algorithm has more resources (more capital on both sides) than an optimal offline algorithm.

3.1 Randomized Algorithms Against Adaptive Adversaries

As shown in recent previous work [3], there is no competitive deterministic algorithm against adaptive adversaries. In this work, we extend this result and prove there is no competitive randomized algorithm against adaptive adversaries.

Theorem 1. *There is no competitive randomized algorithm against adaptive online adversaries.*

Proof. Assume that there is a competitive randomized algorithm against any adaptive online adversary. Theorem 2.1 and Theorem 2.2 in [4] state that the existence of a randomized c-competitive algorithm against any adaptive online adversary and the existence of a randomized d-competitive algorithm against any oblivious adversary imply the existence of a cd-competitive randomized algorithm against any adaptive offline adversary. Additionally, this implies that there is a cd-competitive deterministic algorithm. Since a c-competitive algorithm against any adaptive online adversary is c-competitive against any oblivious adversary, the existence of a c-competitive randomized algorithm against any adaptive online adversary implies the existence of a c^2-competitive deterministic algorithm. From Theorem 19 in [3] we know that there are no deterministic online algorithms against adaptive online adversaries which contradicts our assumption and proves that there is no competitive randomized algorithm. □

From here on, we discuss algorithms against oblivious adversaries since there are no algorithms against stronger adversaries.

3.2 Algorithms with Advice

In this subsection, we examine algorithms with advice. We refer to an algorithm as *optimal*, if the competitive ratio is one, hence it performs as well as the optimal offline algorithm. Next, we study optimal deterministic algorithms with advice, i.e. we examine how many advice bits are necessary for an optimal deterministic algorithm. We prove a tight lower bound of $n - 2$ advice bits for optimal deterministic algorithms. Later, we study the relation between the necessary number of advice bits and the competitiveness of an algorithm.

Theorem 2. *There exists an optimal deterministic algorithm with $n - 2$ advice bits.*

Proof. The advice (oracle) gives the strategy for the first $n - 2$ transactions. For the last two transactions, the algorithm proceeds greedily. If the optimal offline algorithm accepts none of the last two transactions, none of them can be accepted. If one or two transactions can be accepted, our algorithm will accept and reject them accordingly such that it is optimal. □

Theorem 3. *There is no optimal deterministic algorithm with less than $n - 2$ advice bits.*

Proof. Assume an algorithm ALG reading at most $n - 2$ advice bits exists. Then, ALG reads no advice bits for $n = 3$. We construct following sequences of transactions with initial capital distribution $\langle 2; 1 \rangle$:

1. $(l, r, 2)$, $(l, r, 1)$, $(l, r, 1)$
2. $(l, r, 2)$, $(r, l, 3)$, $(r, l, 3)$

For the second sequence of transactions, the optimal offline algorithm accepts the first transaction, whereas, for the first sequence, the first transaction is rejected. Since ALG is deterministic, the decision made on the first transaction will always be the same. Consequently, ALG cannot be optimal. □

We notice that the number of advice bits necessary for optimal algorithms is very high. Therefore, we consider, next, algorithms with competitive ratio greater than one, and show a relation between the competitive ratio and the required advice bits of an algorithm.

Theorem 4. *An algorithm with strictly less than $f(n)$ advice bits has a competitive ratio of $\Omega(\frac{n}{f(n)})$.*

Proof. Let the initial capital distribution be $\langle 2^{f(n)}; 1 \rangle$ and $f(n)$ be a function such that $f(n) \in o(n)$. *loop* denotes transactions that move the total capital from left to right until there are n transactions. We define a sequence of transactions as

$$I_x = (l, r, 1), \ldots, (l, r, 2^{f(n)-1}), (r, l, x + 1), loop$$

such that the set of instances is defined as

$$\mathcal{I} = \{I_x \mid x \in \{0, \ldots, 2^{f(n)} - 1\}\}$$

Similar to the proofs before, an algorithm must accept *loop*, otherwise the competitive ratio is $c \geq \frac{n-f(n)}{f(n)+1}$. Assume that there is an algorithm A reading at most $k < f(n)$ advice bits. For instance I_x, A must transfer x capital with the transactions of the prefix to the right side such that it can accept *loop*. The number of strategies the advice can describe is smaller than the number of instances in the constructed input set. Thus, we know that for two different inputs the same strategy is used. We conclude, A cannot accept *loop* for one element in $\mathcal{I}$ and has competitive ratio $c \geq \frac{n-f(n)}{f(n)+1}$. □

3.3 Randomized Algorithms Against Oblivious Adversaries

In this subsection, we study the competitive ratio of randomized algorithms. To that end, we use the results from Sect. 3.2 to provide lower bounds on the competitive ratio of randomized algorithms, since algorithms with advice and randomized algorithms are closely related. According to [5], Yao's Principle [29] bounds the expected competitive ratio of randomized algorithms from below with

$$R \geq max \left(min_i \frac{\mathbb{E}_{y(j)}[OPT(\sigma_j)]}{\mathbb{E}_{y(j)}[ALG_i(\sigma_j)]}, min_i \frac{1}{\mathbb{E}_{y(j)}[\frac{ALG_i(\sigma_j)}{OPT(\sigma_j)}]} \right)$$

where both arguments in the maximum function are proven lower bounds. $\mathbb{E}[X]$ is the expectation of a random variable X, OPT is the optimal offline algorithm and ALG_i is the optimal strategy for the input sequence σ_i. Moreover, $y(j)$ denotes the distribution of j and σ_j is an input sequence for all j. We denote by c the expected competitive ratio and by I_x the input sequence. We change the other variables accordingly.

Theorem 5. *Every randomized algorithm is* $\Omega(\frac{n}{f(n)+\frac{n}{2^{f(n)}}})$*-competitive.*

Proof. Let $\langle 2^{f(n)}; 1 \rangle$ be the initial capital distribution and

$$\mathcal{I} = \{I_x \mid x \in \{0, \ldots, 2^{f(n)} - 1\}\}$$

the set of sequences of transactions for $f(n) \in o(n)$ where

$$I_x = p, (r, l, x + 1), loop$$

and $p = (l, r, 1), (l, r, 2), \ldots, (l, r, 2^{f(n)-1})$. $loop$ is the movement of the capital back and forth. Then, we define the set of strategies as $\mathcal{A} = \{A_x \mid x \in \{0, \ldots, 2^{f(n)} - 1\}\}$. A non-optimal strategy accepts at most $f(n)$ transactions, whereas the optimal strategy accepts at most n and at least $n - f(n)$ transactions. We will refer to this in the following calculation as $(*)$.

$$c \geq min_i \frac{1}{\mathbb{E}_{y(j)}[\frac{A_i(I_j)}{OPT(I_j)}]} \qquad \text{Yao's principle}$$

$$= \frac{1}{max_i \mathbb{E}_{y(j)}[\frac{A_i(I_j)}{OPT(I_j)}]}$$

$$\geq \frac{1}{max_i \mathbb{E}_{y(j)}[\frac{A_i(I_j)}{n-f(n)}]} \qquad OPT(I_j) \geq n - f(n)$$

$$= \frac{n - f(n)}{max_i \mathbb{E}_{y(j)}[A_i(I_j)]}$$

$$\geq \frac{n - f(n)}{\sum_{k=0, k\neq i}^{2^{f(n)}-1} \left(\frac{1}{2^{f(n)}} (f(n) + 1) \right) + \frac{n}{2^{f(n)}}} \qquad y(j) \text{ uniform and } (*)$$

$$= \frac{n - f(n)}{(2^{f(n)} - 1)\left(\frac{f(n)+1}{2^{f(n)}}\right) + \frac{n}{2^{f(n)}}}$$

$$= \frac{n - f(n)}{f(n) + 1 - \frac{f(n)+1}{2^{f(n)}} + \frac{n}{2^{f(n)}}}$$

$$\geq \frac{n - f(n)}{2\left(f(n) + \frac{n}{2^{f(n)}}\right)}$$

Thus, we have shown that $c \in \Omega\left(\frac{n}{f(n) + \frac{n}{2^{f(n)}}}\right)$. $\qquad\square$

Corollary 1. *There is no competitive randomized algorithm against oblivious adversaries.*

Proof. Follows immediately from Theorem 5 for $f(n) = o(n)$. $\qquad\square$

3.4 Resource Augmentation

In this subsection, we approach the problem from another perspective, and consider a different analysis approach called resource augmentation as described in [18]. In this approach, an online algorithm may have more resources than the optimal offline algorithm in order to improve the performance of an algorithm against an adversary. In our setting, we allow the online algorithm to have $h \geq 1$ times more capital on both sides. An online algorithm starts with $\langle hC_l; hC_r \rangle$ if the optimal offline algorithm starts with the initial capital distribution $\langle C_l; C_r \rangle$. In this section, we discuss the existence of deterministic algorithms for different values of h.

Theorem 6. *There is no competitive deterministic algorithm for $h < 2$.*

Proof. Let the initial capital distribution for the online algorithm be $\langle hC; 0 \rangle = \langle 2C - \epsilon; 0 \rangle$ such that $h = 2 - \frac{\epsilon}{C}$. Then, we define the sequences

$$\left(l, r, C - \frac{\epsilon}{2}\right), (r, l, 1), (r, l, 1), \ldots, (r, l, 1)$$

$$\left(l, r, C - \frac{\epsilon}{2}\right), (l, r, C), (r, l, C), \ldots, (l, r, C)$$

Assume there is a competitive algorithm that rejects the first transaction. Then, none of the consecutive transactions with low value (first sequence) can be accepted. This contradicts the assumption and implies that if a competitive deterministic algorithm exists, the first transaction must be accepted. Suppose there is a competitive algorithm that accepts the first transaction. Then, the capital distribution after accepting the first transaction is $\langle C - \frac{\epsilon}{2}; C - \frac{\epsilon}{2} \rangle$. Then, the following transactions in the second sequence cannot be accepted, since the movement of C back and forth is not possible anymore. This contradicts the assumption that there is a competitive deterministic algorithm that accepts the first transaction. We conclude, that there is no deterministic algorithm for $h = 2 - \frac{\epsilon}{C}$. For $C = n$, $C = 2^n$, or even bigger C, h goes to 2 from below. Thus, there is no competitive deterministic online algorithm for any $h < 2$. $\qquad\square$

3.5 Minimizing the Number of Rejected Transactions

So far, we defined the competitiveness as the ratio between the number of accepted transactions of the optimal offline algorithm and an online algorithm, thus as an online maximisation problem. In this section, we examine the problem from another point of view; defining a minimization problem. To this end, we define the competitive ratio of an algorithm, denoted by c, as the value that upper bounds the ratio between the cost of the online algorithm, $ALG(I_x)$, and the cost of the optimal offline solution, $OPT(I_x)$, for any input sequence I_x, i.e.,

$$\forall x, \ c \geq \frac{ALG(I_x)}{OPT(I_x)}$$

The cost, in both cases, represents the number of rejected transactions.

Similarly to Sects. 3.2 and 3.3, we first lower bound the competitive ratio for a given upper bound on the advice bits and then use this result to show there is no competitive randomized algorithm against oblivious adversaries.

Theorem 7. *An algorithm with strictly less than $f(n)$ advice bits has a competitive ratio of $\Omega(\frac{n}{f(n)})$.*

Proof. Let $\langle 2^{f(n)}; 1 \rangle$ be the initial capital distribution. We define

$$\mathcal{I} = \{I_x \mid x \in \{0, \ldots, 2^{f(n)} - 1\}\}$$

where

$$I_x = (l, r, 1), (l, r, 2), \ldots (l, r, 2^{f(n)-1}), (r, l, x+1), loop$$

for $f(n) \in o(n)$ and *loop* being a sequence of transactions moving all the capital from left to right back and forth. An algorithm that does not accept *loop* has a competitive ratio of at least $c \geq \frac{n-f(n)-1}{f(n)}$. Assume there is an algorithm reading $k < f(n)$ advice bits with a better competitive ratio. Since there are $2^{f(n)}$ different instances in $\mathcal{I}$ and fewer strategies are expressible by the advice, one strategy is used for two elements of the input set. In the previous section, we showed that transferring i is the only strategy accepting *loop* for an instance I_i. Then, for I_j where $j \neq i$ the strategy that transfers i with the transactions of the prefix cannot accept *loop*. As a result, the competitive ratio is at least $c \geq \frac{n-f(n)-1}{f(n)}$ which contradicts the assumption that there is an algorithm with a better competitive ratio reading $k < f(n)$ advice bits. $\square$

As stated in [5], Yao's Principle [29] bounds the expected competitive ratio for minimisation problems from below with

$$\bar{R} \geq max \left(min_i \frac{\mathbb{E}_{y(j)}[ALG_i(\sigma_j)]}{\mathbb{E}_{y(j)}[OPT(\sigma_j)]}, min_i \mathbb{E}_{y(j)} \left[\frac{ALG_i(\sigma_j)}{OPT(\sigma_j)} \right] \right)$$

where both arguments in the maximum function are proven lower bounds. The notation is the same as in Sect. 3.3.

Theorem 8. *Every randomised algorithm (minimising the number of rejected transactions) is $\Omega(\frac{n}{f(n)})$-competitive.*

Proof. We construct the same set of instances as in the proof of Theorem 7 with the same initial capital distribution. Accordingly, we define the set of strategies to be $\mathcal{A} = \{A_x \mid x \in \{0, \dots, 2^{f(n)} - 1\}\}$ where A_x is the optimal strategy for I_x. As discussed in previous proofs, choosing a non-optimal strategy for an input results in the rejection of at least all of the $n - f(n) - 1$ transactions in *loop*. We know that the optimal offline algorithm rejects at most $f(n)$ and at least no transactions.

$$
\begin{aligned}
c &\geq \min_i \frac{\mathbb{E}_{y(j)}[A_i(I_j)]}{\mathbb{E}_{y(j)}[OPT(I_j)]} && \text{Yao's Principle} \\
&\geq \min_i \frac{\mathbb{E}_{y(j)}[A_i(I_j)]}{f(n)} \\
&= \min_i \frac{\sum_{j=0, j\neq i}^{2^{f(n)}-1}(A_i(I_j)) + OPT(I_j)}{2^{f(n)} f(n)} && y(j) \text{ uniform} \\
&\geq \frac{(2^{f(n)} - 1)(n - f(n) - 1)}{2^{f(n)} f(n)} \\
&= \frac{\left(1 - \frac{1}{2^{f(n)}}\right)(n - f(n) - 1)}{f(n)} \\
&= \left(\frac{n}{f(n)} - 1 - \frac{1}{f(n)}\right)\left(1 - \frac{1}{2^{f(n)}}\right) \in \Omega\left(\frac{n}{f(n)}\right)
\end{aligned}
$$

$\square$

Corollary 2. *There is no competitive randomised algorithm (that minimises the number of rejected transactions) against oblivious adversaries.*

Proof. Follows from Theorem 8 for $f(n) = 1$. $\square$

4   Related Work

This work builds on the definitions and results of Avarikioti et al. [3], where a framework to approach the design of payment channel networks from an algorithmic perspective was originally introduced. For a single channel, they showed that maximising the profit with given capital assignments is NP-hard and presented a fully polynomial time approximation scheme in the offline setting, i.e. when the sequence of future transactions is known upfront. Moreover, they studied the online case, where no prior information is known about the future transactions. In particular, they showed that there is no competitive (deterministic) online algorithm and presented an $O(\log(C))$-competitive algorithm that constructs a payment hub that accepts all transactions. In this paper, we extend their work for the online setting and consider randomized algorithms, algorithms with advice and resource augmentation algorithms.

Additionally, the design of payment networks with fees from the viewpoint of a payment service provider who wants to maximise the profit is discussed in [1]. In the contrary to this work where we assume constant fee for every transaction, in [1], each channel requires a different fee, much like the tolls on a road network. Despite the different assumptions, both works share the same objective, to maximize the profit for the network operator (and designer).

Payment channels were originally introduced by Spilman [27] as a solution for the limited transaction throughput of Bitcoin [22]. Spilman channels allowed two parties to transact off-chain as long as the direction of the capital movement was only in one direction. Later, bidirectional channels were introduced simultaneously by Poon et al. [24] and by Decker and Wattenhofer [10]. Many recent constructions of payment or state channels have been proposed addressing different aspects and needs (e.g. state channels handle smart contracts) of various cryptocurrencies [2,7,9,20]. Although these works propose different constructions for payment (or state) channels, they all result in a decentralized payment network and thus require complex routing algorithms and high capital availability on behalf of the users of the network. Therefore, the results of this work apply to all these proposed payment channel solutions.

Multiple works exist that focus on the routing problem of payment networks. Prihodko et al. propose Flare [25], a proposal for path discovery by gathering information about the Lightning network topology. However, Flare raised privacy concerns which were later addressed by SilentWhispers [21] and SpeedyMurmurs [26]. In contrast with these works, we assume a central authority, a payment service provider, that designs the network, and thus has complete knowledge on the network structure and capital capacity of each channel. Our objective is to design the optimal network structure to maximize the profit for the service provider.

On a different direction, Dziembowski et al. proposed Perun [11], a virtual channel hub that allows the users connected to the hub to directly interact off-chain via virtual channels establish through the virtual hub. In the same line of work, Heilman et al. presented Tumblebit [14], a payment channel hub compatible with Bitcoin that guarantees anonymity and security even though the hub is an untrusted intermediary. Similarly, Green and Miers presented Bolt [12], another channel construction that requires smart contracts but offers stronger privacy and security guarantees. Moreover, Khalil et al. introduced Nocust [16] whereas Poon and Buterin introduced Plasma [23], which are layer-2 commit-chains, i.e., off-chain payment hubs. Despite the fact that all these work also assume a central coordinator that enables the off-chain payments through a centralized network, they mainly focus on the construction of the payment hub. In contrast, we focus on the algorithmic perspective of the problem, and address more primitive questions: how can the central coordinator profit the most, and which transactions should he facilitate through the network given a capital constraint.

5   Conclusions

We studied the online problem of a single channel, where a Payment Service Provider (PSP) is called to decide which transactions to accept or reject. The objective is to maximize the number or accepted transactions and thus maximize the PSP's profit.

We showed there is no randomized online algorithm against any adaptive online adversary. Furthermore, we considered deterministic algorithms with advice and proved that there is no competitive randomized algorithm against an oblivious adversary. In addition, we examined resource augmentation, and showed that even with twice as much capital there is no competitive deterministic algorithm against an oblivious adversary. Finally, we considered the complementary minimization problem - minimizing the number of rejected transactions - and similarly proved there is no competitive randomized algorithm against oblivious adversaries.

Given that the single channel case is merely a simple sub-case of the general problem and the flurry of negative results we presented in this work, we conclude that the online channel design is a demanding problem with interesting future work.

References

1. Avarikioti, G., Janssen, G., Wang, Y., Wattenhofer, R.: Payment network design with fees. In: Garcia-Alfaro, J., Herrera-Joancomartí, J., Livraga, G., Rios, R. (eds.) DPM/CBT 2018. LNCS, vol. 11025, pp. 76–84. Springer, Cham (2018). https://doi.org/10.1007/978-3-030-00305-0_6
2. Avarikioti, G., Kogias, E.K., Wattenhofer, R.: Brick: asynchronous state channels. arXiv preprint: arXiv:1905.11360 (2019)
3. Avarikioti, G., Wang, Y., Wattenhofer, R.: Algorithmic channel design. In: 29th International Symposium on Algorithms and Computation (ISAAC), Jiaoxi, Yilan County, Taiwan, December 2018
4. Ben-David, S., Borodin, A., Karp, R., Tardos, G., Wigderson, A.: On the power of randomization in online algorithms. In: Algorithmica, pp. 379–386 (1990)
5. Borodin, A., El-Yaniv, R.: Online Computation and Competitive Analysis. Cambridge University Press, New York (1998)
6. Buterin, V.: Ethereum: a next-generation smart contract and decentralized application platform (2013). https://github.com/ethereum/wiki/wiki/White-Paper
7. Coleman, J., Horne, L., Xuanji, L.: Counterfactual: generalized state channels (2018)
8. Croman, K., et al.: On scaling decentralized blockchains. In: Clark, J., Meiklejohn, S., Ryan, P.Y.A., Wallach, D., Brenner, M., Rohloff, K. (eds.) FC 2016. LNCS, vol. 9604, pp. 106–125. Springer, Heidelberg (2016). https://doi.org/10.1007/978-3-662-53357-4_8
9. Decker, C., Russell, R., Osuntokun, O.: eltoo: a simple Layer2 protocol for Bitcoin (2018)
10. Decker, C., Wattenhofer, R.: A fast and scalable payment network with Bitcoin duplex micropayment channels. In: Pelc, A., Schwarzmann, A.A. (eds.) SSS 2015. LNCS, vol. 9212, pp. 3–18. Springer, Cham (2015). https://doi.org/10.1007/978-3-319-21741-3_1

11. Dziembowski, S., Eckey, L., Faust, S., Malinowski, D.: PERUN: virtual payment channels over cryptographic currencies. IACR Cryptology ePrint Archive 2017, 635 (2017)
12. Green, M., Miers, I.: Bolt: anonymous payment channels for decentralized currencies, October 2017
13. Gudgeon, L., Moreno-Sanchez, P., Roos, S., McCorry, P., Gervais, A.: SoK: off the chain transactions. IACR Cryptology ePrint Archive 2019, 360 (2019). https://eprint.iacr.org/2019/360
14. Heilman, E., Alshenibr, L., Baldimtsi, F., Scafuro, A., Goldberg, S.: TumbleBit: an untrusted Bitcoin-compatible anonymous payment hub. In: Network and Distributed System Security Symposium (2017)
15. Kalyanasundaram, B., Pruhs, K.: Speed is as powerful as clairvoyance. J. ACM **47**(4), 617–643 (2000)
16. Khalil, R., Gervais, A., Felley, G.: Nocust - a securely scalable commit-chain. Cryptology ePrint Archive, Report 2018/642 (2018). https://eprint.iacr.org/2018/642
17. Kokoris-Kogias, E., Jovanovic, P., Gasser, L., Gailly, N., Syta, E., Ford, B.: OmniLedger: a secure, scale-out, decentralized ledger via sharding. In: 2018 IEEE Symposium on Security and Privacy (SP), pp. 583–598. IEEE (2018)
18. Komm, D.: An Introduction to Online Computation: Determinism, Randomization, Advice, 1st edn. Springer, Cham (2016). https://doi.org/10.1007/978-3-319-42749-2
19. Luu, L., Narayanan, V., Zheng, C., Baweja, K., Gilbert, S., Saxena, P.: A secure sharding protocol for open blockchains. In: ACM Conference on Computer and Communications Security (2016)
20. Miller, A., Bentov, I., Kumaresan, R., Cordi, C., McCorry, P.: Sprites and state channels: payment networks that go faster than lightning. arXiv preprint arXiv:1702.05812 (2017)
21. Moreno-Sanchez, P., Kate, A., Maffei, M.: SilentWhispers: enforcing security and privacy in decentralized credit networks (2017)
22. Nakamoto, S.: Bitcoin: a peer-to-peer electronic cash system (2018). http://bitcoin.org/bitcoin.pdf
23. Poon, J., Buterin, V.: Plasma: scalable autonomous smart contracts (2017)
24. Poon, J., Dryja, T.: The Bitcoin lightning network: scalable off-chain instant payments (2015). https://lightning.network
25. Prihodko, P., Zhigulin, S., Sahno, M., Ostrovskiy, A., Osuntokun, O.: Flare: an approach to routing in lightning network white paper (2016)
26. Roos, S., Moreno-Sanchez, P., Kate, A., Goldberg, I.: Settling payments fast and private: efficient decentralized routing for path-based transactions. arXiv preprint arXiv:1709.05748 (2017)
27. Spilman, J.: Anti DoS for TX replacement. https://lists.linuxfoundation.org/pipermail/bitcoin-dev/2013-April/002433.html. Accessed 17 Apr 2019
28. Visa Inc.: Fact sheet - visa (2018). https://usa.visa.com/dam/VCOM/download/corporate/media/visanet-technology/aboutvisafactsheet.pdf. Accessed 10 Apr 2019
29. Yao, A.C.C.: Probabilistic computations: toward a unified measure of complexity. In: 18th Annual Symposium on Foundations of Computer Science (SFCS 1977), pp. 222–227. IEEE (1977)

A Multi-protocol Payment System to Facilitate Financial Inclusion

Kazım Rıfat Özyılmaz[1]([✉])(iD), Nazmi Berhan Kongel[2], Ali Erhat Nalbant[2],
and Ahmet Özcan[2]

[1] Bogazici University, Bebek, 34342 İstanbul, Turkey
kazim@monolytic.com
[2] Arf Labs OÜ, 11412 Tallinn, Estonia
{berhan,alierhat,ahmet}@arf.one
https://www.arf.one

Abstract. In 2017, there were already 250 million unbanked people with a smartphone and internet access out of a total 1,7 billion, while only in 2018, the volume of migrant remittance inflows for low-and middle-income countries was $529 billion. Due to financial crisis leading to devaluation and hyperinflation, 12 countries (other than the USA) with 40,5 million population already adopted the U.S. dollar as their native currency. Alternatively, Bitcoin usage soared among people in developing countries who are looking for dependable ways to store and transfer value. Bitcoin, with its estimated 25 million global users and ten thousand globally distributed nodes, is by far the most widely used, best understood and battle-tested cryptocurrency in the world to reach this financially excluded demographic.

Our goal in this paper is to demonstrate a technical solution that utilizes Bitcoin by making it (a) instantly transferable in a user-friendly way, (b) with minimal (preferably zero) transaction fees and (c) without sacrificing users' security (i.e. by not controlling their private keys).

In our design, instant transactions are enabled by using 2-of-2 multisignature wallet addresses (split between the user and the custodial) and transaction fees are minimized by aggregating unspent transaction outputs (UTXOs). On top of these two pure Bitcoin-based improvements, a fast and low-cost Bitcoin sending mechanism is proposed by creating a "Bitcoin anchor" on Stellar platform, effectively using Stellar as a "Layer 2" solution. Lastly, mapping Bitcoin addresses to user-friendly identifiers (such as email addresses) has been evaluated to increase adoption. By combining these new techniques, we aim to overcome the limitations of Bitcoin and related solutions like off-chain (Lightning Networks) or federated side-chain (Blockstream Liquid) systems.

Keywords: Bitcoin · Multisignature · UTXO aggregation ·
Fee reduction · Stellar · Stellar anchor · Stellar asset

© Springer Nature Switzerland AG 2019
C. Pérez-Solà et al. (Eds.): DPM 2019/CBT 2019, LNCS 11737, pp. 321–335, 2019.
https://doi.org/10.1007/978-3-030-31500-9_21

1   Introduction

Today, more than 1.7 billion people around the world do not have access to basic financial services like sending and receiving money or getting loans from safe institutions. Nearly half of these unbanked people live in developing economies like China (225 million), India (190 million), Pakistan (100 million) and Indonesia (95 million) [34]. Unfortunately, the number of unbanked and underbanked households are not only significant in developing countries but also in developed ones. For example, based on research conducted in 2017, Federal Deposit Insurance Corporation (FDIC) stated that 18.7% of U.S. households (24.2 million) is underbanked, meaning that the household had a checking or savings account but at the same time, obtained financial products and services outside of the banking system [17].

On the other hand, there are billions of people continuously transacting on the internet, who need instant, affordable, easy-to-use and global financial services that match the speed of their online interactions. We are living in a world where Twitch streamers and Instagram shop owners basically "operate" globally. With instant purchase links, subscriptions, donations and tips, a new kind of borderless economy is growing rapidly. For example, Streamlabs, a monetization solution for Twitch with 70% market adoption, paid $400 million in total and $141 million in the last quarter of 2018 to its users just for the tips they received [33].

Finally, for both of these groups the cost of remittance plays a vital role. According to Remittance Prices Report (Q1 2019) compiled by Worldbank, the global average remittance commission is 6.294% where it is 6.66% for G8 countries and 7.07% for G20 countries. It is important to state that it would be possible to save up to $16 billion a year by cutting 5 % points in remittance commissions [36].

The term "financial inclusion", which describes the scope of these improvable activities, is defined by Worldbank as: "... (it) means that individuals and businesses have access to useful and affordable financial products and services that meet their needs - transactions, payments, savings, credit and insurance - delivered in a responsible and sustainable way" [35]. These two aforementioned groups of people, albeit very different in terms of income and geographical location, have similar needs with different goals when it comes to financial transactions.

Once the need is established, it would be possible to define the "means" for a global, affordable and inclusive solution. Bitcoin, with its estimated 25 million holders and ten thousand globally distributed nodes, established itself as the go-to cryptocurrency brand for anyone looking to use an alternative financial solution [6,11,30]. Unfortunately, Bitcoin has certain characteristics (or design choices) that makes it partly unfeasible for certain real-life use cases. For example, instant transactions are impossible due to the Proof-of-Work (PoW) limitations and lack of consensus finality. Similarly, Bitcoin's game-theoretic approach to transaction fees does not play well when the network is under pressure. In those cases, due to high demand in Bitcoin transactions, transaction fees become highly volatile [24]. Last but not least, transaction throughput of Bitcoin, measured in transactions per second (tps), is yet another important limiting factor, where the Bitcoin blockchain's 7 transactions per second today is dwarfed by the VISA network's 150 million transactions every day (1.736 transactions per second) and potentially 24.000 transactions per second at peak [37].

In recent years, many blockchain protocols that claim high transaction throughput rates have emerged. Some of those coins even managed to accumulate significant trade volume. However, these high-throughput protocols are still small in terms of trading volume (biggest one is less than one tenth) compared to Bitcoin [15]. Besides adoption, decentralization is another important aspect that should be considered when designing a financial inclusion solution. For example, Ripple [14] is highly involved with banks and it has features like "freezing funds" that can be induced by gateways in the Ripple system. Similarly, EOS's Delegated Proof-of-Stake (dPoS) [16] and NEO's Delegated Byzantine Fault Tolerant (dBFT) [25] use only a small set of validators, forcing the whole network to trust a limited number of nodes. Another example may be Facebook's new project Libra [20], which promises high adoption globally. However, Libra is a permissioned system and a significant amount of money should be paid in order to become a validator. In short, based on adoption, decentralization and brand recognition, we believe Bitcoin should be the central infrastructure to address the mainstream users, despite its shortcomings.

Bitcoin ecosystem responded back to those emerging protocols with major infrastructure projects, known as "Layer 2". Projects like Lightning Network [28], offering instant transactions on so-called off-chain payment channels, or Blockstream Liquid [2], facilitating an inter-exchange settlement network started to get recognized.

In this paper, we propose a Bitcoin and Stellar based technical solution involving a platform (third party) that (a) makes instant transfers possible, (b) eliminates friction for users in terms of ease of onboarding and operation, (c) operates with minimal (preferably zero) transaction fees and (d) does not manage its users' private keys, i.e. being non-custodial. Similar to Lightning Network and Liquid, the proposed solution will interact with Bitcoin but introduce a chain of operations to ensure instant and minimal-fee Bitcoin transfers without compromising users' security.

The organization of this paper is as follows: in the next section "Design" (Sect. 2), an overview of the core concepts involving the usage of Bitcoin and Stellar are presented. In "Implementation" (Sect. 3), the technical details of the implementation are provided. Then, in "Competition" (Sect. 4), current solutions that work together with Bitcoin and addressing the requirements of financial inclusion are evaluated. Finally, a brief summary of the paper is provided in "Conclusion" (Sect. 5).

2 Design

In order to realize the proposed features (instant and cheap transactions), a three step model is used to describe the flow of any given transaction: *Request,*

Credit and *Settlement*. A transaction starts with the user triggering a transfer to another party (Request), continues with the instant action taken by the platform that confirms the payment (Credit) and ends with the final on-chain action (Settlement). In the proposed system, *Request* is done using multisignature Bitcoin or regular Stellar addresses, *Credit* is realized by issuing and transferring assets on Stellar network and (optimized) *Settlement* is executed on Bitcoin network with UTXO aggregation. The platform in the proposed solution will be:

1. the counter-party in a 2-of-2 multisignature Bitcoin address when a user creates an account,
2. the enabler of instant Bitcoin transactions by being the guarantor for the receiving party once the sender signed the initial transaction,
3. the aggregator of partially-signed unspent transaction outputs (UTXOs) and merge them into cheaper on-chain Bitcoin transactions in terms of fees (satoshis per byte),
4. the address book generator that maps the email addresses to Bitcoin and Stellar addresses and notify users.

With the aforementioned Bitcoin-based improvements, once the user is in the proposed system, the transactions will be instantly confirmed and cheaper due to UTXO aggregation. To be specific, partially-signed UTXOs may be merged together if certain properties of Bitcoin script language is used at signing. As a result, one-way, user-to-merchant transactions will be significantly affordable. However, there is also the real-life scenario of opening up a "credit channel" between the user and the merchant, or between two users, where consecutive transactions take place without final settlement. In order to address this aspect, the platform utilizes another public blockchain, namely *Stellar*. For these special peer-to-peer Bitcoin transactions, the platform:

1. will create an *Anchor* on Stellar public blockchain and issue a Bitcoin-pegged asset, that is called *ABTC* (as in "Anchor Bitcoin"). Quoting Stellar Guide [31]: "Anchors are entities that people trust to hold their deposits and issue credits into the Stellar network for those deposits. All money transactions in the Stellar network (except lumens) occur in the form of credit issued by anchors, so anchors act as a bridge between existing currencies and the Stellar network",
2. will convert any BTC transfer to an ABTC transfer on public Stellar blockchain, effectively rendering it off-chain on Bitcoin blockchain but on-chain on Stellar blockchain,
3. will credit the user with ABTC, once a peer-to-peer, off-chain Bitcoin transaction is triggered,
4. will not interfere with any ABTC transfer until the user requested to convert it back to BTC,
5. should convert ABTC (Stellar blockchain) to BTC (Bitcoin blockchain) and vice versa.

In essence, a stablecoin-like, Bitcoin-pegged asset is created on Stellar blockchain, to effectively use Stellar as a "Layer 2" network. Zamyatin et al.

proposed a method to create Bitcoin-pegged assets on Ethereum blockchain with similar goals [38]. The reason to select BTC instead of XLM, USDT or DAI is as follows: Although transaction fees and confirmation times are lower on Stellar network, the stability of the exchange rate and the liquidity of XLM are low compared to BTC throughout the world. USDT is not selected due to changing collateral structure and possible legal troubles ahead. DAI, on the other hand, forces high collateralization which creates a major disadvantage in this case.

Finally, the most important design choice in this paper is the addition of a third party to the system, namely the platform, which will raise questions about "trust". It is important to understand that:

- the platform does not control the users' private keys and is unable to create any transaction by itself that is not signed by the user first, i.e. non-custodial,
- users will not experience any loss of funds in case either the platform or the users' system got hacked,
- if users lose their private keys, they will be unable to recover their funds,
- there are no operational risks for users like in Lightning Networks, namely, possible loss of funds in case of getting offline or a crashed hard drive,
- the platform utilizes another public blockchain, namely Stellar, to ensure transparent, peer-to-peer transfer.

3   Implementation

In this section, the choices and the flows of the proposed design will be explained. The mechanics of the proposal will be inspected in terms of features like: instant Bitcoin transactions (Request step), off-chain transactions (Credit step), minimal Bitcoin transaction fees (Settlement step), user experience and complexity.

"Instant Bitcoin Transactions" (Sect. 3.1) section will cover the implementations on top of Bitcoin to specifically address its shortcomings. "Off-chain Peer-to-Peer Transactions" (Sect. 3.2) details a flow, where public Stellar blockchain is used like a "Layer 2" solution via issuing Bitcoin-pegged assets. "Minimal Bitcoin Transaction Fees" (Sect. 3.3) section details the UTXO aggregation process and presents the improvements.

3.1   Instant Bitcoin Transactions (Request)

In order to enable instant transactions with Bitcoin, an off-chain mechanism should be introduced to finalize the transaction without committing final state to Bitcoin chain. In the proposed design, users will create an account on the platform which is presented as a wallet application. During that process, a 2-of-2 multisignature Native Pay-to-Witness-Script-Hash (P2WSH) address is created using the public keys of the user and the platform. After that point, users may deposit to or withdraw from that specific multisignature address. This mechanism is similar to Lightning Network's *Funding Transaction* to open payment channels [28,39], or Green Address wallet creation [18]. Once the multisignature address is successfully funded by the user, they may spend their Bitcoin (e.g. create transactions) via signing their UTXOs and sending it to the platform for the final signature.

The whole account creation and spending process will work as follows:

1. user will create a random seed in the wallet application and the first private and public key is created using the "BIP-32: Hierarchical Deterministic Wallets" method [3]. Both seed and keys will never leave the (mobile) application,
2. user will send its public key to the platform,
3. platform will create a 2-of-2 multisignature Native Pay-to-Witness-Script-Hash (P2WSH) address using the users and its own public key and share that address with the user,
4. user will fund that address with Bitcoin,
5. user will query their Bitcoin balance and UTXOs,
6. user will create a raw Bitcoin transaction using the required amount of UTXOs as inputs and receiver addresses and amounts as outputs,
7. user will sign the raw transaction with signature hash type *(SIGHASH_NONE |SIGHASH_ANYONECANPAY)* (where single input is signed and all the other inputs and outputs are modifiable),
8. user will send the partially-signed raw transaction to the platform to be signed and sent to the Bitcoin network,
9. platform will receive the partially-signed raw transaction, verify it and queue it for aggregation (which will be covered in the next Subsect. 3.3),
10. platform will signal the receiving party (merchant or user of the platform) instantly about payment completion and credit that user in the system in an off-chain way,
11. platform will finalize the aggregated transaction, add the required transaction fee based on network conditions, sign it with *SIGHASH_ALL* and broadcast it to the network.

As seen in the flow above, the platform has the capability to signal the completion of payment to the receiver, once the sender has signed the initial transaction. However, besides all the improvements, the proposed system introduces two disadvantages. The first one is, due to the use of multisignature addresses the transaction sizes are bigger than the regular Bitcoin transactions. Roughly, a single signature is 70 bytes and a compressed public key in hexadecimal format is 33 bytes, so every additional signature (which is one in our case) adds up 100 bytes to the transaction [8]. The second disadvantage is about internal risk. The platform notifies the receiver about payment completion however that state may not be reflected on-chain at that instant. Basically, the platform is carrying this internal risk until the settlement is complete.

Luckily, both of these disadvantages can either be eliminated or significantly reduced. Bitcoin is on the verge of adopting *Schnorr signatures*, that will reduce the multisignature size overhead drastically [22]. Instead of storing all the signatures for every required party separately, Schnorr signature scheme makes it possible to use the space for just one signature, independent of the number of required signatures.

About the second disadvantage, it would be possible for the platform to manage its internal risk by sending transactions more frequently. The platform may utilize two metrics: *"total accumulated Bitcoin size in pending transactions"* and *"passed time since the last sent transaction"* to dynamically reduce its risk. In addition, if Bitcoin is sent as a peer-to-peer, off-chain transaction (which we will cover in Sect. 3.2), the Bitcoin amount is already converted to ABTC (Anchor Bitcoin) and tracked on public Stellar blockchain.

3.2 Off-Chain Peer-to-Peer Transactions (Credit)

This section will cover an off-chain method in terms of Bitcoin blockchain, however, the transaction will actually be executed on public Stellar blockchain via a Bitcoin-pegged asset. In its simplest form, the flow will be in parallel to the crediting step (item 10.) in the previous subsection (Sect. 3.1) and work as follows:

1. a Stellar address will be created for the user at wallet initialization and it is shared with the platform
2. a *trustline* between the platform and user will be built in order to send and receive ABTC (Anchor Bitcoin). In Stellar, a trustline is an explicit statement on the peer's part that shows trust to a token. Every account must make a "trustline" to the ABTC asset before owning any [32]
3. user will start an off-chain, peer-to-peer Bitcoin transaction
4. the platform will get the partially-signed UTXO and verify its signature
5. the platform will send ABTC (Anchor Bitcoin) and minimal amount of XLM to the user to cover transaction fees on Stellar blockchain
6. user will sign and send the ABTC (Anchor Bitcoin) to the other party on Stellar
7. receiving party may execute another peer-to-peer transfer or may claim Bitcoin by sending ABTC (Anchor Bitcoin) to the platform's wallet address on Stellar (ABTC issuer address)

Due to the fact that, Stellar transaction fees are fixed to 0.00001 XLM, it is very cheap to send transactions considering XLM was $0.85 USD at its highest.

On a side note, using a public blockchain as a "Layer 2" mechanism also gives the users the flexibility to use atomic swaps. Outside of the proposed protocol, the platform may give users the capability the directly receive ABTC on Stellar for deposits made on Bitcoin. From the user's perspective, it is possible to swap BTC (Bitcoin) with ABTC (Stellar) using the platforms wallet addresses and vice versa. This way, a parallel flow for off-chain and peer-to-peer transactions may be created. Experimental software for such swaps is already in prototype stage [19].

3.3   Minimal Bitcoin Transaction Fees (Settlement)

Bitcoin transaction fee is a game-theoretic construct that is measured in *satoshis per byte* and fluctuates depending on the congestion of the Bitcoin network in terms of pending transactions (i.e. size of mempool). Highest historic daily average Bitcoin transaction fee is estimated as 985 satoshis per byte on the 12th of December 2017, right in the middle of the Bitcoin price spike [10]. Even today, with all the custodial exchange wallets and Lightning Network, spikes in the exchange rate still trigger jumps in transaction fee unit prices. For example, on the 20th May 2019 average transaction fee price jumped to 212 satoshis per byte [10].

In a nutshell, there are only two parameters that can be used to decrease the transaction fee: space and time. Currently, the most cost effective scheme to create transactions is using native SegWit (bech32) addresses [4]. Table 1 presents the fee savings for various transaction types, going as high as 49% in our 2-of-2 multisignature address case [9]. On the other hand, it is possible to reduce transaction fees by just being "patient". If transaction confirmation is not urgent, it is possible to wait confirmation for a couple days and pay up to 92% less (Table 2) [9].

Table 1. Fee reduction for transaction types [9]

Type	Legacy (vbytes)	P2SH-wrapped SegWit (vbytes)	Native SegWit (vbytes)
Single signature	226	167 26% saving	141 37% saving
2-of-2 multisig	335	197 41% saving	169 49% saving

Table 2. Fee reduction for wait time [9]

Wait	Savings
2 h	5%
6 h	18%
12 h	39%
24 h (1 day)	39%
48 h (2 days)	47%
72 h (3 days)	52%
96 h (4 days)	92%

The proposed platform not only utilizes both of these techniques (using bech32 addresses and patient spending via targeting 12 to 24 h on-chain settlement) but also implements additional optimizations that are only possible with the proposed design. The unique opportunities for optimization are:

1. aggregate and spend only completely consumed UTXOs, therefore saving up one output per payment attempt, per user (i.e. do not ever create and send to change addresses [7])
2. aggregate payments to same addresses together (i.e. *(SIGHASH_NONE |SIGHASH_ANYONECANPAY)* makes it possible to modify outputs)
3. use the benefit of sending big transactions, rich in transaction fees. Aggregated transactions will be relatively big (over dozens of inputs and outputs) and even though the satoshi per byte unit price is slightly lower compared to the other pending transactions, the higher mining fee alone may be attractive for adding that single, big transaction to the blockchain

Just like in the previous section, these design choices come with a disadvantage: there are no change addresses created for the user and until the whole single UTXO is consumed (or withdrawn by the user) the final state will not be visible on-chain.

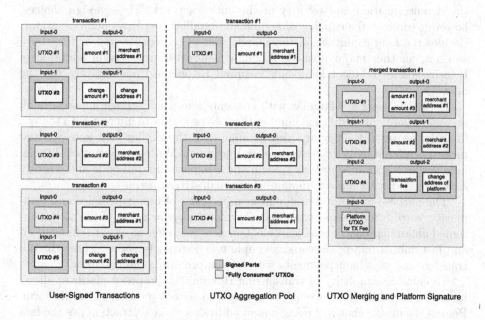

Fig. 1. UTXO aggregation example (3-to-1)

In Fig. 1, a simple demonstration of UTXO aggregation is given. First, the platform will receive user-signed transactions from external devices and analyzes them for possible optimizations. If any change address and amount is given in the transaction output, the platform will check whether the whole UTXO is consumed. Eligible UTXOs, which are fully consumed, are transferred to the *aggregation pool* for possible merges. Once the platform decides to send a transaction to the Bitcoin network, based on a predetermined criteria (again time or space), it will go through the aggregation pool and (a) transfer UTXOs to a new transaction and (b) merge amounts sent to the same addresses (i.e. same script-PubKey). If additional funds are required (e.g. transaction fees), the platform will use its own funds to offset the required amount. The platform may charge its merchants (instantly) or its users. Nonetheless, *Platform UTXO* should be seen as the operating capital of the platform.

In the end, the final transaction will be significantly smaller then the total of the three separate transactions acquired directly from the users. Roughly, 2-input 2-output Segregated Witness (SegWit) transactions are 265 vbytes, 1-input 1-output transactions are 138 vbytes and 4-input 3-output transactions are 485 vbytes [13]. This shows that UTXO aggregation saves 28% space in this particular example.

3.4   User Experience and Complexity

There are several design choices within the Bitcoin that are critical to ensure the decentralization and security of the entire network. These design choices, however, come at the expense of simple and familiar user flows and operations. Besides the long confirmation times and transaction fees covered in the above sections, another major obstacle standing before the mainstream adoption of Bitcoin is the technical know-how required to onboard and actively use the network.

Everyday users are familiar with concepts like email, phone number, bank account number or credit card number that are necessary to interact with today's financial ecosystem. They are also familiar with the similar user journeys and processes offered by third party financial products. With the advent of Bitcoin, however, new concepts such as Bitcoin address, hashrate, full node and custody have emerged, that are very unfamiliar to mainstream users.

The purpose of the proposed platform, in terms of user experience and complexity, is to introduce more familiar abstractions for as many of the aforementioned unfamiliar concepts as possible. Our goal is to offer an easy-to-understand and fast onboarding process and a simple user journey for peer-to-peer Bitcoin transfers and merchant payments without compromising fund security.

To enhance simplicity of transferring Bitcoin, the proposed platform allows its users to send Bitcoin via email to people who are not yet users of the platform. Behind the design choice of using e-mail addresses as an abstraction lies the fact that emails are unique, easy to verify, familiar, ubiquitous and platform agnostic. The transfer is completed automatically once the non-user recipient onboards the platform by verifying their email address using the mobile application. The transfer can be cancelled any time by the sender until such verification occurs. For example, such an "address book" feature may also be implemented as a smart contract on Ethereum platform, making the proposed system more decentralized.

4   Competition

In this section, an overview of the competing technologies will be given, starting with Bitcoin. Every project and feature in this section is evaluated in terms of instant transactions, minimal transaction fees, user experience and complexity.

4.1   Bitcoin

Bitcoin is the cryptocurrency that single handedly changed how people see financial institutions and launched a whole new industry. However, as said in the previous sections, fundamental design decisions that make Bitcoin trustless and decentralized, also affected the transaction times and costs in a negative way. Today, the correct way of receiving a Bitcoin transaction is waiting at least two block confirmations, which lasts around 20 to 30 min. Similarly, Bitcoin transaction fee characteristics have been discussed in the previous section. In essence, our proposed solution is building a logical abstraction around Bitcoin to overcome these limitations.

Partially Signed Bitcoin Transactions (PSBT) aim to define a standard, extensible format that multiple parties may process and pass around a transaction for "co-signing" [5]. By using PSBT, it would be possible to collectively create transactions, therefore, (in theory) gaining the ability to aggregate UTXOs. The disadvantage of PSBT compared to the proposed solution is, PSBT transactions must be created with all input UTXOs filled in first. The users may pass it around and sign it independently, only after creating that raw, skeleton transaction. Due to the fact that transaction signing can not happen interactively, it would be relatively hard to collectively create and sign a transaction effectively, especially if the number of participants are high. It is important to emphasize that, a PSBT transaction will not be valid until all the signatures are complete. In contrast, no extra network overhead is present in the proposed solution, all transaction aggregation is done by the platform.

4.2   Lightning Network

Lighting Network is a payment solution built on top of Bitcoin, that promises an instant, trustless and affordable way of making transactions. It is a peer-to-peer network, where peers are able to "lock" their Bitcoin on chain and able to transfer it to other parties via "channels". It is designed to create a network of micropayment channels that will address the scalability problem of the Bitcoin network. Lightning Network offers instant transactions on its off-chain payment channels, where on-chain transactions are needed when one of the two parties wants to send funds outside of the channel.

The routing capability of the Lightning Network is determined by the funds distributed to channels by the peers. In theory, for a seamless routing, both the sending and receiving end of a given channel should be funded. Unfortunately, in real life, payments are mostly uni-directional (e.g. customers to merchants). As a result, Lightning Network will continuously experience liquidation shortage unless there are major liquidity providers in the system keeping up channels open to merchants. Obviously, a major liquidity provider will hurt decentralization of the network which is one of the design pillars of Lightning. Hopefully, instead of single path routing, the proposed multi path payments may help the network fighting this problem [26].

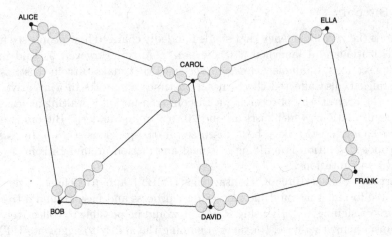

Fig. 2. Sample lightning channel setup [29]

Figure 2 [29] illustrates the operation of Lightning channels and demonstrates a routing limitation. In this example, Alice can only send one coin to Frank in a single Lightning Network transaction due to the current state of the channels (over Bob and David or over Carol and Ella). Basically, Carol is unable to send coins to David due to the depleted channel at her side and this forces the system to route over Bob or Ella in the end. However, if multi path payment feature were active, it would be possible to send two coins (one coin over each route at the same time).

Similarly, a Lightning channel must be funded at least by the amount of the transfer for a seamless one-directional routing. As said, this brings a liquidation shortage if a Bitcoin amount m must be routed through n number of channels. In such a case, not only must the entire route be funded with the amount of $m * n$ but also each of the locked funds on sending ends of the nodes must be greater than or equal to the amount of transfer m. This brings an "opportunity cost" since these funds cannot be put in use anywhere else during the idle times of the node and the corresponding channel.

It is preferred that all peers in the Lightning Network maintain a full Bitcoin node besides the Lightning daemon. This requirement is to prevent the so-called fraudulent channel close where one peer (online) broadcasts an outdated channel state to the Bitcoin network without the knowledge of the other (offline) [28]. It is also a major issue for user onboarding, since opening, maintaining and funding a Lightning Network node requires significant amount of technical know-how.

To mitigate this issue, researchers are working on so-called "Watchtowers" that Lightning peers may trust with their channel states [1, 23, 27].

4.3 Liquid Network

Liquid Network is an inter-exchange settlement network, designed to be a side-chain of Bitcoin. Liquid Network is run by federation of exchanges and financial institutions using special hardware [12]. Bitcoin is transferred to Liquid Network via a peg-in process and after locking funds on Bitcoin main chain, a 1-to-1 asset called L-BTC is issued inside the network [2]. Another important feature of Liquid Network is having the capability to execute *Confidential Transactions* that conceal the amount value that is transferred among peers [21].

With its federated design, Liquid Network is an enterprise solution for institutions that need the speed and flexibility while operating with cryptocurrency-based assets. In short, it can execute transactions very fast with minimal fees however it is not possible to on-board regular users due to the requirements (or limitations) of the federation.

5 Conclusion

We believe that equal opportunity for financial self-sufficiency and independence is now more reachable than ever in human history via blockchain and cryptocurrencies. Bitcoin, with its estimated 25 million global users and ten thousand globally distributed nodes, is closer than any other cryptocurrency to get mainstream adoption to reach this goal.

As presented throughout the paper, the proposed design improves Bitcoin transactions in terms of cost and speed via UTXO aggregation and utilizes another public blockchain, Stellar, just like a "Layer 2" solution. Since the platform manages transaction aggregation by itself, the design works better compared to Partially Signed Bitcoin Transactions (PSBT) which has high peer-to-peer communication overhead. It is superior than the Lightning Network in terms of complexity (e.g. payment routing) and efficiency (e.g. locking of funds) and is also open to any participant unlike Liquid Network. Our final goal is to complement Bitcoin and utilize Stellar to create instant, affordable transactions and improve the whole user experience at least for the everyday user without compromising security.

References

1. Avarikioti, G., Laufenberg, F., Sliwinski, J., Wang, Y., Wattenhofer, R.: Towards secure and efficient payment channels. arXiv preprint arXiv:1811.12740 (2018)
2. Back, A., et al.: Enabling blockchain innovations with pegged sidechains, p. 72 (2014). http://www.opensciencereview.com/papers/123/enablingblockchain-innovations-with-pegged-sidechains

3. Bitcoin: Hierarchical Deterministic Wallets (2013). https://github.com/bitcoin/bips/blob/master/bip-0032.mediawiki
4. Bitcoin: Base32 address format for native v0-16 witness outputs, March 2017, https://github.com/bitcoin/bips/blob/master/bip-0173.mediawiki
5. Bitcoin: Partially signed bitcoin transaction format, July 2017. https://github.com/bitcoin/bips/blob/master/bip-0174.mediawiki
6. Bitcoin Market Journal: How many people use Bitcoin in 2019? February 2019. https://www.bitcoinmarketjournal.com/how-many-people-use-bitcoin
7. Bitcoin Wiki: Change (2017). https://en.bitcoin.it/wiki/Change
8. Bitcoin Wiki: Elliptic curve digital signature algorithm (2019). https://en.bitcoin.it/wiki/Elliptic_Curve_Digital_Signature_Algorithm
9. Bitcoin Wiki: Techniques to reduce transaction fees (2019). https://en.bitcoin.it/wiki/Techniques_to_reduce_transaction_fees
10. bitcoinfees.info: Bitcoin transaction fees (2019). https://bitcoinfees.info
11. Bitnodes: Bitcoin nodes worldwide, May 2019. https://bitnodes.earn.com
12. Blockstream: Liquid (2018). https://blockstream.com/liquid
13. Buy Bitcoin Worldwide: Bitcoin fee calculator & estimator (2019). https://estimatefee.com
14. Chase, B., MacBrough, E.: Analysis of the XRP ledger consensus protocol. arXiv preprint arXiv:1802.07242 (2018)
15. Coinmarketcap: Monthly volume rankings (currency) (2019). https://coinmarketcap.com/currencies/volume/monthly
16. EOS.IO: Eos.io technical white paper (2017). https://github.com/EOSIO/Documentation
17. FDIC: FDIC national survey of unbanked and underbanked households, October 2018. https://www.fdic.gov/householdsurvey
18. Green Address: Frequently asked questions (2018). https://greenaddress.it
19. Hatch, C.: XCAT: cross chain atomic trades protocol and API for trades between Stellar and Ethereum (2018). https://github.com/chatch/xcat
20. Libra Association Members: An introduction to Libra (2019). https://libra.org/en-US/white-paper
21. Maxwell, G.: Confidential transactions (2015). https://people.xiph.org/~greg/confidential_values.txt
22. Maxwell, G., Poelstra, A., Seurin, Y., Wuille, P.: Simple Schnorr multi-signatures with applications to Bitcoin. Des. Codes Cryptogr. 1–26 (2018)
23. McCorry, P., Bakshi, S., Bentov, I., Miller, A., Meiklejohn, S.: Pisa: arbitration outsourcing for state channels. IACR Cryptology ePrint Archive 2018, 582 (2018)
24. Nakamoto, S.: Bitcoin: a peer-to-peer electronic cash system (2008). https://bitcoin.org/bitcoin.pdf
25. NEO: Neo technical white paper (2017). https://docs.neo.org/en-us/whitepaper.html
26. Osuntokun, O.: AMP: Atomic multi-path payments over lightning, February 2018 . https://lists.linuxfoundation.org/pipermail/lightning-dev/2018-February/000993.html
27. Osuntokun, O.: Hardening lightning, January 2018. https://cyber.stanford.edu/sites/g/files/sbiybj9936/f/hardening_lightning_updated.pdf
28. Poon, J., Dryja, T.: The Bitcoin lightning network: scalable off-chain instant payments (2016). https://lightning.network/lightning-network-paper.pdf
29. Rizun, P.: Lightning quiz (2019). https://medium.com/@peter_r/the-answer-to-yesterdays-lightning-network-quiz-was-a-1-coin-d8125c056758

30. Statista: Number of blockchain wallet users worldwide, April 2019. https://www.statista.com/statistics/647374/worldwide-blockchain-wallet-users
31. Stellar: Architecture (2019). https://www.stellar.org/developers/guides/anchor
32. Stellar: Assets (2019). https://www.stellar.org/developers/guides/concepts/assets.html
33. Streamlabs: Live streaming Q4'18 report, January 2019. https://blog.streamlabs.com
34. The World Bank: The global Findex database, May 2017. https://globalfindex.worldbank.org
35. The World Bank: Financial inclusion, October 2018. http://www.worldbank.org/en/topic/financialinclusion/overview
36. The World Bank: Remittance prices worldwide, March 2019. https://remittanceprices.worldbank.org/en
37. VISA: Visa transaction throughput, August 2010. https://usa.visa.com/run-your-business/small-business-tools/retail.html
38. Zamyatin, A., Harz, D., Lind, J., Panayiotou, P., Gervais, A., Knottenbelt, W.: XCLAIM: trustless, interoperable, cryptocurrency-backed assets. IEEE Secur. Priv. (2019)
39. ZmnSCPxj: Funding transactions as a generalized design pattern for offchain protocols (2017). https://zmnscpxj.github.io/offchain/generalized.html

Simulation Extractability in Groth's zk-SNARK

Shahla Atapoor and Karim Baghery[✉]

University of Tartu, Tartu, Estonia
karim.baghery@ut.ee

Abstract. A Simulation Extractable (SE) zk-SNARK enables a prover to prove that she knows a witness for an instance in a way that the proof: (1) is succinct and can be verified very efficiently; (2) does not leak information about the witness; (3) is simulation-extractable -an adversary cannot come out with a new valid proof unless it knows a witness, even if it has already seen arbitrary number of simulated proofs. Non-malleable succinct proofs and very efficient verification make SE zk-SNARKs an elegant tool in various privacy-preserving applications such as cryptocurrencies, smart contracts and etc. In Eurocrypt 2016, Groth proposed the most efficient pairing-based zk-SNARK in the CRS model, but its proof is vulnerable to the malleability attacks. In this paper, we show that one can efficiently achieve simulation extractability in Groth's zk-SNARK by some changes in the underlying language using an OR construction. Analysis and implementations show that in practical cases overload has minimal effects on the efficiency of original scheme which currently is the most efficient zk-SNARK. In new construction, proof size is extended with one element from $\mathbb{G}_1$, one element from $\mathbb{G}_2$, plus a bit string that totally is less than 256 bytes for 128-bit security. Its verification is dominated with 4 pairings which is the most efficient verification among current SE zk-SNARKs.

Keywords: ZK proofs · SNARKs · Simulation extractability · CRS model

1 Introduction

Non-Interactive Zero-Knowledge (NIZK) proofs are one of the central design tools in cryptographically secure systems, allowing one to verify the veracity of statements without leaking extra information. Technically speaking, a NIZK allows a prover to prove that, for a public statement x she knows a witness w which hold in a relation $\mathbf{R}$, $(x, w) \in \mathbf{R}$, without leaking any information about her witness w. In the Common Reference String (CRS) model [BFM88], a NIZK is a three-party protocol that works as the following. First, there exists a trusted party K (a.k.a. CRS generator) who takes security parameter λ as an input and generates CRS elements $\mathsf{crs} := (\mathsf{crs}_P, \mathsf{crs}_V)$ which later will be used by prover and verifier for proof generation and proof verification, respectively. Then the

© Springer Nature Switzerland AG 2019
C. Pérez-Solà et al. (Eds.): DPM 2019/CBT 2019, LNCS 11737, pp. 336–354, 2019.
https://doi.org/10.1007/978-3-030-31500-9_22

prover P gets **crs**$_P$, the statement x and her witness w and generates a proof π, attesting that for the statement x, I know a witness w s.t. $(x, w) \in \mathbf{R}$. Finally, a verifier V takes **crs**$_V$, the statement x and the proof π and returns either accept (if proof is valid) or reject (if verification failed). If V does not need any secret information to verify the proof π, the proof is called *publicly verifiable* that can be verified by many public verifiers (e.g. by nodes of a distributed network).

Generally, a NIZK argument satisfies three properties known as *completeness*, *soundness* and *zero-knowledge*. *Completeness* guarantees that an honest P always convinces an honest V. The *soundness* ensures that a malicious P cannot convince the honest V except with negligible probability. *Zero-knowledge property* assures that the proof generated by P does not leak any information about the witness w. Formal definitions will be given later in Sect. 2.1.

During last few years, a very efficient family of NIZK proof systems are developed which are known as zero-knowledge Succinct Non-interactive Argument of Knowledge (zk-SNARK) [Gro10, Lip12, PHGR13, BCTV13, Gro16, GM17]. A zk-SNARK generates a *succinct* proof that allows a computationally weak verifier to efficiently verify the proof. Differently from a standard NIZK, an SNARK guarantees *knowledge-soundness* that is a stronger notion in comparison with standard soundness. Knowledge-soundness (more precisely non-black-box knowledge soundness) guarantees that if an adversarial prover manages to come out with an acceptable proof, there exists an efficient extractor which given source code and random coins of the adversary can efficiently extract the witness. Knowledge-soundness of zk-SNARKs is non-black-box and achieved under knowledge assumptions [Dam92][1]. Impossibility result of Gentry and Wichs [GW11] also confirms that extraction in zk-SNARKs should be based on non-falsifiable assumptions (e.g. knowledge assumptions). By the date, the most efficient zk-SNARK is proposed by Groth [Gro16] in Eurocrypt 2016, that is constructed for Quadratic Arithmetic Programs (QAPs) and works in a bilinear group. The proof in Groth's zk-SNARK consists of 2 elements in $\mathbb{G}_1$ and 1 element in $\mathbb{G}_2$, and V needs to check one equation that is dominated with 3 pairings.

In practice, however knowledge-soundness is an amplified notion in comparison with standard soundness but still a knowledge-sound proof is vulnerable to the man-in-the-middle attacks[2]. In other words, knowledge soundness only guarantees that a successful prover knows the witness, and it does not guarantee non-malleability of proofs. Due to this fact, zk-SNARKs that only guarantee knowledge-soundness cannot be deployed in many of practical applications straightforwardly [BCG+14, KMS+16, JKS16, Bag19]. For instance,

---

[1] In non-black-box extraction, extractor ext$_A$ needs to get full access to the source code and random coins of adversary $\mathcal{A}$ to be able to extract the witness. But in black-box extraction, one can extract the witnesses straightforwardly from the proof using CRS trapdoors [Bag19].

[2] For instance, in the verification equations that have paring structure such as $a \bullet b = c$, where a and b are proof elements from $\mathbb{G}_1$ and $\mathbb{G}_2$ with prime orders, one can see that such verification equation will be satisfied also for new proof elements such as $a' = a^r$ and $b' = b^{1/r}$, for arbitrary $r \leftarrow \mathbb{Z}_p$.

privacy-preserving crypocurrencies such as Zerocash that uses zk-SNARKs [BCTV13, Gro16] as a subroutine, takes extra steps to prevent malleability attacks in the SNARK proofs for *pour* transactions [BCG+14]. Similarly, privacy-preserving smart contract systems [KMS+16, JKS16] show that knowledge-soundness of zk-SNARKs is not enough for their systems. *Simulation-knowledge soundness* which also is known as *simulation extractability*, is an amplified version of knowledge-soundness that is proposed to achieve extractability and non-malleable proofs. A Simulation-Extractable (SE) zk-SNARK guarantees that the proof is succinct, zero-knowledge and *simulation-extractable*. Simulation extractability implies that an adversary cannot generate a new proof unless he knows a witness, even if he has seen arbitrary number of simulated proofs.

In Crypto 2017, Groth and Maller [GM17] proposed the first SE zk-SNARK in the CRS model that allows to generate non-malleable proofs (referred as GM zk-SNARK). They also proved that a SE zk-SNARK requires at least two verification equations. Their scheme is constructed in the bilinear groups for Square Arithmetic Programs (SAPs) and achieves the lower bound in the number of verification equations. To verify a proof, V needs to check two equations that are dominated with 5 pairings [GM17]. To guarantee non-malleability in proofs, their scheme removes one of the bilinear group generators from the CRS, which might create some different challenges in some practical cases (e.g. in CRS generation by multi-party computation protocols [ABL+19], or in achieving subversion security [ABLZ17]). Above all, GM zk-SNARK is constructed for SAPs that require twice number of multiplication (MUL) gates, as $ab = ((a+b)^2 - (a-b)^2)/4$. Implementations also approves that for a particular arithmetic circuit, Groth's zk-SNARK [Gro16] considerably has better efficiency than GM zk-SNARK [GM17], but Gorth's scheme does not achieve simulation extractability, which makes its proofs vulnerable to the malleability attacks. For a Rank-1 Constraint System (R1CS) instance, efficiency metrics of both schemes are compared in Table 1.

Table 1. Performance of zk-SNARKs proposed by Groth and Maller [GM17], Groth [Gro16] and this work for arithmetic circuit satisfiability with an R1CS instance with 10^6 constraints and 10^6 variables, where 10 are input variables. SE: Simulation Extractable, KS: Knowledge Soundness, BS: λ-Bit String.

SNARK	CRS size, run time	Proof size, P run time	Verification, V time	Security
[GM17]	376 MB, 103 s	$2\mathbb{G}_1 + 1\mathbb{G}_2$, 120 s	5 pairings, 2.3 ms	SE
[Gro16]	196 MB, 75 s	$2\mathbb{G}_1 + 1\mathbb{G}_2$, 83 s	3 pairings, 1.4 ms	KS
This work	205 MB, 80.5 s	$3\mathbb{G}_1 + 2\mathbb{G}_2 + 1$BS, 90 s	4 pairings, 2.0 ms	SE

Problem Statement. By reminding that currently Groth's zk-SNARK [Gro16] has the best efficiency but only achieves knowledge-soundness, and the fact that GM zk-SNARK [GM17] ensures simulation extractability but with less efficiency and only one group generator in the CRS, a research question can be raised as if we can achieve simulation extractability in Groth's scheme efficiently? Such that, new scheme will (1) work for QAPs (2) have both generators of bilinear groups in the CRS (3) have a comparable or even better efficiency than GM zk-SNARK.

Our Contribution. In this paper, we address the questions discussed above and propose a variation of Groth's zk-SNARK that can achieve simulation extractability with minimal efficiency loss in practical cases. To this end, we use the known OR technique and define a new language L' based on the language L in Groth's zk-SNARK that is inspired by the works of De Santis et al. [DDO+01] and Kosba et al. [KZM+15].

Defining new language based on original language results some changes in algorithms of the original scheme. Evaluations show that in practical cases, new changes have minimal affects on the efficiency of original scheme which currently is the state-of-the-art. Strictly speaking, the verification of new scheme has two equations as the optimal case, and only adds 1 pairing to the verification of Groth's scheme. As a result, totally verification of new scheme is dominated with 4 pairings which is less than 5 pairings in GM SE zk-SNARK [GM17]. Empirical analysis show that, for the considered instance in Table 1, verification of new scheme takes 2.0 ms. In the proposed variation, the proof size will be extended by one element from $\mathbb{G}_1$, one element from $\mathbb{G}_2$ plus a 256-bit string, that totally will be 3 elements from $\mathbb{G}_1$, 2 elements from $\mathbb{G}_2$ and one 256-bit string, which for 128-bit security still it is less than 256 bytes. The prover should give a proof for a new circuit that has around 50×10^3 gates more than before, where in practical scenarios the overload is very small. I.e. Zerocash uses zk-SNARKs to give a proof for a circuit with approximately 2×10^6 MUL gates[3]. In comparison with P running times in Table 1, prover of new scheme requires 90 sec to generate a proof; particularly with smaller CRS in comparison with GM [GM17] scheme (with 205 MB, instead of 376 MB). Efficiency of the proposed variation is summarized in Table 1.

Discussion and Related Works. Among different NIZK arguments, zk-SNARKs are the most practically-interested ones; because of their succinct proofs and very efficient verifications. But as majority of them guarantee knowledge-soundness by default, that is vulnerable to the man-in-the-middle attacks, so they cannot be deployed directly in practical systems. Actually, in constructing large cryptographic systems, this issue can make some challenges for non-expert users. To address this, recently constructing efficient SE zk-SNARKs, that by default can guarantee non-malleability of proofs, has gotten more attention [GM17,BG18,KLO19,Lip19]. In [BG18], Bowe and Ariel also proposed a variation of Groth's scheme that achieves simulation extractability but in the Random Oracle (RO) model. In their variation, the proof consists of 3 elements from $\mathbb{G}_1$ and 2 elements from $\mathbb{G}_2$, and verification is dominated with 5 pairings. A good point about their case is that they keep the language as original one, and

---

[3] Their initial circuit had $\approx 4 \times 10^6$ gates, but recently they optimized the system and reduced the number of gates to $\approx 2 \times 10^6$, but still it is very larger than $\approx 50 \times 10^3$.

add some computations to the proof generation and verification with relying on a random oracle that returns group elements[4]. Implementing such random oracle might cause some challenges in practice. Since Groth's zk-SNARK is constructed and proven in the CRS model, so we aim to achieve simulation extractability in the same model using more practical cryptographic primitives.

The rest of paper is organized as follows; Sect. 2 introduces notations and preliminaries. A simulation-extractable version of Groth's zk-SNARK is presented in Sect. 3. In Sect. 4, we discuss about instantiation and efficiency of the proposed construction. Finally we conclude the paper in Sect. 5.

2   Preliminaries

Let PPT denote probabilistic polynomial-time, and NUPPT denote non-uniform PPT. Let $\lambda \in \mathbb{N}$ be the information-theoretic security parameter, say $\lambda = 128$. All adversaries will be stateful. For an algorithm $\mathcal{A}$, let $\text{im}(\mathcal{A})$ be the image of $\mathcal{A}$, i.e. the set of valid outputs of $\mathcal{A}$, let $\text{RND}(\mathcal{A})$ denote the random tape of $\mathcal{A}$, and let $r \leftarrow \text{RND}(\mathcal{A})$ denote sampling of a randomizer r of sufficient length for $\mathcal{A}$'s needs. By $y \leftarrow \mathcal{A}(x; r)$ we denote the fact that $\mathcal{A}$, given an input x and a randomizer r, outputs y. For algorithms $\mathcal{A}$ and $\text{ext}_{\mathcal{A}}$, we write $(y \,\|\, y') \leftarrow (\mathcal{A} \,\|\, \text{ext}_{\mathcal{A}})(x; r)$ as a shorthand for "$y \leftarrow \mathcal{A}(x; r)$, $y' \leftarrow \text{ext}_{\mathcal{A}}(x; r)$". We denote by $\text{negl}(\lambda)$ an arbitrary negligible function in λ. For distributions A and B, $A \approx_c B$ means that they are computationally indistinguishable.

In pairing-based groups, we use additive notation together with the bracket notation, i.e., in group $\mathbb{G}_{\mu}$, $[a]_{\mu} = a[1]_{\mu}$, where $[1]_{\mu}$ is a fixed generator of $\mathbb{G}_{\mu}$. A *bilinear group generator* $\text{BGgen}(1^{\lambda})$ returns $(p, \mathbb{G}_1, \mathbb{G}_2, \mathbb{G}_T, \hat{e}, [1]_1, [1]_2)$, where p (a large prime) is the order of cyclic abelian groups $\mathbb{G}_1$, $\mathbb{G}_2$, and $\mathbb{G}_T$. Finally, $\hat{e} : \mathbb{G}_1 \times \mathbb{G}_2 \to \mathbb{G}_T$ is an efficient non-degenerate bilinear pairing, s.t. $\hat{e}([a]_1, [b]_2) = [ab]_T$. Denote $[a]_1 \bullet [b]_2 = \hat{e}([a]_1, [b]_2)$. The current recommendation is to use an optimal (asymmetric) Ate pairing [HSV06] over Barreto-Naehrig curves [BN05]. In that case, at security level of $\lambda = 99$, an element of $\mathbb{G}_1/\mathbb{G}_2/\mathbb{G}_T$ can be represented in respectively 256/512/3072 bits.[5]

Next we review QAPs that defines NP-complete language specified by a quadratic equation over polynomials and have reduction from the language CIRCUIT-SAT [GGPR13, Gro16].

Quadratic Arithmetic Programs. QAP was introduced by Gennaro *et al.* [GGPR13] as a language where for an input x and witness w, (x, w) $\in$ **R** can be verified by using a parallel quadratic check. Consequently, any efficient simulation-extractable zk-SNARK for QAP results in an efficient simulation-extractable zk-SNARK for CIRCUIT-SAT. An QAP instance $\mathcal{Q}_p$ is specified by

---

[4] Intuitively, some part of their changes play the role of a one-time secure signature scheme, but add two pairings to the verification of original scheme.

[5] The value $\lambda = 99$ takes account recent cryptanalysis of the Barreto-Naehrig curves by Kim and Barbulescu [KB16, BD17]. One can use different settings for 128-bit security. Since we use the library `libsnark` [BCTV13] that currently offers the mentioned security level, we just refer the reader to [KB16, BD17] for more discussion.

the so defined $(\mathbb{Z}_p, m_0, \{u_j, v_j, w_j\}_{j=0}^m, \ell(X))$. This instance defines the following relation, where we assume that $A_0 = 1$:

$$\mathbf{R} = \left\{ \begin{array}{l} (\mathsf{x}, \mathsf{w}) : \mathsf{x} = (A_1, \ldots, A_{m_0})^\top \wedge \mathsf{w} = (A_{m_0+1}, \ldots, A_m)^\top \wedge \\ \left(\sum_{j=0}^m A_j u_j(X)\right) \left(\sum_{j=0}^m A_j v_j(X)\right) \equiv \sum_{j=0}^m A_j w_j(X) \pmod{\ell(X)} \end{array} \right\}.$$

Alternatively, $(\mathsf{x}, \mathsf{w}) \in \mathbf{R}$ if there exists a (degree $\leq n - 2$) polynomial $h(X)$, s.t. $\left(\sum_{j=0}^m A_j u_j(X)\right) \left(\sum_{j=0}^m A_j v_j(X)\right) - \sum_{j=0}^m A_j w_j(X) = h(X)\ell(X)$, where $\ell(X) = \prod_{i=1}^n (X - \omega^{i-1})$ is a polynomial related to Lagrange interpolation, and ω is an n-th primitive root of unity modulo p. Roughly speaking, the goal of the prover of a zk-SNARK for QAP [GGPR13] is to prove that for public statement $(A_1, \ldots, A_{m_0})$ and $A_0 = 1$, she knows the witnesses $(A_{m_0+1}, \ldots, A_m)$ and a degree $\leq n - 2$ polynomial $h(X)$, such that above equation holds.

One-time Signature Schemes [Lam79]. A one-time signature (OTS) scheme is a digital signature scheme that can be used to sign one message per key pair. An OTS scheme is made up three PPT algorithms (KGen, Sign, SigVerify), for key generation, signing, and verification, respectively. A signature scheme is *complete* if an honesty generated signature by Sign always successfully passes the verifications by SigVerify. We say that a signature scheme is *strong unforgeability under a one-time message attack* (SUF-1CMA) if all PPT adversaries have at most negligible advantage in the following experiment.

$\mathsf{EXP}_{\mathsf{SUF-1CMA}}$:

- *Setup:* The challenger C runs $\mathsf{KGen}(\lambda)$ to generate a signing-verification key pair $(\mathsf{sk}, \mathsf{pk})$ and gives pk to the adversary $\mathcal{A}$,
- *Signing Query:* $\mathcal{A}$ selects a message m from message space and gives it to challenger C. Challenger C computes $\sigma = \mathsf{Sign}(\mathsf{sk}, m)$ and sends it to $\mathcal{A}$,
- *Forgery:* $\mathcal{A}$ outputs a message-signature pair (m^*, σ^*),

where adversary's advantage in above experiment is defined as $\mathsf{Adv}_\mathcal{A}(\lambda) = \Pr[\mathsf{SigVerify}(\mathsf{pk}, m^*, \sigma^*) = 1 \wedge (m^*, \sigma^*) \neq (m, \sigma)]$.

2.1 Definitions

We use the definitions of NIZK arguments from [Gro16, GM17]. Let $\mathcal{R}$ be a relation generator, such that $\mathcal{R}(1^\lambda)$ returns a polynomial-time decidable binary relation $\mathbf{R} = \{(\mathsf{x}, \mathsf{w})\}$. Here, x is the statement and w is the witness. Security parameter λ can be deduced from the description of $\mathbf{R}$. The relation generator also outputs auxiliary information ξ that will be given to the honest parties and the adversary. As in [Gro16, ABLZ17], ξ is the value returned by $\mathsf{BGgen}(1^\lambda)$, so ξ is given as an input to the honest parties; if needed, one can include an additional auxiliary input to the adversary. Let $\mathbf{L_R} = \{\mathsf{x} : \exists \mathsf{w}, (\mathsf{x}, \mathsf{w}) \in \mathbf{R}\}$ be an NP-language. A *NIZK argument system* Ψ for $\mathcal{R}$ consists of tuple of PPT algorithms $(\mathsf{K}, \mathsf{P}, \mathsf{V}, \mathsf{Sim})$, such that:

CRS generator: K is a PPT algorithm that, given $(\mathbf{R}, \xi)$ where $(\mathbf{R}, \xi) \in$ $\mathrm{im}(\mathcal{R}(1^\lambda))$, outputs $\mathbf{crs} := (\mathbf{crs}_P, \mathbf{crs}_V)$ and stores trapdoors of $\mathbf{crs}$ as $\mathbf{ts}$. We distinguish $\mathbf{crs}_P$ (needed by the prover) from $\mathbf{crs}_V$ (needed by the verifier).

Prover: P is a PPT algorithm that, given $(\mathbf{R}, \xi, \mathbf{crs}_P, x, w)$, where $(x, w) \in \mathbf{R}$, outputs an argument π. Otherwise, it outputs $\perp$.

Verifier: V is a PPT algorithm that, given $(\mathbf{R}, \xi, \mathbf{crs}_V, x, \pi)$, returns either 0 (reject) or 1 (accept).

Simulator: Sim is a PPT algorithm that, given $(\mathbf{R}, \xi, \mathbf{crs}, \mathbf{ts}, x)$, outputs a simulated argument π.

A zk-SNARK system is required to be complete, knowledge-sound, ZK, and succinct as in the following definitions.

Definition 1 (Perfect Completeness). A non-interactive argument Ψ is *perfectly complete for* $\mathcal{R}$, if for all λ, all $(\mathbf{R}, \xi) \in \mathrm{im}(\mathcal{R}(1^\lambda))$, and $(x, w) \in \mathbf{R}$, $\Pr\left[\, \mathbf{crs} \leftarrow \mathsf{K}(\mathbf{R}, \xi),\ \pi \leftarrow \mathsf{P}(\mathbf{R}, \xi, \mathbf{crs}_P, x, w) : \mathsf{V}(\mathbf{R}, \xi, \mathbf{crs}_V, x, \pi) = 1 \,\right] = 1$.

Definition 2 (Computationally Knowledge-Soundness [Gro16]). A non-interactive argument Ψ is computationally (adaptively) *knowledge-sound for* $\mathcal{R}$, if for every NUPPT $\mathcal{A}$, there exists a NUPPT extractor $\mathsf{ext}_\mathcal{A}$, s.t. for all λ,

$$\Pr\left[\begin{array}{l} \mathbf{crs} \leftarrow \mathsf{K}(\mathbf{R}, \xi), r \leftarrow \mathsf{RND}(\mathcal{A}), ((x, \pi) \,\|\, w) \leftarrow (\mathcal{A} \,\|\, \mathsf{ext}_\mathcal{A})(\mathbf{R}, \xi, \mathbf{crs}; r) : \\ (x, w) \notin \mathbf{R} \wedge \mathsf{V}(\mathbf{R}, \xi, \mathbf{crs}_V, x, \pi) = 1 \end{array}\right] = \mathsf{negl}(\lambda).$$

Here, ξ can be seen as a common auxiliary input to $\mathcal{A}$ and $\mathsf{ext}_\mathcal{A}$ that is generated by using a benign [BCPR14] relation generator.

Definition 3 (Computationally Zero-Knowledge (ZK) [Gro16]). A non-interactive argument Ψ is *computationally ZK for* $\mathcal{R}$, if for all λ, all $(\mathbf{R}, \xi) \in \mathrm{im}(\mathcal{R}(1^\lambda))$, and for all NUPPT $\mathcal{A}$, $\varepsilon_0 \approx_c \varepsilon_1$, where

$$\varepsilon_b = \Pr[(\mathbf{crs} \,\|\, \mathbf{ts}) \leftarrow \mathsf{K}(\mathbf{R}, \xi) : \mathcal{A}^{\mathsf{O}_b(\cdot, \cdot)}(\mathbf{R}, \xi, \mathbf{crs}) = 1] \ .$$

Here, the oracle $\mathsf{O}_0(x, w)$ returns $\perp$ (reject) if $(x, w) \notin \mathbf{R}$, and otherwise it returns $\mathsf{P}(\mathbf{R}, \xi, \mathbf{crs}_P, x, w)$. Similarly, $\mathsf{O}_1(x, w)$ returns $\perp$ (reject) if $(x, w) \notin \mathbf{R}$, otherwise it returns $\mathsf{Sim}(\mathbf{R}, \xi, \mathbf{crs}, \mathbf{ts}, x)$. Ψ is *perfect ZK for* $\mathcal{R}$ if one requires that $\varepsilon_0 = \varepsilon_1$.

Definition 4 (Succinctness [GM17]). A non-interactive argument Ψ is *succinct* if the proof size is polynomial in λ and the verifier's computation time is polynomial in security parameter λ and the size of instance x.

In the rest, we recall the definition of (non-black-box) simulation extractability that we aim to achieve in a variation of Groth's zk-SNARK.

Definition 5 ((Non-Black-Box) Simulation Extractability [GM17]). A non-interactive argument Ψ is *(non-black-box) simulation-extractable for* $\mathcal{R}$, if for any NUPPT $\mathcal{A}$, there exists a NUPPT extractor $\mathsf{ext}_\mathcal{A}$ s.t. for all λ,

$$\Pr\left[\begin{array}{l} \mathbf{crs} \leftarrow \mathsf{K}(\mathbf{R}, \xi), r \leftarrow \mathsf{RND}(\mathcal{A}), ((x, \pi) \,\|\, w) \leftarrow (\mathcal{A}^{\mathsf{O}(\cdot)} \,\|\, \mathsf{ext}_\mathcal{A})(\mathbf{R}, \xi, \mathbf{crs}; r) : \\ (x, \pi) \notin Q \wedge (x, w) \notin \mathbf{R} \wedge \mathsf{V}(\mathbf{R}, \xi, \mathbf{crs}_V, x, \pi) = 1 \end{array}\right] = \mathsf{negl}(\lambda).$$

Here, Q is the set of (x, π)-pairs generated by the adversary's queries to $O(.)$. Note that *(non-black-box) simulation extractability* implies *knowledge-soundness*.

3   A Variation of Groth's zk-SNARK

As briefly discussed in the introduction, Groth's zk-SNARK [Gro16] guarantees knowledge-soundness (defined in Definition 2) which is weaker than simulation extractability. Technically speaking, knowledge-sound proofs are not secure against man-in-the-middle attacks. In this section, we present a variation of Groth's zk-SNARK which can achieve (non-black-box) simulation extractability, defined in Definition 5, that can guarantee non-malleability of the proofs.

3.1   New Construction

In construction of new variation, we define a new language $\mathbf{L}'$, using an OR technique [DDO+01, KZM+15], which combines original language $\mathbf{L}$ in Groth's zk-SNARK with a commitment scheme which commits to a secret randomness as a key for a pseudorandom function. Let (KGen, Sign, SigVerify) be a one-time signature scheme and (Com, ComVerify) be a perfectly binding commitment scheme.

Given the language $\mathbf{L}$ with the corresponding NP relation $\mathbf{R_L}$, we define a new language $\mathbf{L}'$ such that $((x, \mu, \mathsf{pk}_{\mathsf{Sign}}, \rho), (w, s, r)) \in \mathbf{R_{L'}}$ iff:

$$\left((x, w) \in \mathbf{R_L} \vee (\mu = f_s(\mathsf{pk}_{\mathsf{Sign}}) \wedge \rho = \mathsf{Com}(s, r))\right),$$

where $\{f_s : \{0, 1\}^* \to \{0, 1\}^\lambda\}_{s \in \{0,1\}^\lambda}$ is a pseudo-random function family. The intuition for a pseudo-random function $f_s(\cdot)$ is that without the knowledge of the key s, $f_s(\cdot)$ behaves like a true random function. However, given s, one can compute $f_s(\cdot)$ easily. In new language $\mathbf{L}'$, for a statement-witness pair to be valid, either a witness for $\mathbf{R_L}$ is provided (by honest prover) or an opening to ρ together with the value of $\mu = f_s(\mathsf{pk}_{\mathsf{Sign}})$ is provided (by simulator), where s is the open value of ρ (in CRS). One may note that in order for a statement to pass the verification without a valid witness, the prover must generate $f_s(\mathsf{pk}_{\mathsf{Sign}})$ without the knowledge of s (thus breaking the pseudo-random function f_s). By considering new language $\mathbf{L}'$, zk-SNARK of Groth for the relation $\mathbf{R}$ constructed from PPT algorithms (K, P, V, Sim) can be lifted to a simulation-extractable zk-SNARK Ψ' with PPT algorithms (K', P', V', Sim') as described in Fig. 1. To simplify the description, we assume Com takes exactly λ random bits as randomness and that the witness for original language $\mathbf{L}$ is exactly λ bits; it is straight forward to adapt the proof when they are of different lengths [KZM+15]. Note that in the simulation-extractable zk-SNARK Ψ', the algorithms of original scheme will be executed for a new arithmetic circuit which encodes new language $\mathbf{L}'$ and has slightly larger number of gates. Namely, CRS generation algorithm of Groth's zk-SNARK will be executed with a new QAP instance that has larger parameters; $(\mathsf{crs} \parallel \mathsf{ts}) \leftarrow \mathsf{K}(\mathbf{R_{L'}}, \xi)$. Similarly prover of new variation will execute prover of Groth's zk-SNARK with a new arithmetic circuit that has larger number of gates; namely $\pi \leftarrow \mathsf{P}(\mathbf{R_{L'}}, \xi, \mathsf{crs}, (x, z_0, \mathsf{pk}_{\mathsf{Sign}}, \rho), (w, z_1, z_2))$, where z_1 and z_2 play the role of witnesses s and r for prover.

CRS generator $K'(\mathbf{R_L}, \xi)$: Sample $(\mathbf{crs} \,\|\, \mathsf{ts}) \leftarrow K(\mathbf{R_{L'}}, \xi)$; $s, r \leftarrow \{0,1\}^\lambda$; $\rho := \mathsf{Com}(s, r)$; and output $(\mathbf{crs'} \,\|\, \mathsf{ts'}) := ((\mathbf{crs}, \rho) \,\|\, (\mathsf{ts}, (s, r)))$; where $\mathsf{ts'}$ is new simulation trapdoor.

Prover $P'(\mathbf{R_L}, \xi, \mathbf{crs'}, \mathsf{x}, \mathsf{w})$: Parse $\mathbf{crs'} := (\mathbf{crs}, \rho)$; abort if $(\mathsf{x}, \mathsf{w}) \notin \mathbf{R_L}$; generate $(\mathsf{pk_{Sign}}, \mathsf{sk_{Sign}}) \leftarrow \mathsf{KGen}(1^\lambda)$; sample $z_0, z_1, z_2 \leftarrow \{0,1\}^\lambda$; generate $\pi \leftarrow P(\mathbf{R_{L'}}, \xi, \mathbf{crs}, (\mathsf{x}, z_0, \mathsf{pk_{Sign}}, \rho), (\mathsf{w}, z_1, z_2))$ using the prover of Groth's scheme; sign $\sigma \leftarrow \mathsf{Sign}(\mathsf{sk_{Sign}}, (\mathsf{x}, z_0, \pi))$; and return $\pi' := (z_0, \pi, \mathsf{pk_{Sign}}, \sigma)$.

Verifier $V'(\mathbf{R_L}, \xi, \mathbf{crs'}, \mathsf{x}, \pi')$: Parse $\mathbf{crs'} := (\mathbf{crs}, \rho)$ and $\pi' := (z_0, \pi, \mathsf{pk_{Sign}}, \sigma)$; abort if $\mathsf{SigVerify}(\mathsf{pk_{Sign}}, (\mathsf{x}, z_0, \pi), \sigma) = 0$; call the verifier of Groth's scheme $V(\mathbf{R_{L'}}, \xi, \mathbf{crs}, (\mathsf{x}, z_0, \mathsf{pk_{Sign}}, \rho), \pi)$ and abort if it outputs 0.

Simulator $\mathsf{Sim}'(\mathbf{R_L}, \xi, \mathbf{crs'}, \mathsf{ts'}, \mathsf{x})$: Parse $\mathbf{crs'} := (\mathbf{crs}, \rho)$ and $\mathsf{ts'} := (\mathsf{ts}, (s, r))$; generate $(\mathsf{pk_{Sign}}, \mathsf{sk_{Sign}}) \leftarrow \mathsf{KGen}(1^\lambda)$; set $\mu = f_s(\mathsf{pk_{Sign}})$; generate $\pi \leftarrow \mathsf{Sim}(\mathbf{R_{L'}}, \xi, \mathbf{crs}, (\mathsf{x}, \mu, \mathsf{pk_{Sign}}, \rho), (\mathsf{ts} \,\|\, (s, r)))$; sign $\sigma \leftarrow \mathsf{Sign}(\mathsf{sk_{Sign}}, (\mathsf{x}, \mu, \pi))$; and output $\pi' := (\mu, \pi, \mathsf{pk_{Sign}}, \sigma)$.

Fig. 1. A variation of Groth's zk-SNARK.

3.2   Security Proofs

In the rest we present security proofs of the proposed scheme.

Theorem 1 (Completeness). *The variation of Groth's zk-SNARK described in Sect. 3.1, guarantees completeness.*

Proof. In new scheme internal computations of P and V are the same as original one, except few extra efficient computations. Precisely, P needs to do the computation for a new instance that has slightly larger size (e.g. $n = n_{old} + n_{new}$, where n_{new} is the number of MUL gates added to the old circuit) and sign the proof and statement with a one-time signature scheme. So following the completeness of original scheme, and the fact that the deployed signature scheme is complete, which means $\mathsf{SigVerify}(\mathsf{pk_{Sign}}, m, \mathsf{Sign}(m, \mathsf{sk_{Sign}})) = 1$, one can conclude that the modified construction satisfies *completeness*.   □

Theorem 2 (Zero-Knowledge). *Assume that Groth's zk-SNARK satisfies computational zero-knowledge, that the pseudo-random function family is secure, that the commitment scheme is perfectly binding and computational hiding, and that the one-time signature scheme is unforgeable, then the presented SNARK described in Sect. 3.1, guarantees computational zero-knowledge.*

Proof. We write a series of hybrid experiments which start from an experiment with the simulator and ends with an experiment that uses the real prover. We show that all experiments are two-by-two indistinguishable. Changes between successive experiments are shown with highlights. Recall that Groth's zk-SNARK guarantees perfect zero-knowledge and the simulator of the modified scheme is expressed in Fig. 1. Now consider the following experiments,

$\underline{\mathsf{EXP}_1}$(simulator):

- *Setup:* Sample $(\mathbf{crs} \,\|\, \mathsf{ts}) \leftarrow \mathsf{K}(\mathbf{R_{L'}}, \xi)$; $s, r \leftarrow \{0,1\}^\lambda$; $\rho := \mathsf{Com}(s, r)$; and output $(\mathbf{crs}' \,\|\, \mathsf{ts}') := ((\mathbf{crs}, \rho) \,\|\, (\mathsf{ts}, (s, r)))$; where ts' is simulation trapdoor.
- *Define function* $\mathsf{O}(\mathsf{x}, \mathsf{w})$: Abort if $(\mathsf{x}, \mathsf{w}) \notin \mathbf{R_L}$; $(\mathsf{pk}_{\mathsf{Sign}}, \mathsf{sk}_{\mathsf{Sign}}) \leftarrow \mathsf{KGen}(1^\lambda)$; set $\mu = f_s(\mathsf{pk}_{\mathsf{Sign}})$; generate $\pi \leftarrow \mathsf{Sim}(\mathbf{R_{L'}}, \xi, \mathbf{crs}, (\mathsf{x}, \mu, \mathsf{pk}_{\mathsf{Sign}}, \rho), \mathsf{ts}')$; sign $\sigma \leftarrow \mathsf{Sign}(\mathsf{sk}_{\mathsf{Sign}}, (\mathsf{x}, \mu, \pi))$; return $\pi' := (\mu, \pi, \mathsf{pk}_{\mathsf{Sign}}, \sigma)$.
- $b \leftarrow \mathcal{A}^{\mathsf{O}(\mathsf{x}, \mathsf{w})}(\mathbf{crs}')$

$\underline{\mathsf{EXP}_2}$(separate secret key of pseudo random function):

- *Setup:* Sample $(\mathbf{crs} \,\|\, \mathsf{ts}) \leftarrow \mathsf{K}(\mathbf{R_{L'}}, \xi)$; $s', s, r \leftarrow \{0,1\}^\lambda$; $\rho := \mathsf{Com}(s', r)$; and output $(\mathbf{crs}' \,\|\, \mathsf{ts}') := ((\mathbf{crs}, \rho) \,\|\, (\mathsf{ts}, (s, s', r)))$; where ts' are new trapdoors.
- *Define function* $\mathsf{O}(\mathsf{x}, \mathsf{w})$: Abort if $(\mathsf{x}, \mathsf{w}) \notin \mathbf{R_L}$; generate $(\mathsf{pk}_{\mathsf{Sign}}, \mathsf{sk}_{\mathsf{Sign}}) \leftarrow \mathsf{KGen}(1^\lambda)$; set $\mu = f_s(\mathsf{pk}_{\mathsf{Sign}})$; generate $\pi \leftarrow \mathsf{Sim}(\mathbf{R_{L'}}, \xi, \mathbf{crs}, (\mathsf{x}, \mu, \mathsf{pk}_{\mathsf{Sign}}, \rho), (\mathsf{ts} \,\|\, (s, r)))$; sign $\sigma \leftarrow \mathsf{Sign}(\mathsf{sk}_{\mathsf{Sign}}, (\mathsf{x}, \mu, \pi))$; return $\pi' := (\mu, \pi, \mathsf{pk}_{\mathsf{Sign}}, \sigma)$.
- $b \leftarrow \mathcal{A}^{\mathsf{O}(\mathsf{x}, \mathsf{w})}(\mathbf{crs}')$

Lemma 1. *If the underlying commitment scheme is computationally hiding, then for two experiments* EXP_2 *and* EXP_1 *we have* $\Pr[\mathsf{EXP}_2] \approx_c \Pr[\mathsf{EXP}_1]$.

Proof. Computationally hiding property of a commitment scheme implies that $\mathsf{Com}(m_1, r)$ is computationally indistinguishable from $\mathsf{Com}(m_2, r)$. So this property straightforwardly results the lemma. $\qquad\square$

$\underline{\mathsf{EXP}_3}$(replace pseudo random function):

- *Setup:* Sample $(\mathbf{crs} \,\|\, \mathsf{ts}) \leftarrow \mathsf{K}(\mathbf{R_{L'}}, \xi)$; s', $\not{s}$, $r \leftarrow \{0,1\}^\lambda$; $\rho := \mathsf{Com}(s', r)$; and output $(\mathbf{crs}' \,\|\, \mathsf{ts}') := ((\mathbf{crs}, \rho) \,\|\, (\mathsf{ts}, (\not{s}, s', r)))$; where ts' is simulation trapdoor and barred characters such as $\not{s}$ are removed characters.
- *Define function* $\mathsf{O}(\mathsf{x}, \mathsf{w})$: Abort if $(\mathsf{x}, \mathsf{w}) \notin \mathbf{R_L}$; $(\mathsf{pk}_{\mathsf{Sign}}, \mathsf{sk}_{\mathsf{Sign}}) \leftarrow \mathsf{KGen}(1^\lambda)$; set $\mu \leftarrow \{0,1\}^\lambda$; generate $\pi \leftarrow \mathsf{Sim}(\mathbf{R_{L'}}, \xi, \mathbf{crs}, (\mathsf{x}, \mu, \mathsf{pk}_{\mathsf{Sign}}, \rho), (\mathsf{ts} \,\|\, (s', r)))$; sign $\sigma \leftarrow \mathsf{Sign}(\mathsf{sk}_{\mathsf{Sign}}, (\mathsf{x}, \mu, \pi))$; return $\pi' := (\mu, \pi, \mathsf{pk}_{\mathsf{Sign}}, \sigma)$.
- $b \leftarrow \mathcal{A}^{\mathsf{O}(\mathsf{x}, \mathsf{w})}(\mathbf{crs}')$

Lemma 2. *If the pseudo random function* $f_s(\cdot)$ *is secure and the underlying one-time signature scheme is unforgeable, we have* $\Pr[\mathsf{EXP}_3] \approx_c \Pr[\mathsf{EXP}_2]$.

Proof. By considering that the signature scheme is secure, we note that the generated $\mathsf{pk}_{\mathsf{Sign}}$ is unique with overwhelming probability. Additionally, we can replace the pseudo random function $f_s(\cdot)$ with a true random function that will result EXP_4. By considering unique values of $\mathsf{pk}_{\mathsf{Sign}}$ and indistinguishability of output of $f_s(\cdot)$ and truly random function, one can conclude the claim. $\qquad\square$

$\underline{\mathsf{EXP}_4}$(prover):

- *Setup:* Sample $(\mathbf{crs} \,\|\, \mathsf{ts}) \leftarrow \mathsf{K}(\mathbf{R_{L'}}, \xi)$; $s', r \leftarrow \{0,1\}^\lambda$; $\rho := \mathsf{Com}(s', r)$; and output $(\mathbf{crs'} \,\|\, \mathsf{ts'}) := ((\mathbf{crs}, \rho) \,\|\, (\mathsf{ts}, (s', r)))$; where $\mathsf{ts'}$ is simulation trapdoor.
- *Define function* $\mathsf{O}(\mathsf{x}, \mathsf{w})$: Abort if $(\mathsf{x}, \mathsf{w}) \notin \mathbf{R_L}$; $(\mathsf{pk}_{\mathsf{Sign}}, \mathsf{sk}_{\mathsf{Sign}}) \leftarrow \mathsf{KGen}(1^\lambda)$; set $\mu \leftarrow \{0,1\}^\lambda$ (μ plays the role of z_0 in Fig. 1); sample $z_1, z_2 \leftarrow \{0,1\}^\lambda$; generate $\pi \leftarrow \mathsf{P}(\mathbf{R_{L'}}, \xi, \mathbf{crs}, (\mathsf{x}, \mu, \mathsf{pk}_{\mathsf{Sign}}, \rho), (\mathsf{w}, z_1, z_2))$; sign $\sigma \leftarrow \mathsf{Sign}(\mathsf{sk}_{\mathsf{Sign}}, (\mathsf{x}, \mu, \pi))$; return $\pi' := (\mu, \pi, \mathsf{pk}_{\mathsf{Sign}}, \sigma)$.
- $b \leftarrow \mathcal{A}^{\mathsf{O}(\mathsf{x}, \mathsf{w})}(\mathbf{crs'})$

Lemma 3. *If Groth's zk-SNARK guarantees zero-knowledge, then we have* $\Pr[\mathsf{EXP}_4] \approx_c \Pr[\mathsf{EXP}_3]$.

Proof. The last experiment exactly models the real prover of construction in Fig. 1, and as already Groth's scheme guarantees zero-knowledge, so one can conclude that the real proof in experiment EXP_4 is indistinguishable from the simulated proof in EXP_3. Intuitively this is because all new elements added to the new construction are chosen randomly and independently. ☐

This concludes the main theorem. ☐

Theorem 3 ((Non-Black-Box) Simulation Extractability). *Assume that Groth's zk-SNARK satisfies knowledge soundness and computational zero-knowledge, that the pseudo-random function family is secure, that the commitment scheme is perfectly binding and computational hiding, and that the one-time signature scheme is unforgeable, then the presented SNARK described in Sect. 3.1, guarantees (non-black-box) simulation extractability.*

Proof. Similarly, we write a sequence of hybrid experiences and finally show that the success probability in the last game is negligible. Recall that Groth's scheme is proven to achieve knowledge-soundness (defined in Definition 2). Now consider the following game,

$\underline{\mathsf{EXP}_1}$(main experiment):

- *Setup:* Sample $(\mathbf{crs} \,\|\, \mathsf{ts}) \leftarrow \mathsf{K}(\mathbf{R_{L'}}, \xi)$; $s, r \leftarrow \{0,1\}^\lambda$; $\rho := \mathsf{Com}(s, r)$; and output $(\mathbf{crs'} \,\|\, \mathsf{ts'}) := ((\mathbf{crs}, \rho) \,\|\, (\mathsf{ts}, (s, r)))$; where $\mathsf{ts'}$ is new CRS trapdoor.
- *Define function* $\mathsf{O}(\mathsf{x})$: $(\mathsf{pk}_{\mathsf{Sign}}, \mathsf{sk}_{\mathsf{Sign}}) \leftarrow \mathsf{KGen}(1^\lambda)$; set $\mu = f_s(\mathsf{pk}_{\mathsf{Sign}})$; generate $\pi \leftarrow \mathsf{P}(\mathbf{R_{L'}}, \xi, \mathbf{crs}, (\mathsf{x}, \mu, \mathsf{pk}_{\mathsf{Sign}}, \rho), (\mathsf{w}, (s, r)))$; sign $\sigma \leftarrow \mathsf{Sign}(\mathsf{sk}_{\mathsf{Sign}}, (\mathsf{x}, \mu, \pi))$; return $\pi' := (\mu, \pi, \mathsf{pk}_{\mathsf{Sign}}, \sigma)$.
- $(\mathsf{x}, \pi') \leftarrow \mathcal{A}^{\mathsf{O}(\mathsf{x})}(\mathbf{crs'})$.
- Parse $\pi' := (\mu, \pi, \mathsf{pk}_{\mathsf{Sign}}, \sigma)$; $\mathsf{w} \leftarrow \mathsf{ext}_{\mathcal{A}}(\mathbf{crs'}, \mathsf{x}, \pi, \xi)$.
- Return 1 iff $((\mathsf{x}, \pi') \notin Q) \wedge (\mathsf{V'}(\mathbf{R_L}, \xi, \mathbf{crs'}, \mathsf{x}, \pi') = 1) \wedge ((\mathsf{x}, \mathsf{w}) \notin \mathbf{R_L})$; where Q shows the set of statement-proof pairs generated by $\mathsf{O}(\mathsf{x})$.

$\underline{\mathsf{EXP}_2}$(relaxing the return checking):

- *Setup:* Sample $(\mathbf{crs} \,\|\, \mathsf{ts}) \leftarrow \mathsf{K}(\mathbf{R_{L'}}, \xi)$; $s, r \leftarrow \{0,1\}^\lambda$; $\rho := \mathsf{Com}(s, r)$; and output $(\mathbf{crs'} \,\|\, \mathsf{ts'}) := ((\mathbf{crs}, \rho) \,\|\, (\mathsf{ts}, (s, r)))$; where $\mathsf{ts'}$ is new CRS trapdoor.

- *Define function* $O(x)$: $(\mathsf{pk_{Sign}}, \mathsf{sk_{Sign}}) \leftarrow \mathsf{KGen}(1^\lambda)$; set $\mu = f_s(\mathsf{pk_{Sign}})$; generate $\pi \leftarrow P(\mathbf{R_{L'}}, \xi, \mathbf{crs}, (x, \mu, \mathsf{pk_{Sign}}, \rho), (w, (s, r)))$; sign $\sigma \leftarrow \mathsf{Sign}(\mathsf{sk_{Sign}}, (x, \mu, \pi))$; return $\pi' := (\mu, \pi, \mathsf{pk_{Sign}}, \sigma)$.
- $(x, \pi') \leftarrow \mathcal{A}^{O(x)}(\mathbf{crs'})$.
- Parse $\pi' := (\mu, \pi, \mathsf{pk_{Sign}}, \sigma)$; $w \leftarrow \mathsf{ext}_\mathcal{A}(\mathbf{crs'}, x, \pi, \xi)$.
- Return 1 iff $((x, \pi') \notin Q) \wedge (V'(\mathbf{R_L}, \xi, \mathbf{crs'}, x, \pi') = 1) \wedge (\mathsf{pk_{Sign}} \notin \mathcal{PK}) \wedge (\mu = f_s(\mathsf{pk_{Sign}}))$; where Q is the set of statement-proof pairs and $\mathcal{PK}$ is the set of signature verification keys, both generated by $O(x)$.

Lemma 4. *If the one-time signature scheme is unforgeable, and Groth's scheme guarantees knowledge-soundness, then* $\Pr[\mathsf{EXP_2}] \leq \Pr[\mathsf{EXP_1}] + \mathsf{negl}(\lambda)$.

Proof. We note that if $(x, \pi') \notin Q$ and "$\mathsf{pk_{Sign}}$ from (x, π'), has been generated by $O(\cdot)$", then the (x, μ, π) is a valid message/signature pairs. Therefore by unforgeability of the signature scheme, we know that $(x, \pi) \notin Q$ and "$\mathsf{pk_{Sign}}$ has been generated by $O(\cdot)$" happens with negligible probability, which allows us to focus on $\mathsf{pk_{Sign}} \notin \mathcal{PK}$.

Now, due to knowledge-soundness of the original scheme (there is an extractor $\mathsf{ext}_\mathcal{A}$ where can extract witness from $\mathcal{A}$), if some witness is valid for $\mathbf{L'}$ and $(x, w) \notin \mathbf{R_L}$, so we conclude it must be the case that there exists some s', such that ρ is valid commitment of s' and $\mu = f_{s'}(\mathsf{pk_{Sign}})$, which by perfectly binding property of the commitment scheme, it implies $\mu = f_s(\mathsf{pk_{Sign}})$. $\square$

$\underline{\mathsf{EXP_3}}$(simulator):

- *Setup:* Sample $(\mathbf{crs} \| \mathsf{ts}) \leftarrow K(\mathbf{R_{L'}}, \xi)$; $s, r \leftarrow \{0, 1\}^\lambda$; $\rho := \mathsf{Com}(s, r)$; and output $(\mathbf{crs'} \| \mathsf{ts'}) := ((\mathbf{crs}, \rho) \| (\mathsf{ts}, (s, r)))$; where $\mathsf{ts'}$ is new CRS trapdoor.
- *Define function* $O(x)$: $(\mathsf{pk_{Sign}}, \mathsf{sk_{Sign}}) \leftarrow \mathsf{KGen}(1^\lambda)$; set $\mu = f_s(\mathsf{pk_{Sign}})$; generate $\pi \leftarrow \mathsf{Sim}(\mathbf{R_{L'}}, \xi, \mathbf{crs}, (x, \mu, \mathsf{pk_{Sign}}, \rho), (\mathsf{ts} \| (s, r)))$; sign $\sigma \leftarrow \mathsf{Sign}(\mathsf{sk_{Sign}}, (x, \mu, \pi))$; return $\pi' := (\mu, \pi, \mathsf{pk_{Sign}}, \sigma)$.
- $(x, \pi') \leftarrow \mathcal{A}^{O(x)}(\mathbf{crs'})$.
- Parse $\pi' := (\mu, \pi, \mathsf{pk_{Sign}}, \sigma)$; $w \leftarrow \mathsf{ext}_\mathcal{A}(\mathbf{crs'}, x, \pi, \xi)$.
- Return 1 iff $((x, \pi') \notin Q) \wedge (V'(\mathbf{R_L}, \xi, \mathbf{crs'}, x, \pi') = 1) \wedge (\mathsf{pk_{Sign}} \notin \mathcal{PK}) \wedge (\mu = f_s(\mathsf{pk_{Sign}}))$; where Q is the set of statement-proof pairs and $\mathcal{PK}$ is the set of signature verification keys, both generated by $O(x)$.

Lemma 5. *If Groth's SNARK guarantees zero-knowledge, then for two experiments* $\mathsf{EXP_3}$ *and* $\mathsf{EXP_2}$, *we have* $\Pr[\mathsf{EXP_3}] \leq \Pr[\mathsf{EXP_2}] + \mathsf{negl}(\lambda)$.

Proof. As the original scheme ensures (perfect) zero-knowledge, so it implies no polynomial time adversary can distinguish a proof generated by the simulator from a proof that is generated by the prover. So, as we are running in polynomial time, thus two experiments are indistinguishable. $\square$

$\underline{\mathsf{EXP_4}}$(separating secret key of pseudo random function):

- *Setup:* Sample $(\mathbf{crs} \| \mathsf{ts}) \leftarrow K(\mathbf{R_{L'}}, \xi)$; $s', s, r \leftarrow \{0, 1\}^\lambda$; $\rho := \mathsf{Com}(s', r)$; and output $(\mathbf{crs'} \| \mathsf{ts'}) := ((\mathbf{crs}, \rho) \| (\mathsf{ts}, (s, s', r)))$; where $\mathsf{ts'}$ is new CRS trapdoor.

- *Define function* $O(x)$: $(pk_{Sign}, sk_{Sign}) \leftarrow KGen(1^\lambda)$; set $\mu = f_s(pk_{Sign})$; generate $\pi \leftarrow Sim(\mathbf{R_{L'}}, \xi, \mathbf{crs}, (x, \mu, pk_{Sign}, \rho), (ts \| (s, r)))$; sign $\sigma \leftarrow Sign(sk_{Sign}, (x, \mu, \pi))$; return $\pi' := (\mu, \pi, pk_{Sign}, \sigma)$.
- $(x, \pi') \leftarrow \mathcal{A}^{O(x)}(\mathbf{crs'})$.
- Parse $\pi' := (\mu, \pi, pk_{Sign}, \sigma)$; $w \leftarrow ext_{\mathcal{A}}(\mathbf{crs'}, x, \pi, \xi)$.
- Return 1 iff $((x, \pi') \notin Q) \wedge (V'(\mathbf{R_L}, \xi, \mathbf{crs'}, x, \pi') = 1) \wedge (pk_{Sign} \notin \mathcal{PK}) \wedge (\mu = f_s(pk_{Sign}))$; where Q is the set of statement-proof pairs and $\mathcal{PK}$ is the set of signature verification keys, both generated by $O(x)$.

Lemma 6. *If the commitment scheme used in the CRS generation is computationally hiding, then* $\Pr[EXP_4] \leq \Pr[EXP_3] + negl(\lambda)$.

Proof. Computationally hiding of a commitment scheme implies that $Com(m_1, r)$ and $Com(m_2, r)$ are computationally indistinguishable, as in this lemma. □

$\underline{EXP_5}$(replace pseudo random function $f_s(\cdot)$ with true random function $F(\cdot)$):

- *Setup:* Sample $(\mathbf{crs} \| ts) \leftarrow K(\mathbf{R_{L'}}, \xi)$; s', $\sharp$, $r \leftarrow \{0, 1\}^\lambda$; $\rho := Com(s', r)$; and output $(\mathbf{crs'} \| ts') := ((\mathbf{crs}, \rho) \| (ts, (\sharp, s', r)))$; where ts' is simulation trapdoor.
- *Define function* $O(x)$: $(pk_{Sign}, sk_{Sign}) \leftarrow KGen(1^\lambda)$; set $\mu = F(pk_{Sign})$; generate $\pi \leftarrow Sim(\mathbf{R_{L'}}, \xi, \mathbf{crs}, (x, \mu, pk_{Sign}, \rho), (ts \| (s, r)))$; sign $\sigma \leftarrow Sign(sk_{Sign}, (x, \mu, \pi))$; return $\pi' := (\mu, \pi, pk_{Sign}, \sigma)$.
- $(x, \pi') \leftarrow \mathcal{A}^{O(x)}(\mathbf{crs'})$.
- Parse $\pi' := (\mu, \pi, pk_{Sign}, \sigma)$; $w \leftarrow ext_{\mathcal{A}}(\mathbf{crs'}, x, \pi, \xi)$.
- Return 1 iff $((x, \pi') \notin Q) \wedge (V'(\mathbf{R_L}, \xi, \mathbf{crs'}, x, \pi') = 1) \wedge (pk_{Sign} \notin \mathcal{PK}) \wedge (\mu = F(pk_{Sign}))$; where Q is the set of statement-proof pairs and $\mathcal{PK}$ is the set of signature verification keys, both generated by $O(x)$.

Lemma 7. *If the truly random function is secure, then* $\Pr[EXP_4] \leq \Pr[EXP_5]$.

Proof. By assuming function $F(\cdot)$ is secure, we can conclude no polynomial time adversary can distinguish an output of the true random function $F(\cdot)$ from an output of the pseudo random function $f_s(\cdot)$. Indeed, experiment EXP_5 can be converted to an adversary for the game of a *true random function*. □

Claim. For experiment EXP_5, we have $\Pr[EXP_5] \leq 2^{-\lambda}$.

Proof. From verification we know $pk_{Sign} \notin \mathcal{PK}$, therefore $F(pk_{Sign})$ has not been queried already. Thus, we will see $F(pk_{Sign})$ as a newly generated random string independent from μ, which implies adversary only can guess. □

This completes proof of the main theorem. □

4   Instantiation and Efficiency Evaluation

We observed in Sect. 3.1 that defining the new language $\mathbf{L}'$ led to some changes in the algorithms of original scheme. In this section, we discuss how efficient can be such changes (described in Fig. 1). We first discuss how the used primitives can be instantiated and then evaluate efficiency of the whole protocol.

Recall that in result of new changes, one needed a pseudo random function, a commitment scheme and a one-time secure signature scheme. In similar practical cases, both pseudo random function and commitment scheme are instantiated using an efficient SHA-256 circuit that has around $\approx 25 \times 10^3$ MUL gates for one block (512-bit input) [BCG+14, KMS+16][6].

The next primitive that we need to instantiate is the digital signature that should be one-time signature scheme and unforgeable. As Groth's zk-SNARK is paring-based and is constructed with bilinear groups, so we instantiate the signature scheme with Boneh and Boyen's signature [BB08] where works in bilinear groups and has very efficient verification; it requires only one pairing and one multi-exponentiation. Their scheme is proven to guarantee unforgeability under chosen message attack and consequently unforgeability under *one-time* chosen message attack. The key generation, signing and verification of Boneh and Boyen's signature scheme [BB08] for message m is summarized below.

- **Key Generation**, $(\mathsf{pk}_{\mathsf{Sign}}, \mathsf{sk}_{\mathsf{Sign}}) \leftarrow \mathsf{KGen}(1^\lambda)$: Given system parameters for a prime-order bilinear group $(p, \mathbb{G}_1, \mathbb{G}_2, \mathbb{G}_T, \hat{e}, [1]_1, [1]_2)$, randomly selects $\mathsf{sk} \leftarrow \mathbb{Z}_p^*$, and computes $\mathsf{sk} \cdot [1]_1$ and returns $(\mathsf{pk}_{\mathsf{Sign}}, \mathsf{sk}_{\mathsf{Sign}}) := ([\mathsf{sk}]_1, \mathsf{sk})$.
- **Signing**, $[\sigma]_2 \leftarrow \mathsf{Sign}(\mathsf{sk}_{\mathsf{Sign}}, m)$: Given system parameters, a secret key $\mathsf{sk}_{\mathsf{Sign}}$, and a message m, computes $[\sigma]_2 = [1/(m + \mathsf{sk})]_2$ and returns $[\sigma]_2$ as the signature.
- **Verification**, $\{1, 0\} \leftarrow \mathsf{SigVerify}(\mathsf{pk}_{\mathsf{Sign}}, [\sigma]_2)$: Given a public key $\mathsf{pk}_{\mathsf{Sign}}$, a message m, and a signature $[\sigma]_2$, verifies if $[m + \mathsf{sk}]_1 \bullet [1/(m + \mathsf{sk})]_2 = [1]_T$; if so, it returns 1; otherwise it returns 0,

where $\bullet$ denotes the paring operation. In our case, we use the same bilinear group as in the original zk-SNARK and m would be the hash (e.g. with SHA224 or SHA256) of concatenations of the proof elements with the statement, i.e. $m := H(\mathsf{x}\|z_0\|\pi)$[7]. As it can be seen, the scheme generates a single-element signature from $\mathbb{G}_2$, its public key is an element from $\mathbb{G}_1$, and above all its verification only requires one paring. Note that $[1]_T$ can be preprocessed and shared in the CRS.

So by considering the above instantiation, new proof $\pi' = (\mu, \pi, \mathsf{pk}_{\mathsf{Sign}}, [\sigma]_2)$ will be as $\pi' = (\mu, \pi, [\mathsf{sk}]_1, [1/(m + \mathsf{sk})]_2)$ where from original scheme

---

[6] It has 25.538 gates in the xjsnark library, https://github.com/akosba/xjsnark.

[7] As shown in [BB08], by taking hash of input message the signature scheme can be used to sign arbitrary messages in $\{0,1\}^*$. To do so, a collision resistant hash function $H : \{0,1\}^* \rightarrow \{0, \dots, 2^b\}$ such that $2^b < p$ is sufficient [BB08]. By considering recent analysis on Barreto-Naehrig curves by Kim and Barbulescu [BD17], one can use different settings for various security levels which would need to use different hash functions for signing arbitrary messages in [BB08] signature scheme.

$\pi = ([a]_1, [b]_2, [c]_1)$, and μ is an output of the pseudo random function $f_s(\cdot)$, which is instantiated with SHA-256 hash function [KMS+16]. As a result, the proof in new scheme will be 3 elements from $\mathbb{G}_1$, 2 elements from $\mathbb{G}_2$ and one 256-bit string. Consequently, new changes add only one paring to the verification of original scheme. To the best of our knowledge, this is the first simulation-extractable zk-SNARK in the CRS model which its verification is dominated with 4 pairings.

Next, we empirically analyse efficiency of the proposed scheme from different perspectives. Table 2 summarizes asymptotic and empirical performance of new scheme and two zk-SNARKs proposed by Groth's [Gro16] and GM [GM17]. Implementations of Groth's [Gro16] and GM [GM17] zk-SNARKs are available in `libsnark` library [BCTV13][8], so similarly implementation of new scheme is done in the same library.

Table 2. An efficiency comparison of new scheme with Groth's [Gro16] and GM [GM17] zk-SNARKs for arithmetic circuit satisfiability with m_0 elements instance, m wires, n MUL gates. In [GM17], n MUL gates translate to $2n$ squaring gates. Implementations (Implem.) are done on a Laptop with 2.50 GHz Intel Core i5-7200U CPU, with 16 GB RAM, in single-threaded mode, for an R1CS instance with $n = 10^6$ constraints and $m = 10^6$ variables, of which $m_0 = 10$ are input variables. $\mathbb{G}_1$ and $\mathbb{G}_2$: group elements, E: exponentiations, P: pairings. In the new scheme, the statement contains $(x, \mu, pk_{Sign}, \rho)$ which has 3 new elements (μ, pk_{Sign}, ρ), so $m'_0 = m_0 + 3$. All asymptotic analysis of new scheme are done based on our particular instantiation of commitment and pseudo random function. So, as new changes add $\approx 50 \times 10^3$ MUL gates to n and m, so $n' = n + 50.000$ and $m' = m + 50.000$.

SNARK	CRS size & gen. time	Proof size	Comp. & time of P	V	Sec
[GM17] & Implem.	$m + 4n + 5\mathbb{G}_1$	$2\,\mathbb{G}_1$	$m + 4n - m_0\,E_1$	$m_0\,E_1$	SE
	$2n + 3\,\mathbb{G}_2$	$1\,\mathbb{G}_2$	$2n\,E_2$	$5\,P$	
	376 MB, 103 s	127 bytes	120 s	2.3 ms	
[Gro16] & Implem.	$m + 2n - m_0\,\mathbb{G}_1$	$2\,\mathbb{G}_1$	$m + 3n - m_0 + 3\,E_1$	$m_0\,E_1$	KS
	$n + 3\,\mathbb{G}_2$	$1\,\mathbb{G}_2$	$n + 1\,E_2$	$3\,P$	
	196 MB, 75 s	127 bytes	83 s	1.4 ms	
Sect. 3.1 & Implem.	$m' + 2n' - m'_0 + 5\,\mathbb{G}_1$	$3\,\mathbb{G}_1 + 2\,\mathbb{G}_2$	$m' + 3n' - m'_0 + 4\,E_1$	$m'_0 + 1\,E_1$	SE
	$n' + 3\,\mathbb{G}_2$	1 bit string	$n' + 2\,E_2$	$4\,P$	
	205 MB, 80.5 s	254 bytes	90.1 s	2.0 ms	

In Table 2, all implementation results are reported for the same R1CS instance. In the rest of analysis, we evaluate efficiency of new scheme for different R1CS instances and compare with *knowledge sound* scheme of Groth [Gro16] and simulation-extractable scheme of GM [GM17]. Strictly speaking, in top plots of Fig. 2, we compare CRS size and CRS generation time of new scheme with the mentioned zk-SNARKs for R1CS instances with 100 input variables and different number of MUL gates, from range 25×10^3 till 2×10^6 gates. Similarly, in bottom left of Fig. 2, we plot prover's running time in three zk-SNARKs for

---

[8] Available on https://github.com/scipr-lab/libsnark.

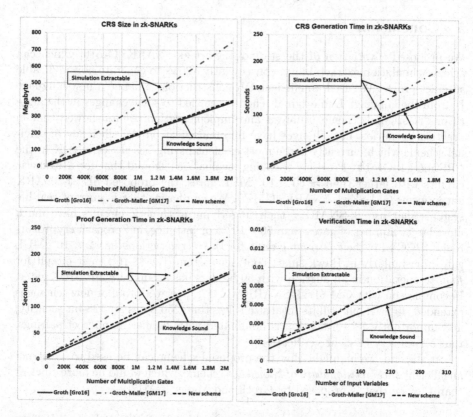

Fig. 2. A comparison of various efficiency metrics in zk-SNARKs of Groth [Gro16], Groth-Maller [GM17] and the proposed variation. Except the plot of verification time in zk-SNARKs (bottom right), all plots are drawn for R1CS instances with 10 input variables and various number of constraints (multiplication gates). In the plot of verification time (bottom right), we draw running time of verifiers in all three zk-SNARKs for R1CS instances with 10^5 constraints and different number of inputs.

various R1CS instances with 100 input variables and different number of MUL gates. Finally, the plot in bottom right of Fig. 2, compares the verification time of new SE zk-SNARK for various R1CS instances with 10^5 constraints and various number of input variables.

From the comparisons in Table 2 and empirical analysis in Fig. 2, one can observe that in order to give non-malleable proofs for an arithmetic circuit satisfiability in circuits with larger than 50×10^3 MUL gates, the proposed SE zk-SNARK can outperform GM SE zk-SNARK considerably. Note that, however new construction has larger proof size than GM zk-SNARK, 254 bytes in comparison with 127 bytes, but still its verification requires smaller number of pairings, and in the worst cases it is as efficient as verification of GM SE zk-SNARK [GM17].

5   Conclusion

We proposed a variation of the state-of-the-art zk-SNARK [Gro16] which can achieve simulation extractability; consequently allows to generate non-malleable succinct proofs. We used an efficient OR construction to define a new language L' from the language L in original scheme, that led to some changes in the algorithms of original scheme. Analysis and implementation results showed that in practical scenarios, new changes have minimal effect on the efficiency of original scheme which currently is the most efficient paring-based zk-SNARK in the CRS model [Gro16]. Precisely speaking, evaluations showed that for arithmetic circuits with larger than $\approx 50 \times 10^3$ MUL gates, the proposed SE zk-SNARK outperforms GM SE zk-SNARK [GM17]. We emphasize that in current real-life systems that use zk-SNARKs, their underlying arithmetic circuits have much more larger number of gates than 50×10^3. For instance, in Zerocash cryptocurrency [BCG+14] their current circuit for *pour* transactions has 2×10^6 MUL gates; or similarly in Hawk smart contract system [KMS+16], their circuit for *finalize* operation in an auction with 50 bidders has around 4×10^6 MUL gates. In comparison with GM SE zk-SNARK [GM17], however proof of new scheme is extended slightly, but still its total size is less than 256 bytes for 128-bit security; and importantly its verification is dominated with smaller number of pairings, that allows very efficient verification.

At the end, we highlight that the proposed scheme can be used to construct an efficient *succinct signature of knowledge* scheme, which would be more efficient than the one that is proposed by Groth and Maller [GM17].

Acknowledgments. The authors were supported by the European Union's Horizon 2020 research and innovation programme under grant agreements No 780477 (project PRIViLEDGE), and by the Estonian Research Council grant (PRG49).

References

[ABL+19] Abdolmaleki, B., Baghery, K., Lipmaa, H., Siim, J., Zając, M.: UC-secure CRS generation for SNARKs. In: Buchmann, J., Nitaj, A., Rachidi, T. (eds.) AFRICACRYPT 2019. LNCS, vol. 11627, pp. 99–117. Springer, Cham (2019). https://doi.org/10.1007/978-3-030-23696-0_6

[ABLZ17] Abdolmaleki, B., Baghery, K., Lipmaa, H., Zając, M.: A subversion-resistant SNARK. In: Takagi, T., Peyrin, T. (eds.) ASIACRYPT 2017. LNCS, vol. 10626, pp. 3–33. Springer, Cham (2017). https://doi.org/10.1007/978-3-319-70700-6_1

[Bag19] Baghery, K.: On the efficiency of privacy-preserving smart contract systems. In: Buchmann, J., Nitaj, A., Rachidi, T. (eds.) AFRICACRYPT 2019. LNCS, vol. 11627, pp. 118–136. Springer, Cham (2019). https://doi.org/10.1007/978-3-030-23696-0_7

[BB08] Boneh, D., Boyen, X.: Short signatures without random oracles and the SDH assumption in bilinear groups. J. Cryptol. **21**(2), 149–177 (2008)

[BCG+14] Ben-Sasson, E., et al.: Zerocash: decentralized anonymous payments from bitcoin. In: 2014 IEEE Symposium on Security and Privacy, pp. 459–474. IEEE Computer Society Press, May 2014

[BCPR14] Bitansky, N., Canetti, R., Paneth, O., Rosen, A.: On the existence of extractable one-way functions. In: Shmoys, D.B. (ed.) 46th ACM STOC, pp. 505–514. ACM Press, May/June 2014

[BCTV13] Ben-Sasson, E., Chiesa, A., Tromer, E., Virza, M.: Succinct non-interactive arguments for a von neumann architecture. Cryptology ePrint Archive, Report 2013/879 (2013). http://eprint.iacr.org/2013/879

[BD17] Barbulescu, R., Duquesne, S.: Updating key size estimations for pairings. Cryptology ePrint Archive, Report 2017/334 (2017). http://eprint.iacr.org/2017/334

[BFM88] Blum, M., Feldman, P., Micali, S.: Non-interactive zero-knowledge and its applications (extended abstract). In: 20th ACM STOC, pp. 103–112. ACM Press, May 1988

[BG18] Bowe, S., Gabizon, A.: Making groth's zk-snark simulation extractable in the random oracle model. IACR Cryptol. ePrint Arch. **2018**, 187 (2018)

[BN05] Barreto, P.S.L.M., Naehrig, M.: Pairing-friendly elliptic curves of prime order. Cryptology ePrint Archive, Report 2005/133 (2005). http://eprint.iacr.org/2005/133

[Dam92] Damgård, I.: Towards practical public key systems secure against chosen ciphertext attacks. In: Feigenbaum, J. (ed.) CRYPTO 1991. LNCS, vol. 576, pp. 445–456. Springer, Heidelberg (1992). https://doi.org/10.1007/3-540-46766-1_36

[DDO+01] De Santis, A., Di Crescenzo, G., Ostrovsky, R., Persiano, G., Sahai, A.: Robust non-interactive zero knowledge. In: Kilian, J. (ed.) CRYPTO 2001. LNCS, vol. 2139, pp. 566–598. Springer, Heidelberg (2001). https://doi.org/10.1007/3-540-44647-8_33

[GGPR13] Gennaro, R., Gentry, C., Parno, B., Raykova, M.: Quadratic span programs and succinct NIZKs without PCPs. In: Johansson, T., Nguyen, P.Q. (eds.) EUROCRYPT 2013. LNCS, vol. 7881, pp. 626–645. Springer, Heidelberg (2013). https://doi.org/10.1007/978-3-642-38348-9_37

[GM17] Groth, J., Maller, M.: Snarky signatures: minimal signatures of knowledge from simulation-extractable SNARKs. In: Katz, J., Shacham, H. (eds.) CRYPTO 2017. LNCS, vol. 10402, pp. 581–612. Springer, Cham (2017). https://doi.org/10.1007/978-3-319-63715-0_20

[Gro10] Groth, J.: Short pairing-based non-interactive zero-knowledge arguments. In: Abe, M. (ed.) ASIACRYPT 2010. LNCS, vol. 6477, pp. 321–340. Springer, Heidelberg (2010). https://doi.org/10.1007/978-3-642-17373-8_19

[Gro16] Groth, J.: On the size of pairing-based non-interactive arguments. In: Fischlin, M., Coron, J.-S. (eds.) EUROCRYPT 2016. LNCS, vol. 9666, pp. 305–326. Springer, Heidelberg (2016). https://doi.org/10.1007/978-3-662-49896-5_11

[GW11] Gentry, C., Wichs, D.: Separating succinct non-interactive arguments from all falsifiable assumptions. In: Fortnow, L., Vadhan, S.P. (eds.) 43rd ACM STOC, pp. 99–108. ACM Press, June 2011

[HSV06] Hess, F., Smart, N.P., Vercauteren, F.: The eta pairing revisited. Cryptology ePrint Archive, Report 2006/110 (2006). http://eprint.iacr.org/2006/110

[JKS16] Juels, A., Kosba, A.E., Shi, E.: The ring of gyges: investigating the future of criminal smart contracts. In: Weippl, E.R., Katzenbeisser, S., Kruegel, C., Myers, A.C., Halevi, S. (eds.) ACM CCS 16, pp. 283–295. ACM Press, October 2016

[KB16]   Kim, T., Barbulescu, R.: Extended tower number field sieve: a new complexity for the medium prime case. In: Robshaw, M., Katz, J. (eds.) CRYPTO 2016. LNCS, vol. 9814, pp. 543–571. Springer, Heidelberg (2016). https://doi.org/10.1007/978-3-662-53018-4_20

[KLO19]   Kim, J., Lee, J., Oh, H.: QAP-based simulation-extractable SNARK with a single verification. Cryptology ePrint Archive, Report 2019/586 (2019). https://eprint.iacr.org/2019/586

[KMS+16]   Kosba, A.E., Miller, A., Shi, E., Wen, Z., Papamanthou, C.: Hawk: the blockchain model of cryptography and privacy-preserving smart contracts. In: 2016 IEEE Symposium on Security and Privacy, pp. 839–858. IEEE Computer Society Press, May 2016

[KZM+15]   Kosba, A.E., et al.: C∅C∅: A Framework for Building Composable Zero-Knowledge Proofs. Technical report 2015/1093, 10 November 2015. http://eprint.iacr.org/2015/1093. Accessed 9 Apr 2017

[Lam79]   Lamport, L.: Constructing digital signatures from a one-way function. Technical report SRI-CSL-98, SRI International Computer Science Laboratory, October 1979

[Lip12]   Lipmaa, H.: Progression-free sets and sublinear pairing-based non-interactive zero-knowledge arguments. In: Cramer, R. (ed.) TCC 2012. LNCS, vol. 7194, pp. 169–189. Springer, Heidelberg (2012). https://doi.org/10.1007/978-3-642-28914-9_10

[Lip19]   Lipmaa, H.: Simulation-extractable SNARKs revisited. Cryptology ePrint Archive, Report 2019/612 (2019). http://eprint.iacr.org/2019/612

[PHGR13]   Parno, B., Howell, J., Gentry, C., Raykova, M.: Pinocchio: nearly practical verifiable computation. In: 2013 IEEE Symposium on Security and Privacy, pp. 238–252. IEEE Computer Society Press, May 2013

Auditable Credential Anonymity Revocation Based on Privacy-Preserving Smart Contracts

Rujia Li[1,2], David Galindo[2,3], and Qi Wang[1(✉)]

[1] Southern University of Science and Technology, Shenzhen, China
wangqi@sustech.edu.cn
[2] University of Birmingham, Birmingham, UK
{rxl635,d.galindo}@cs.bham.ac.uk
[3] Fetch.AI, Cambridge, UK

Abstract. Anonymity revocation is an essential component of credential issuing systems since unconditional anonymity is incompatible with pursuing and sanctioning credential misuse. However, current anonymity revocation approaches have shortcomings with respect to the auditability of the revocation process. In this paper, we propose a novel anonymity revocation approach based on privacy-preserving blockchain-based smart contracts, where the code self-execution property ensures availability and public ledger immutability provides auditability. We describe an instantiation of this approach, provide an implementation thereof and conduct a series of evaluations in terms of running time, gas cost and latency. The results show that our scheme is feasible and efficient.

Keywords: Anonymity revocation · Auditability · Smart contract · Privacy preserving

1 Introduction

Anonymity revocation was first discussed by von Solms and Naccache [32], as they pointed out that Chaum's blind signatures [10] could potentially lead to nonpunishable crime. Subsequently, anonymity revocation has been studied comprehensively, especially in e-cash systems designed to combat money laundering and blackmailing [2,7,9]. The idea of adding anonymity revocation to anonymous credential systems was first proposed by Camenisch and Lysyanskaya [6], where they offered an optional anonymity tracing approach to find the identity of pseudonymous tokens involved in suspicious transactions. In general, anonymity revocation in a credential system allows an issuer to find out who the owner of an anonymous credential is.

The blindness issuance property of an anonymous credential system prevents an issuer from completing the task of anonymity revocation by themselves. The party who helps the issuer to reveal the identity is referred to as *revelator*. Intuitively, there are two parties that can act as the revelator: the user (credential

R. Li and Q. Wang were supported by the National Science Foundation of China under Grant No. 11601220.

C. Pérez-Solà et al. (Eds.): DPM 2019/CBT 2019, LNCS 11737, pp. 355–371, 2019.
https://doi.org/10.1007/978-3-030-31500-9_23

holder) and the judge (trusted third party). Voluntary anonymity revocation by the user is usually straightforward. The issuer cannot link the identity, the message and the resulting signature together unless the user does. One typical example is Microsoft's U-Prove [29]. In such a system, the issuing protocol and the showing protocol are unlinkable. Even if the issuer colludes with the verifier, it cannot associate the message with the resulting signature. The only possibility is that the user chooses to lift the anonymity. Meanwhile, lifting anonymity by a judge, which is inspired by fair blind signature scheme [15], is widely used in systems such as [8,12,30,31]. Taking ABC4Trust [31] as an example, it introduced an inspector to uncover the user who created a presentation token to prevent abuse.

However, some weaknesses in the mentioned anonymity revocation approaches still remain. Firstly, revealing anonymity through the credential holder relies too heavily on the user's will, which ultimately leads to the nonavailability problem. This means if a user behaves maliciously and rejects to cooperate with the issuer, the issuer would never learn the relationship between the identity and the credential. Furthermore, even if the user is honest, they may be offline, resulting in the failure of blindness removal. Meanwhile, in the majority of previous proposals revealing anonymity through the judge lacks transparency, which raises some security concerns: (1) even without the user's consent, the issuer and the judge may conspire to map the credential to the real identity of that user; (2) the judge is a single point of failure. More importantly, the user has no auxiliary information to detect whether the judge has been compromised or not. These challenges lead to the following question:

Is it possible to build an anonymity revocation mechanism that satisfies the requirements: (1) the process of lifting anonymity is transparent and auditable; (2) the revelator always accept revealing the anonymity if necessary?

In this paper, we give a positive answer to this question. Instead of using a trusted third party, we use a neutral and transparent privacy-preserving smart contract as the revelator (to revoke the blindness). The self-execution property of the smart contract ensures the availability of the revelator. This means the neutral blockchain is always honest and is willing to revoke the anonymity whenever it is needed by the issuer. Meanwhile, our privacy-preserving smart contract-based approach allows anonymity revocation in an auditable manner. More precisely, the anonymity tracing must interact with the privacy-preserving smart contract that "lives" on the blockchain and automatically renders the progress auditable. Such revocation progress is recorded in a blockchain transaction which is publicly visible. This auditability provided by the smart contract calling records avoids the misuse of revocation and reduces potential collusion problems to a great extent. Furthermore, the transparent contract calling records provide the user with auxiliary information to detect whether the issuer has been compromised.

In addition, our scheme brings the benefit of greater availability. The service of blockchain is maintained by a large group of nodes [27], which avoids the offline revelator problem. Alternatively, the high-availability blockchain service, being continuously online, provides greater actualization of blindness disclosure and the tracer could trace the identity or credential at any time.

In summary, the contributions of this paper can be summarized as follows:

- We propose a new auditable blind credential system based on privacy-preserving smart contracts, which provides a powerful auditability and neutrality for credential anonymity revocation
- We give an instantiation of our construction and provide a proof of concept implementation. The performance evaluation shows that our scheme is feasible and efficient.

The rest of the paper is structured as follows: further related work is discussed in Sect. 2. Notation and cryptographic building blocks are presented in Sect. 3. An overview of our construction is given in Sect. 4, followed by an instantiation in Sect. 5. The implementation and evaluation of the instantiation are detailed out in Sect. 6. Some example applications are given in Sect. 7. Finally, Sect. 8 concludes with some future work.

2 Related Work

In this section, we first survey current anonymity revocation approaches and make a comparison with our solution. Then, we give some background on blockchain and privacy-preserving smart contracts.

In the last few decades, a series of works [1,15,21,34], have been proposed in the field of anonymity revocation, especially in e-cash systems. Brickell et al. [3] introduced the first trustee-based tracing electronic cash system, in which the coin owner can be revealed by several publicly appointed trustees. Camenisch et al. [7] proposed an anonymous digital payment system with a passive anonymity-revoking trustee. In their system, the trustee only needs to be involved in the anonymity-revoking progress rather than the regular transactions such as opening a new account. Jakobsson and Yung [18] presented an e-money system that makes the value of funds and user anonymity revocable with the consumer rights organisations, even given an extreme condition that an active attacker gets the bank's key or forces the bank to release the money.

In 1995, a fair blind signature scheme was first proposed by Stadler et al. [34]. It involved a judge and allowed this judge to deliver information to the signer to link the issuing session and the resulting message-signature pair. Later, Jakobsson and Yung [19] pointed out that the reused session identifier may make the anonymity revocation invalid, and proposed a fair blind signature scheme that guarantees the one-to-one mapping revocability between the issuing session and the resulting signature. Thereafter, Hufschmitt et al. [17] presented a formal security model for fair blind signatures in the random oracle model. Then, based on Hufschmitt's model, Fuchsbauer et al. [15] proposed a fair blind signature scheme that is not based on the random oracle model. To the best of our knowledge, Camenisch and Lysyanskaya [6] was the first to use anonymity revocation in the credential system. They offered an optional approach to trace the identity of the pseudonymous token for some transactions. After that, some practical systems like IBM's Identity Mixer [8], ABC4Trust [31] started to consider the anonymity revocation. An interesting revocation approach is traceable anonymous certificate [23]. It allows one sub-issuer to verify the ownership of a user and another sub-issuer to validate the contents. Then, these two issuers collaborate to map the certificate to its real identity.

However, the aforementioned anonymity revocation approaches have some drawbacks: the repudiation and the lack of auditability in the revocation progress. The assumptions that the revelator always remains honest is unrealistic. The revelator may be offline when it is needed, or may conspire with the issuer to seek profits, or even be entirely controlled by an attacker. Our scheme is the first to use a privacy-preserving smart contract as the revelator to solve the above problems. The self-executing nature of the contract ensures the neutrality of the revelator. The transparent contract calling records guarantee that the revelator's revocation progress is auditable. The continuous blockchain service keeps the high-availability of the revelator.

Privacy-Preserving Smart Contracts. The concept of a *smart contract*, as a primary application of blockchain technologies [27], was first proposed by Szabo [35]. It is originally defined as a set of digital protocols within which the parties abide by some pre-agreed commitments. In the blockchain system, the smart contract is designed as a self-executing protocol that can verify or execute the fulfilment for the shared instruction code. The smart contract is generally made up of two parts: the instruction code and the executed status. The smart contract in the traditional blockchain systems such as Bitcoin and Ethereum [24] lacks privacy since the instruction code and executed status are publicly shared and visible among all the participants (nodes) in the network. Recently, a new line of work [5, 20, 22, 25] claimed that they had solved these privacy issues by proposing the privacy-preserving smart contract platforms. To verify the feasibility of our scheme, we selected Ekiden [11] and its implementation Oasis Devnet as our privacy-preserving platform. Ekiden [11] combines trusted execution environments (TEEs) and blockchain to achieve confidentiality as well as decentralisation. It allows replicating the contract execution to TEE-powered nodes, where these TEE-powered nodes guarantee the private state and data of the contracts by encrypting them with crytographic keys only known to them.

3   Preliminaries

In this section, we define the notation and recall a well-known cryptographic building block that will be used in our construction.

3.1   Notation

Let λ be the security parameter, $\Sigma(\mathsf{KeyGen}, \mathsf{Sig}, \mathsf{Vf})$ represent a standard signature scheme, and $\mathcal{SM}.\mathsf{Enc}(\mathsf{KeyGen}, \mathsf{Enc}, \mathsf{Dec})$ stand for symmetric encryption and $\mathcal{ASM}.\mathsf{Enc}(\mathsf{KeyGen}, \mathsf{Enc}, \mathsf{Dec})$ refer to asymmetric encryption.

3.2   Fair Blind Signature

Informally, a fair blind signature is an interactive protocol between three parties: the user, the issuer and the tracer. It is defined by eight probabilistic polynomial-time algorithms Setup, KeyGen, $\mathsf{Issue_{sig}}$, $\mathsf{Verify_{sig}}$, $\mathsf{Trace_{sig}}$, $\mathsf{Trace_{id}}$, $\mathsf{Match_{sig}}$, and $\mathsf{Match_{id}}$ as follows. For a formal functional definition of fair blind signature schemes see for example [1,15].

- Setup is a parameter generation algorithm that takes the security parameter λ, and outputs the common parameters $params$ for the following algorithms; $params \leftarrow \mathsf{Setup}(1^\lambda)$.
- KeyGen is a key generation algorithm that takes the parameter $params$, and outputs a key pair $(\mathsf{sk}, \mathsf{pk})$; $(\mathsf{sk}, \mathsf{pk}) \leftarrow \mathsf{KeyGen}(params)$.
- $\mathsf{Issue_{sig}}$ is an algorithm that takes the message msg and outputs a blind signature; $\Sigma_{mgs} \leftarrow \mathsf{Issue_{sig}}(msg)$.
- $\mathsf{Verify_{sig}}$ is an algorithm to verify the signature Σ_{mgs}. It outputs 1 if sig_m is valid, and 0 otherwise; $0/1 \leftarrow \mathsf{Verify_{sig}}(\Sigma_{mgs})$.
- $\mathsf{Trace_{sig}}$ is a revocation algorithm that generates a resulting signature sig'_m, where this signature is yielded from the target session identifier id_u; $sig'_u \leftarrow \mathsf{Trace_{cred}}(id_u)$.
- $\mathsf{Trace_{id}}$ is a revocation algorithm that generates the session identifier id'_u which has produced target signature sig_u; $id'_u \leftarrow \mathsf{Trace_{id}}(sig_u)$.
- $\mathsf{Match_{sig}}$ is a matching algorithm that examines whether the original signature sig_u matches to the resulting signature sig'_u or not. It outputs 1 if they match, and 0 otherwise; $0/1 \leftarrow \mathsf{Match_{sig}}(sig_u, sig'_u)$.
- $\mathsf{Match_{id}}$ is a matching algorithm that examines whether the original session identifier id_u matches to the resulting identifier id'_u or not. It outputs 1 if they match, and 0 otherwise; $0/1 \leftarrow \mathsf{Match_{id}}(id_u, id'_u)$.

4   Construction Overview

An auditable blind credential system has six participants (see Fig. 1): the issuer, the user, the verifier, the tracer, the inspector and the privacy-preserving smart contract platform. The user is the holder of a credential. The issuer is in charge of blindly issuing a credential. The verifier is responsible for checking the validity of the credential. The tracer is used to reveal the relationship of the credential and its identity. It is noted that, to have a clear understanding, we introduce the concept of tracer and allow both the issuer and the verifier to act as the tracer. The inspector is used to check the suspicious revocation activities and report them. The privacy-preserving smart contract platform is employed as a reve-lator to provide the revocation service. The privacy-preserving smart contract platform includes two types of blockchain nodes: the TEE-powered blockchain nodes and the consensus nodes. The TEE-powered blockchain nodes are com-posed of the contract TEE and the key manager TEE, where contract TEE is used to execute the smart contract and then encrypt the resulting state with the key from the key manager TEE. The consensus nodes are used to achieve the agreement of the encrypted state of the smart contract.

Table 1. A high-level description of anonymity revocation with blockchain

---

System Setup

$params \leftarrow \mathsf{Setup}(1^\lambda)$; the system takes 1^λ and outputs the system parameters $params$.
$(\mathsf{sk}_*, \mathsf{pk}_*) \leftarrow \mathsf{KeyGen}_{\mathsf{entities}}(params)$; the entities (issuer, user, tracer) input $params$ and output their key pair $(\mathsf{sk}_*, \mathsf{pk}_*)$.

Smart Contract Registration

$\widehat{contract} \leftarrow \mathsf{Deploy}_{\mathsf{contract}}(params, code)$; the system takes $params$ and a piece of contract code $code$ and outputs the privacy-preserving smart contract $\widehat{contract}$.
$(\mathsf{sk}_t, \mathsf{pk}_t) \leftarrow \mathsf{KeyGen}_{\mathsf{ppsc}}(params, \widehat{contract})$; given $params$ and $\widehat{contract}$, $\mathsf{KeyGen}_{\mathsf{ppsc}}$ generates the tracing key paris $(\mathsf{sk}_t, \mathsf{pk}_t)$.

Credential Generation

$sig_{attrs} \leftarrow \mathsf{Issue}_{\mathsf{sig}}(attrs, \mathsf{pk}_t, \ldots)$; the issuer inputs the user's attributes $attrs$ and public tracing key pk_t, etc., and outputs the signature of these attributes.
$cred_u \leftarrow \mathsf{FormCred}(attrs, sig_{attrs})$; the issuer inputs the attributes and its signature and outputs a credential.

Credential Verification

$0/1 \leftarrow \mathsf{Verify}_{\mathsf{sig}}(cred_u)$; the verifier checks the signature of the credential $cred_u$ with output 0 or 1.

Credential Tracing

$cred_u \leftarrow \mathsf{Trace}_{\mathsf{cred}}(id'_u)$; $\mathsf{Trace}_{\mathsf{cred}}$ takes the identity id'_u and outputs the credential of that identity.
$tran_{cred} \leftarrow \mathsf{FormTrans}(cred_u, \widehat{contract})$; the tracer invokes $\widehat{contract}$ to obtain the $tran_{cred}$ that contains the encrypted $cred_u$.

Identity Tracing

$id_u \leftarrow \mathsf{Trace}_{\mathsf{identity}}(cred'_u)$; $\mathsf{Trace}_{\mathsf{identity}}$ takes the credential $cred'_u$ and outputs the identity of that credential.
$tran_{id} \leftarrow \mathsf{FormTrans}(id_u, \widehat{contract})$; the tracer invokes $\widehat{contract}$ to obtain the $tran_{id}$ that contains the encrypted id_u.

Tracing Inspection

$views_t \leftarrow \mathsf{Collect}_{\mathsf{trans}}(\mathsf{pk}_t, type)$; the inspector scans the blockchain to collect the tracer's invoking records (represented as the transactions) depending on the type of identity tracing or credential tracing.
$0/1 \leftarrow \mathsf{Inspect}_{\mathsf{trans}}(views_t)$; the inspector takes the $views_t$ and outputs the inspection result.

In general, a basic version of auditable blind credential system works as follows: the system sets up the parameters and prepares for the key pairs for the issuer, the user and the tracer. Then, the system sends a smart contract to a TEE-powered blockchain node to obtain a privacy-preserving contract $\widehat{contract}$, in which the method name, arguments, and return data are externally invisible.

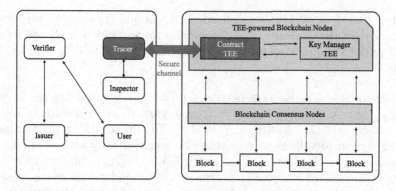

Fig. 1. Overview of our construction

Then, the system invokes the $\widehat{contract}$ through contract TEE to generate the tracing key pair (x_t, y_t). The private key x_t is kept secretly, and only contract TEE can access it internally. Key y_t is public and is used in the issuing protocol. Next, the user authenticates himself to the issuer to obtain an anonymous credential. After that, the user shows the credential to the verifier who wants to check the validity. So far, due to the blind issuance, neither the issuer nor the verifier knows the relationship of the credential and its holder.

In the revocation stage, the tracer firstly builds an encrypted and authenticated channel with the contract TEE (one crucial property of remote attestation [26] in TEE). Then, given the user's identifier or the anonymous credential, the $\widehat{contract}$ lifts the blindness and returns the result to the tracer bearing a transaction. Due to the protection of the encrypted channel, the contents of the transaction including the input and the output data are kept secret. However, the invoking records of the transaction remain visible and become immutable because of the confirmation by the consensus nodes. Alternatively, any entity can see the fact that the tracer is interacting with the contract, but nobody except the tracer knows the exact data in the transaction. Subsequently, the inspector scans the blockchain to collect the tracer's calling records and inspect the suspicious credential tracing activity. We give a high-level description in Table 1.

5   Concrete Instantiation

In this section, we present an instantiation based on Abe's [1] blind signature scheme and the privacy-preserving smart contract platform of Ekiden [11]. For security and efficiency purposes, we slightly modified Abe's [1] scheme by using elliptic curve cryptography. Thus, all the following arithmetic operations are based on addition of points and hereafter unless otherwise noted.

Let $\mathcal{G}$ be a probabilistic polynomial-time algorithm that generates an elliptic curve group $(E(\mathbb{Z}_p), p, q, g, h) \leftarrow \mathcal{G}(1^\lambda)$, where p is a big prime number, q is the order and (g, h) are elements of $E(\mathbb{Z}_p)$. Hash functions $\mathcal{H}_1 : \{0, 1\}^\star \rightarrow \langle g \rangle$, and

$\mathcal{H}_2, \mathcal{H}_3 : \{0,1\}^* \rightarrow \{0,1\}^{|q|}$ are defined. The function $\mathcal{H}_1$ refers to mapping an arbitrary string to an element of the subgroup $\langle g \rangle$ and function $\mathcal{H}_2$ and function $\mathcal{H}_3$ all refer to mapping an arbitrary string to an element of $\mathbb{Z}_q$ with the fixed length.

Key Generation. The issuer generates a public key y and a tag key z, where $x \in \mathbb{Z}_q$, $y = g^x \bmod q$ and $z = \mathcal{H}_1(p, q, g, h, y)$. A user generates a key pair (γ, ξ), where $\gamma \in \mathbb{Z}_q$ and $\xi = g^\gamma \bmod q$. To simplify the instantiation, we use the session identifiers to represent the user's identity and allow one user to generate multiple identities $(\gamma_1, \xi_1), (\gamma_2, \xi_2), \ldots (\gamma_n, \xi_n)$. Similarly, the tracer generates the session key pair (ι, τ), where $\iota \in \mathbb{Z}_q$ and $\tau = g^\tau \bmod q$. It should be noted that the tracer's session key is only used to establish the authenticated channels to the contract TEE.

Contract Registration. The system compiles pieces of code of a smart contract $\widetilde{contract}$ and sends its bytecode to a TEE-powered blockchain node. Then, the TEE-powered blockchain node first loads bytecode into the contract TEE. Then, the contract TEE creates a new contract identifier *ppsc*, obtains a fresh internal contract key pair $(\mathsf{pk}_{cid}^{in}, \mathsf{sk}_{cid}^{in})$ and an internal state key k_{state} from the key manager TEE. Thereafter, the contract TEE outputs an encrypted initial contract state $state_{init} = \mathcal{SM}.\mathsf{Enc}(\mathsf{k}_{state}, \overrightarrow{state_0})$ and an attestation Ω_{cid}, where Ω_{cid} is used to prove the correctness of this initialization. After that, the TEE-powered blockchain node gets a proof π of Ω_{cid} by the attestation service and push the final composition $(\widetilde{contract}, \mathsf{pk}_{cid}^{in}, state_{init}, \Omega_{cid}, \pi)$ to the blockchain consensus nodes. The blockchain consensus nodes would like to accept this smart contract if all the attestations and proofs are verified successfully. As for parameter registration, given the common parameters of $\mathrm{E}(\mathbb{Z}_p)$ and the public key of an issuer, say pk_i, $\widetilde{contract}$ takes a random number x_t under $\mathbb{Z}_q$ as the private tracing key and generates its public key $y_t = g^{x_t} \bmod q$. The private key x_t is held by the secret state, which can only be accessed by the contract TEE internally.

5.1   Blind Issuance

Credential Generation. Credential generation is an interactive protocol that involves only the user and issuer, which means that it runs independently from the privacy-preserving smart contract. The main idea of this protocol is witness indistinguishable [14]. Namely, the issuer owns a key pair (x, y) where $x \in \mathbb{Z}_q, y = g^x \bmod q$, and a "one-time" tag key pair (ω, z), where $\omega \in \mathbb{Z}_q, z = g^\omega$. The signature can only be issued by the real private key x but no one can distinguish which of the two secret keys (x or ω) was used. A full description is as follows: The user firstly computes $z_u = z^{1/\gamma}$ and proves to the issuer that $\log_g \xi$ is equal to $\log_{z_u} z$. Then, the issuer generates random string v, and computes $z_1 = y_t^v$ and $z_2 = z_u/z_1$, and then proves to the user that z_1 is made as it should be. Based on y, z_1, z_2, the issuer and the user engage in an interactive proof protocol, in a witness indistinguishable way, to prove the knowledge of the following two parts:

- **y-side:** proof of knowledge of x of $y = g^x$.
- **z-side:** proof of knowledge of (ω_1, ω_2) of $b_1 = g^{\omega_1}, b_2 = g^{\omega_2}$.

After that, the user blinds (z_1, z_u) into (ξ_1, z) by raising them with the private key λ under the standard diversion technique [28]. The converted proof is eventually transformed to a signature with the Fiat-Shamir technique. Next, the issuer stores ξ^v as the identity of this session. Clearly, ξ^v is easy to map to known ξ which is verified in key generation step. Finally, the user outputs a $cred_u$ with Σ, say $\Sigma = (\zeta_1, \rho, \overline{\omega}, \sigma_1, \sigma_2, \delta, m)$ is the signature for the message m.

Credential Verification. Credential verification, proceeding after credential generation, is another interactive protocol that runs independently from blockchain involving only the user and the verifier. We say a credential (Σ, m) is *valid* if it satisfies:

$$\overline{\omega} + \delta = \mathcal{H}_2(\zeta_1 | g^\rho y^{\overline{\omega}} | g^{\sigma_1} \zeta_1^\delta | h^{\sigma_2} (z/\zeta_1)^\delta | m).$$

5.2 Auditable Revocation

Credential Tracing. Credential Tracing is an interactive protocol that involves the tracer, the TEE-powered blockchain node and blockchain consensus nodes. It covers the following sub-protocols:

1. A tracer first fetches the pk_{cid} of the tracing contract $\widehat{contract}$, and then encrypts the input of the user's identity ξ^v as $inpt_c = \mathcal{ASM}.\mathsf{Enc}(\mathsf{pk}_{cid}, \xi^v)$ and sends the $\widehat{contract}$ within $inpt_c$ to a TEE-powered blockchain node. Obviously, the input of this smart contract remains secret due to encryption.
2. To start the process of the execution, the TEE-powered blockchain node first loads the contract $\widehat{contract}$, the input $inpt_c$ and the previous encrypted state $state_{init}$ into the contract TEE.
3. The contract TEE decrypts inp_c and $state_{init}$ with the keys from the key manager TEE, and starts to execute the anonymity tracing function with output I_{cred} and state $state_t$. Observes that,

$$I_{cred} = (\xi^v)^{x_t} = g^{\gamma v x_t} = y_t^{\gamma v} = \zeta_1. \tag{1}$$

4. The contract TEE obtains a fresh symmetric-key k_{cid}^{out} from the key manager TEE and calculates a new encrypted output $outp_{new}^{TEE} = \mathcal{SM}.\mathsf{Enc}(\mathsf{sk}_{cid}^{out}, I_{cred})$ and a new encrypted state $state_{new}^{TEE} = \mathcal{SM}.\mathsf{Enc}(\mathsf{k}_{state}, state_t)$. Then, it sends $state_{new}^{TEE}$, $outp_{new}^{TEE}$ and the proper attestation to the tracer through a secure channel established by the tracer's session keys (ι, τ).
5. The tracer acknowledges the reception by calling back the TEE-powered blockchain node, which triggers the contract TEE to send the transaction $tran = (\widehat{contract}, outp_{new}^{TEE}, state_{new}^{TEE}, proof)$ to the blockchain. $proof$ is used to protect the integrity of the transaction and the correctness of the $outp_{new}^{TEE}$ and $state_{new}^{TEE}$.

6. Once the consensus nodes confirm $\widehat{contract}$, the contract TEE decrypts $outp_{new}^{TEE}$ and $state_{new}^{TEE}$ as $outp_{new}^t$ and $state_{new}^t$ and then sends them to the tracer through the mentioned secure channel.

7. The tracer parses the $outp_{new}^t$ and $state_{new}^t$ and ultimately learns I_{cred} that is the relationship of the credential and that real owner.

Among all the sub-protocols, we emphasize that the sub-protocols five and six are atomic operations, and we refer to [11] for more details. Also, we highlight two main features. Firstly, $\widehat{contract}$ will be confirmed by the consensus nodes mentioned in sub-protocol six. Thus, the contract invoked eventually becomes immutable and auditable(see Appendix A for security proofs). Second, the output $outp_{new}^t$ and state $state_{new}^t$ are kept secret in the whole life of the execution and transmission.

Identity Tracing. Identity tracing and the credential tracing have the same tracing mechanism. Due to the space limit, we skip its full description. Observes that,

$$I_{id} = \zeta_1^{1/x_t} = z_1^{\lambda/x_t} = y_t^{v\lambda/x_t} = g^{v\lambda} = \xi^v. \tag{2}$$

Since ξ^v is stored or published by the issuer, the tracer can instantly identify the user who issued the credential.

Tracing Inspection. The tracing activities checking is straightforward. Given the inspector type (identity tracing or credential tracing) and the smart contract identifier, the inspector scans the blockchain to collect all the transactions related to this contract. Then, the inspector checks all these transactions to recognise suspicious activities.

6   Implementation and Evaluation

We have implemented a proof of concept of our instantiation. Next we report on our proof of concept and its performance. The corresponding code has been made available open source and is to be found at https://github.com/typex-1/auditable-credential-core.

6.1   Implementation

We focus on implementing the blind issuance protocols and the anonymity revocation smart contracts, and leave the implementation of the menthoned TEE-related protocols to the Oasis Devnet [11]. Specifically, our implementation is divided into two modules: the issuing module and the tracing module. The issuing module covers the protocol of credential generation and credential verification, and it is realised by Python in 168 lines of code. The issuing module is responsible for blindly issuing credentials and verifying the issued ones. Meanwhile, the tracing module which performs the protocol of credential tracing and identity tracing is achieved by Solidity in 449 lines of code and deployed in Oasis Devnet. The tracing module allows the tracer to uncover the identity of a credential or the credential of a specific user.

```
// example code;
mapping (address => uint256) private CredentialTraceResults;
function CredentialRevocation (uint256 upsilon) {
    CredentialTraceResults[upsilon] = power(upsilon, xt, p);
}
```

Two key properties are highlighted in our implementation: the full protection of private state and the auditable anonymity tracing records. The full protection of private state is represented as that the input data and the output data in the contract are kept secret in the full life cycle. For example, as is shown in the example code, the parameter of *CredentialTraceResults* is designed to privately store the relationship of the identity and credential. The other entities can not read them unless through an end-to-end secure channel that has been established with the contract TEE. The auditability of anonymity tracing records is evident in that all the smart contract invoking records are publicly visible and immutable (The Fig. 2 is a smart contract creation and invoking example). In addition, we provide a web-based client to present an interactive process of credential and identity tracing and show the full code in the repository[1].

6.2   Evaluation

Our performance evaluation covers five operations: tracing parameter generation, credential issuing, credential verifying, credential tracing and identity tracing (see Table 2). All experiments are conducted on a Dell precision 3630 tower with 16 GB of RAM and one 3.7 GHz six core i7-8700K processors running Ubuntu 18.04. Experiments are measured in seconds through wall clock run time, where a time difference is obtained between the start and end of the code execution. To have an accurate and fair test result, we repeat the measure for each execution 300 times and calculate its average. Also, to simplify the performance evaluation, we measure the running time of each step and accumulate them together if there are many steps involved. It is noted that, all operations take much less than one second to complete and the credential issuing is the main performance bottleneck. This operation takes more time than others because issuing a new credential requires many interactions between users and issuers. Fortunately, this bottleneck can be ignored in real applications because after all it meets the nature of credential using scenario, which means a credential is issued only once but could be verified or traced multiple times.

We then examine the operating cost. Similar to the performance testing, the cost evaluation covers five mentioned credential operations. Table 2 shows the data size and the cost of these operations in gas under an elliptic curve with 128 bits security level. An analysis of the data size and cost points to some trends. The data size of the operation of parameter generation is the largest since this operation needs to register the group parameters to the smart contract. Surprisingly, the cost of the parameter generation is not the largest as this operation does not cover the

---

[1] https://github.com/typex-1/auditable-credential-core.

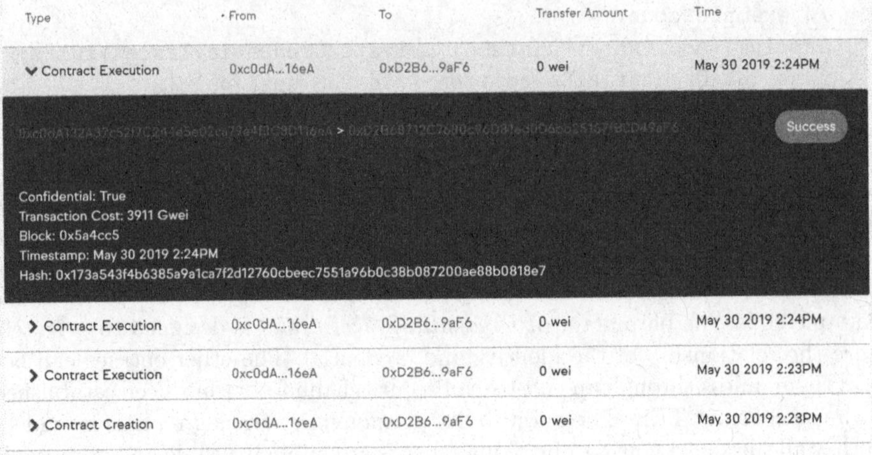

Fig. 2. Credential anonymity revocation records.

complex computations. On the contrary, the data size of the operation of the credential issuing and verifying is zero, and there is no gas cost since these operations are executed independently from the blockchain. Meanwhile, the credential tracing and the identity tracing have static gas cost since the length of input data of these operations is constant, and the data handling procedure is fixed. In our scheme, a one-time elliptic-curve exponentiation (see Eqs. (1) and (2)) is adequate to conduct the complete tracing activity. The gas cost of the one-time computation is quite lower and more easier to adopt by users when compared with some blockchain-based applications such as [4,33], where they have massive elliptic-curve exponentiation operations and significant cost.

Finally, we conduct latency testing as latency is an essential consideration for adopting a system. For our implementation, the latency time includes blockchain confirming time, the network request time and network response time. It is observed that the latency of credential issuing as well as identity verifying is much smaller than other operations. The main reason behind this is that these two operations run independently from blockchain and do not wait for the block to be confirmed. Meanwhile, the average latency of credential tracing and identity tracing is approximately eighteen seconds, which would be a primary drawback

Table 2. The performance, input data size, cost and latency of various operations.

Operation	Performance (seconds)	Size (bytes)	Gas	Latency (seconds)
Parameter generation	0.00084	260	20672	14.781
Credential issuing	0.00740	0	0	1.601
Credential verifying	0.00232	0	0	1.175
Credential tracing	0.00306	132	390261	17.538
Identity tracing	0.00455	132	388944	18.905

of our system. Given these latency constraints, our system, at least built on the current version of Oasis Devnet is not suitable for applications that require fast credential tracing or identity tracing. However, for some privacy-priority applications such as medical record tracing system, our scheme provides a powerful framework to protect patients' privacy.

The main roadblock to the business adoption of blockchain is its low throughput of on-chain transaction. Our system, armed with the blockchain and trusted execution environment, suffers the same scalability issues. Fortunately, the flexible smart contract makes our scheme easier to support batch anonymity revealing. This means a tracer can collect a group of credentials and send them the blockchain once. With such a mechanism, a massive chunk of tracing transactions can be off-loaded from the blockchain, which mitigates the scalability flaw. Meanwhile, some efforts [13,36] have been made to increase the scalability of the blockchain. Our system would benefit from these works.

7   Example Applications

Our scheme has numerous practical applications in some privacy-sensitive scenarios. Two typical use cases are described as follows:

Medical Record Protection. Our scheme may be used for privacy medical record protection specifically for unrestricted research purposes. A medical record is supposed to be very sensitive in some cases such as in HIV and sexually transmitted infections. A hospital might share the medical record with a research institution without patients' permission thereby causing information leakage. Our mechanism allows the hospital to show the real patient records without knowing the patients' real identities, so privacy is respected. In the case of family genetic disorder, patients may disclose their identities to the research institution on their own free will by invoking the privacy-preserving smart contract.

Vehicle Registration Management. Traditionally, the vehicle registration office issues a vehicle plate number knowing all the identities and corresponding car information. If someone with an intent to identify the specific driver colludes with the registration office, privacy invasion occurs. Moreover, the plate number may become a surveillance tool in conjunction with a closed-circuit television camera, which is in wide use almost anywhere. Our scheme allows the issuance of a vehicle plate·number without the vehicle registration office knowing the relationship between the number and the driver's identity. Furthermore, it allows the certificate to be traced in an auditable way when some emergencies such as traffic accidents.

8   Conclusion

Anonymity credentials and anonymity revocation were proposed several decades ago, but they have not yet gained significant adoption. Some potential obstacles are the lack of auditability and neutrality for the revocation process.

In this paper, we proposed a blockchain-powered traceable anonymous credential framework. Our approach allows the issuer to blindly issue a credential, then leverages a privacy-preserving smart contract that acts as a revelator to trace the credential. More importantly, all these tracing activities are auditable due to the immutable smart contract calling records provided by the public ledger. The auditability and neutrality guaranteed by the blockchain avoid misuse of tracing and potential collision problems to a great extent.

Future Work. Even if our scheme provides a powerful approach to trace the anonymity with auditability, in practice, it is still possible that one of a tracer's private keys is stolen or misused. Fortunately, the flexibility of smart contract makes our scheme amenable to support threshold-revealing using well-known multiparty computation techniques.

Acknowledgement. The authors would like to thank Feng Liu, Geyang Wang and Alphea Pagalaran for their constructive suggestions on the manuscript. The authors would also like to thank the anonymous referees for their valuable comments that improved the quality of the paper.

A   Appendix A

This appendix gives an informal security proof of credential tracing and identity tracing for the concrete instantiation described in Sect. 5.

Theorem 1. *The credential tracing of our scheme is auditable under the following assumptions that the TEE is a secure enclave, the DDH assumption holds, the fair blind signature scheme is signature traceable and the blockchain meets the property of liveness.*

Proof. Given a valid identity, say ξ_0^v, there are four possible ways for an adversary to obtain its corresponding credential (signature) Σ_0 without being audited. (i) An adversary has successfully accessed the private key x_t which is stored in the TEE. Thus, the adversary can locally calculate the elliptic-curve exponentiation (see Eqs. (1) and (2)) to conduct the complete tracing activity without interacting with the blockchain. (ii) As mentioned in sub-protocol six of credential tracing, the contract TEE sends the tracing resulting $outp_{new}^t$ and $state_{new}^t$ to the tracer through a DDH-based secure channel. A compromised secure channel makes an adversary free to lift anonymity with auditability. (iii) An adversary has successfully forged a valid signature $\Sigma_0^\star$ independently from the blockchain, where $\Sigma_0^\star$ meets the conditions: $\Sigma_0^\star \neq \Sigma_0$ and $1 \leftarrow \mathsf{Match}_{\mathsf{sig}}(\Sigma_0, \Sigma_0^\star)$. (iv) An adversary called the smart contract $contract$ and successfully hid the invoked transactions from the inspector.

Scenario (i) contradicts our assumption that the TEE provides an isolated secure environment. The proof of Scenario (ii) is done by contradiction. Suppose that there exists an adversary $\mathcal{A}$ that compromised the secure channel with success probability $\mathsf{Adv}_{\mathcal{A}}^{\mathsf{sc}}$, where $\mathsf{Adv}_{\mathcal{A}}^{\mathsf{sc}}$ is not negligible. Then, based on $\mathsf{Adv}_{\mathcal{A}}^{\mathsf{sc}}$ of the adversary $\mathcal{A}$, we can construct another adversary $\mathcal{B}$ to solve DDH problem with non-negligible advantage $\mathsf{Adv}_{\mathcal{A}}^{\mathsf{ddh}}$. However, it contradicts the DDH assumption. Scenario (iii) indicates two properties: an adversary has successfully forged a signature and the forged signature and the original signature can be linked to one identity. These properties violate the unforgeability and signature traceability of fair blind signature scheme, which was proved secure by Abe [1]. Liveness [16] guarantees that the submitted transactions will eventually be included in the ledger. For Scenario (4), if an adversary can successfully hide the invoked transaction, that indicates the transaction does not eventually appear in the ledger, which contradicts our assumption that the blockchain meets the liveness property.

Theorem 2. *The identity tracing of our scheme is auditable under the following assumptions that the TEE is a secure enclave, the DDH assumption holds, the fair blind signature scheme is session traceable and the blockchain meets the property of liveness.*

Proof. Given a valid credential (signature), say Σ_0, there are four possible ways for an adversary to illegally obtain the corresponding identity ξ_0^v without being audited. An adversary has successfully (i) accessed the TEE, (ii) compromised the DDH-based secure channel, (iii) linked one credential to more than one identity and (iv) damaged the liveness of blockchain. Scenario (i), (ii) and (iv) in Theorem 2 are the same as Theorem 1. Thus, this part of proof is omitted here. Scenario (iii) violates the unforgeability and the session traceability of fair blind signature scheme, which was proved secure by Abe [1].

References

1. Abe, M., Ohkub, M.: Provably secure air blind signatures with tight revocation. In: Boyd, C. (ed.) ASIACRYPT 2001. LNCS, vol. 2248, pp. 583–601. Springer, Heidelberg (2001). https://doi.org/10.1007/3-540-45682-1_34
2. Blazy, O., Canard, S., Fuchsbauer, G., Gouget, A., Sibert, H., Traoré, J.: Achieving optimal anonymity in transferable E-cash with a judge. In: Nitaj, A., Pointcheval, D. (eds.) AFRICACRYPT 2011. LNCS, vol. 6737, pp. 206–223. Springer, Heidelberg (2011). https://doi.org/10.1007/978-3-642-21969-6_13
3. Brickell, E.F., Gemmell, P., Kravitz, D.W.: Trustee-based tracing extensions to anonymous cash and the making of anonymous change. In: SODA (1995)
4. Bünz, B., Agrawal, S., Zamani, M., Boneh, D.: Zether: towards privacy in a smart contract world. IACR Cryptology ePrint Archive, p. 191 (2019)
5. Bünz, B., Agrawal, S., Zamani, M., Boneh, D.: Zether: towards privacy in a smart contract world. Cryptology ePrint Archive, Report 2019/191 (2019). https://eprint.iacr.org/2019/191
6. Camenisch, J., Lysyanskaya, A.: An efficient system for non-transferable anonymous credentials with optional anonymity revocation. In: Pfitzmann, B. (ed.) EUROCRYPT 2001. LNCS, vol. 2045, pp. 93–118. Springer, Heidelberg (2001). https://doi.org/10.1007/3-540-44987-6_7

7. Camenisch, J., Maurer, U., Stadler, M.: Digital payment systems with passive anonymity-revoking trustees. In: Bertino, E., Kurth, H., Martella, G., Montolivo, E. (eds.) ESORICS 1996. LNCS, vol. 1146, pp. 33–43. Springer, Heidelberg (1996). https://doi.org/10.1007/3-540-61770-1_26

8. Camenisch, J., Mödersheim, S., Sommer, D.: A formal model of identity mixer. In: Kowalewski, S., Roveri, M. (eds.) FMICS 2010. LNCS, vol. 6371, pp. 198–214. Springer, Heidelberg (2010). https://doi.org/10.1007/978-3-642-15898-8_13

9. Canard, S., Traoré, J.: On fair E-cash systems based on group signature schemes. In: Safavi-Naini, R., Seberry, J. (eds.) ACISP 2003. LNCS, vol. 2727, pp. 237–248. Springer, Heidelberg (2003). https://doi.org/10.1007/3-540-45067-X_21

10. Chaum, D.: Blind signatures for untraceable payments. In: Chaum, D., Rivest, R.L., Sherman, A.T. (eds.) Advances in Cryptology, pp. 199–203. Springer, Boston, MA (1983). https://doi.org/10.1007/978-1-4757-0602-4_18

11. Cheng, R., et al.: Ekiden: a platform for confidentiality-preserving, trustworthy, and performant smart contract execution. arXiv:1804.05141 [cs], April 2018

12. Escala, A., Herranz, J., Morillo, P.: Revocable attribute-based signatures with adaptive security in the standard model. In: Nitaj, A., Pointcheval, D. (eds.) AFRICACRYPT 2011. LNCS, vol. 6737, pp. 224–241. Springer, Heidelberg (2011). https://doi.org/10.1007/978-3-642-21969-6_14

13. Eyal, I., Gencer, A.E., Sirer, E.G., Van Renesse, R.: Bitcoin-NG: a scalable blockchain protocol. In: 13th USENIX Symposium on Networked Systems Design and Implementation NSDI 16, pp. 45–59 (2016)

14. Feige, U., Shamir, A.: Witness indistinguishable and witness hiding protocols. In: STOC (1990)

15. Fuchsbauer, G., Vergnaud, D.: Fair blind signatures without random oracles. In: Bernstein, D.J., Lange, T. (eds.) AFRICACRYPT 2010. LNCS, vol. 6055, pp. 16–33. Springer, Heidelberg (2010). https://doi.org/10.1007/978-3-642-12678-9_2

16. Garay, J., Kiayias, A., Leonardos, N.: The bitcoin backbone protocol: analysis and applications. In: Oswald, E., Fischlin, M. (eds.) EUROCRYPT 2015. LNCS, vol. 9057, pp. 281–310. Springer, Heidelberg (2015). https://doi.org/10.1007/978-3-662-46803-6_10

17. Hufschmitt, E., Traoré, J.: Fair blind signatures revisited. In: Takagi, T., Okamoto, T., Okamoto, E., Okamoto, T. (eds.) Pairing 2007. LNCS, vol. 4575, pp. 268–292. Springer, Heidelberg (2007). https://doi.org/10.1007/978-3-540-73489-5_14

18. Jakobsson, M., Yung, M.: Revokable and versatile electronic money (extended abstract). In: ACM Conference on Computer and Communications Security (1996)

19. Jakobsson, M., Yung, M.: Distributed "magic ink" signatures. In: Fumy, W. (ed.) EUROCRYPT 1997. LNCS, vol. 1233, pp. 450–464. Springer, Heidelberg (1997). https://doi.org/10.1007/3-540-69053-0_31

20. Kalodner, H., Goldfeder, S., Chen, X., Weinberg, S.M., Felten, E.W.: Arbitrum: scalable, private smart contracts. In: 27th USENIX Security Symposium, pp. 1353–1370 (2018)

21. Kiayias, A., Zhou, H.-S.: Concurrent blind signatures without random oracles. In: De Prisco, R., Yung, M. (eds.) SCN 2006. LNCS, vol. 4116, pp. 49–62. Springer, Heidelberg (2006). https://doi.org/10.1007/11832072_4

22. Kosba, A., Miller, A., Shi, E., Wen, Z., Papamanthou, C.: Hawk: the blockchain model of cryptography and privacy-preserving smart contracts. In: 2016 IEEE Symposium on Security and Privacy (SP), pp. 839–858, May 2016

23. Kwon, T.: Privacy preservation with X.509 standard certificates. Inf. Sci. 181(13), 2906–2921 (2011)

24. Luu, L., Chu, D.H., Olickel, H., Saxena, P., Hobor, A.: Making smart contracts smarter. In: Proceedings of the 2016 ACM SIGSAC Conference on Computer and Communications Security, CCS 2016, pp. 254–269. ACM, New York (2016)

25. McCorry, P., Shahandashti, S.F., Hao, F.: A smart contract for boardroom voting with maximum voter privacy. In: Kiayias, A. (ed.) FC 2017. LNCS, vol. 10322, pp. 357–375. Springer, Cham (2017). https://doi.org/10.1007/978-3-319-70972-7_20

26. McKeen, F., et al.: Innovative instructions and software model for isolated execution. In: Proceedings of the 2nd International Workshop on Hardware and Architectural Support for Security and Privacy - HASP 2013, Tel-Aviv, Israel, p. 1. ACM Press (2013)

27. Nakamoto, S.: Bitcoin: a peer-to-peer electronic cash system (2008)

28. Okamoto, T., Ohta, K.: Divertible zero knowledge interactive proofs and commutative random self-reducibility. In: Quisquater, J.-J., Vandewalle, J. (eds.) EUROCRYPT 1989. LNCS, vol. 434, pp. 134–149. Springer, Heidelberg (1990). https://doi.org/10.1007/3-540-46885-4_16

29. Paquin, C., Zaverucha, G.: U-prove cryptographic specification v1. 1. Technical report, Microsoft Corporation (2011)

30. Park, S., Park, H., Won, Y., Lee, J., Kent, S.: Traceable anonymous certificate. Technical report RFC5636, RFC Editor, August 2009. https://doi.org/10.17487/rfc5636, https://www.rfc-editor.org/info/rfc5636

31. Rannenberg, K., Camenisch, J., Sabouri, A.: Attribute-based Credentials for Trust. Identity in the Information Society. Springer, Cham (2015). https://doi.org/10.1007/978-3-319-14439-9

32. von Solms, S., Naccache, D.: On blind signatures and perfect crimes. Comput. Secur. 11(6), 581–583 (1992)

33. Sonnino, A., Al-Bassam, M., Bano, S., Meiklejohn, S., Danezis, G.: Coconut: Threshold issuance selective disclosure credentials with applications to distributed ledgers. arXiv preprint arXiv:1802.07344 (2018)

34. Stadler, M., Piveteau, J.-M., Camenisch, J.: Fair blind signatures. In: Guillou, L.C., Quisquater, J.-J. (eds.) EUROCRYPT 1995. LNCS, vol. 921, pp. 209–219. Springer, Heidelberg (1995). https://doi.org/10.1007/3-540-49264-X_17

35. Szabo, N.: Smart contracts: building blocks for digital markets. EXTROPY J. Transhumanist Thought (16) (1996)

36. Zamfir, V.: Casper the friendly ghost: a correct by construction blockchain consensus protocol. White paper (2017). https://github.com/ethereum/research/blob/master/papers/caspertfg/caspertfg.pdf

Bonded Mining: Difficulty Adjustment by Miner Commitment

George Bissias[1]($\boxtimes$), David Thibodeau[2], and Brian N. Levine[1]

[1] College of Information and Computer Sciences, UMass Amherst, Amherst, USA
{gbiss,levine}@cs.umass.edu
[2] Florida Department of Corrections, Tallahassee, USA
davidpthibodeau@gmail.com

Abstract. Proof-of-work blockchains must implement a difficulty adjustment algorithm (DAA) in order to maintain a consistent inter-arrival time between blocks. Conventional DAAs are essentially feedback controllers, and as such, they are inherently reactive. This approach leaves them susceptible to manipulation and often causes them to either under- or over-correct. We present *Bonded Mining*, a proactive DAA that works by collecting hash rate commitments secured by bond from miners. The difficulty is set directly from the commitments and the bond is used to penalize miners who deviate from their commitment. We devise a statistical test that is capable of detecting hash rate deviations by utilizing only on-blockchain data. The test is sensitive enough to detect a variety of deviations from commitments, while almost never misclassifying honest miners. We demonstrate in simulation that, under reasonable assumptions, Bonded Mining is more effective at maintaining a target block time than the Bitcoin Cash DAA, one of the newest and most dynamic DAAs currently deployed. In this preliminary work, the lowest hash rate miner our approach supports is 1% of the total and we directly consider only two types of fundamental attacks. Future work will address these limitations.

Keywords: Difficulty adjustment · Protocols · Cryptocurrencies

1 Introduction

Blockchain protocols maintain a public ledger of account balances that are updated by authorized transactions. Proof-of-work (PoW) mining is the process of assembling transactions into blocks and earning the right to add the block to a growing chain [24]. PoW mining involves repeatedly cryptographically hashing the assembled block, each time with a different nonce. The hashes are generated uniformly at random from a space with maximum value S. When a hash falls below a *target* t, the corresponding block is said to be mined, and it is added to the blockchain. Closely related to target is the *difficulty*[1] D, which

---

[1] Technically $D = t_0/t$, where $t_0 \approx S$ is the target with highest possible difficulty, but this detail is not important for our analysis.

© Springer Nature Switzerland AG 2019
C. Pérez-Solà et al. (Eds.): DPM 2019/CBT 2019, LNCS 11737, pp. 372–390, 2019.
https://doi.org/10.1007/978-3-030-31500-9_24

is equal to S/t. The expected time required to mine a block is a function of D and the rate that hashes are generated, or *hash rate h*. Hash rate fluctuates (sometimes rapidly) over time, and therefore PoW blockchains must adjust D to ensure that the expected block time remains roughly constant. Currently, all PoW blockchains use a *difficulty adjustment algorithm* (DAA) to adjust D as h fluctuates.

Although implementations vary widely, each DAA is essentially a feedback controller analogous to a thermostat. The DAA uses previous block creation times to detect a change in h, and then it makes an adjustment to D in order to move the expected times toward the desired value T. There are three major limitations to this *reactive* approach.

1. There is a tendency to either over or under correct, which can cause oscillations in block time [29, 30].
2. Contentious hard forks create significant instability in block times for minority hash rate blockchains, which must resort to a backup controller that compensates for swings in miner hash rate allocation preference [31].
3. Most control algorithms can be gamed by miners without consequence in order to extract higher rewards [18, 21], causing fluctuations in block time as a result.
4. Feedback control is inherently reactionary; it only uses historical block time and difficulty data to produce future difficulty values.

Contributions. We present *Bonded Mining*, a protocol that enhances PoW mining with a *proactive* approach to difficulty adjustment so that inter-block times are always near their desired value despite sudden hash rate changes. The idea is to ask miners to *commit* to their individual hash rate and financially bind them to it by holding *bond*. Difficulty is adjusted based on these self-reported commitments. Miners are incentivized to commit to a realistic estimate of their future hash rate and honor their commitment, even if it becomes nominally more profitable to direct their hash rate elsewhere. The protocol is flexible: miner commitments last until they mine their next block; and they can deviate from the commitment (incurring a penalty commensurate with their deviation) as long as they are truthful about the deviation.

For security, we derive a statistical test (using on-blockchain data only) that is capable of detecting both short- and long-term deception from miners. Miners who fail the test suffer a significant financial penalty. The test is sensitive enough to detect a miner who drops to 20% of her commitment for a week or more, and it can also detect when she strays from her commitment by as little as 1% every block over the course of 70 days or more. This sensitivity comes with very little risk for honest miners. Even when a miner deviates from her commitment, if she truthfully reports that deviation, then the probability of failing *any* test over the course of a year is less than 0.3%.

Because of its proactive design and the penalties associated with deception, Bonded Mining is better capable of maintaining the desired expected block times than are conventional, reactive approaches. In Bonded Mining, the extent to which block times remain close to desired time T is the extent to which miners value their bond more than a change in their hash rate. In simulation, we find that even when miners are willing to sacrifice 25% of their bond in order to change their hash rate, Bonded Mining still demonstrates lower amplitude and duration of deviations from T than does the DAA of Bitcoin Cash, the latter of which is one of the newest and most dynamic DAAs currently deployed.

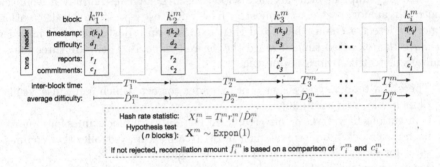

Fig. 1. An overview of Bonded Mining for miner m.

2   Protocol Details

Bonded Mining enhances existing PoW mining schemes by adding a *collateral* requirement to ensure hash rate commitments are honored. Figure 1 illustrates many aspects of the protocol, and a List of Symbols appears in the Appendix. To bootstrap, miner m must post *bond* to the blockchain by paying b coins into a *bond pool* account via a validated *deposit transaction*. The bond is posted prior to mining the first block and is locked until the miner produces his first n blocks, where n is a tunable security parameter depending on the miner's hash rate. As part of the deposit transaction, miner m also stipulates his *commitment* c_1^m for the hash rate he will apply to mining his first block, k_1^m.

Mining. Now consider the next $n - 1$ blocks mined by m: $k_2^m, \ldots, k_n^m$. Define $t(k_i^m)$ as the time that k_i^m is mined. Let $t(k_0^m)$ be defined as the time of the block containing the deposit transaction. When mining block k_i^m, miner m adds a new deposit transaction to the set of transactions being mined, which deposits an additional b coins to the bond pool. In assembling the block, he uses the typical block header, and includes two new pieces of information: *report* r_i^m and commitment c_{i+1}^m. Report r_i^m is an attestation from miner m of his actual average hash rate during the time period between $t(k_{i-1}^m)$ and $t(k_i^m)$. And c_{i+1}^m is m's commitment for the next block k_{i+1}^m. If m wishes to stop after n blocks, then he can issue a divestment transaction (see *Bond state* section below) and c_{n+1}^m is unnecessary; otherwise c_{n+1}^m should contain the commitment for the next block. When block k_i^m is eventually mined by m, the coinbase is immediately transferred to the miner's wallet, but bond is released only upon reconciliation. To be clear: the miner can adjust commitments each block starting from block

k_2^m; but reconciliation begins from block k_n^m onward and evaluates a window of the previous n blocks.

Reconciliation. If i blocks, $i > n$, have been mined by m, then his block k_i^m includes a *reconciliation transaction* that pays himself a *reconciliation payment* $f_i^m \le b$ from the bond pool. f_i^m is the refunded portion of the bond deposited by m while mining block k_{i-n}^m, which is now eligible to be reconciled. (Miners never forfeit coinbase rewards.) Miner m signs and confirms his own reconciliation transaction, but if he repays an inappropriate amount, then the transaction is considered to be invalid by other miners and the entire block is ignored. The value of f_i^m is determined in two stages. The first tests m's *reporting accuracy* via binary hypothesis test Valid($\mathbf{r}^m$), which deterministically rejects or accepts the null hypothesis that the inter-block times from $t(k_{i-n}^m), \dots, t(k_i^m)$ are samples that came from the distribution implied by reports $\mathbf{r}^m = r_{i-n}^m, \dots, r_i^m$. If the null hypothesis is rejected, then $f_i^m = 0$; i.e., m loses all his bond deposited in block k_{i-n}^m. However, if the null hypothesis is not rejected, then in the second stage we evaluate m's *commitment fulfillment* by setting

$$f_i^m = b - b \cdot \mathbf{min} \left\{ 1, \left| \frac{r_i^m - c_i^m}{c_i^m} \right| \right\}. \qquad (1)$$

We tune the hypothesis test so that if m honestly reports his hash rate for each block k_j^m, then he is very likely to pass, and then m will be repaid his bond as the absolute difference between his committed and actual hash rates.

Bond State. We define four distinct states for bonded miners: *Bootstrapping, Fully Bonded, Divested,* and *Abandoned*. A new miner who has not yet deposited nb total bond is bootstrapping. Once she has contributed n mined blocks to the main chain, her total bond reaches nb and the miner is considered fully bonded. Note that the number of blocks required to reach the Fully Bonded state depends on the miner's committed hash rate as a fraction of the total. Accordingly, miners may fluctuate between Bootstrapping and Fully Bonded states if the total hash rate changes. A fully bonded miner is eligible to divest her bond in order to reduce her hash rate on the blockchain to zero. The miner signals this intent by submitting a *divestment transaction*, which can appear in any block, not just one that she mines. The transaction contains reconciliation payments (see above) for each of her n bond deposits that remain in the bond pool. Once the divestment transaction is confirmed, the miner is fully divested: she is committed to zero hash rate and has no remaining bond. In order to begin mining again, a divested miner must proceed through the Bootstrap state. Finally, a miner can reach the Abandoned state from either the Bootstrapping or Fully Bonded states if she fails to generate a block in a reasonable amount of time given her last commitment (see below). If the miner reaches the Abandoned state, all of her bond is lost, and she transitions to the Divested state. (The bond is burned, but could be redistributed as future coinbase.)

Abandonment Detection. The test for abandonment is distinct from the test Valid. It is conducted every block (as opposed to only those generated by the miner). For each miner m in either the Bootstrapping or Fully Bonded state, we have c_i^m, the latest commitment from m. We argue in Sect. 4 that if m honors her commitment, then the inter-arrival time of her ith block, T_i^m, is exponential. Specifically, $T_i^m \sim \text{Expon}(T/c_i^m)$ where T is the target inter-block interval for the network. Now let $Q(p)$ be the quantile function for T_i^m such that $P(T_i^m < Q(p)) < p$. With probability p, we know that T_i^m will be less than $Q(p)$. Therefore, for large p, we can be almost certain that m has abandoned mining at her committed hash rate if no block hash been seen for time $Q(p)$.

For example, consider a miner who commits to 10% of the total hash rate in a protocol like Bitcoin with $T = 10$ min block times. With probability exceeding 99.999%, it will take m no more than 20 h to mine her next block. Thus, by setting the abandonment time to 20 h, we can be highly confident that m has in fact abandoned mining at her committed hash rate.

Difficulty Adjustment. Let c_i and h_i be the total *committed* and *actual* hash rates, respectively, across all miners for the time period in which block k_i will be mined. Bonded Mining makes the assumption that security parameters b and n are tuned so that miners are incentivized to honor their committed hash rate during this time period, i.e. $c_i = h_i$. It is known [25] that the hash target t is related to total hash rate h_i by the equation

$$h_i T = S/t, \tag{2}$$

where T is the expected time to produce a block and S is the size of the hash space, as defined above. Given the equivalences $c_i = h_i$ and $D = S/t$ (defined in Sect. 1), we arrive at the following formula for difficulty.

$$D_i = c_i T. \tag{3}$$

Commitment Constraints. Because the difficulty is derived directly from miner commitments (Eq. 3), it is possible for an attacker to falsely raise the difficulty arbitrarily high at a cost limited to bond b. To prevent this attack, it is important to ensure that every miner's commitment is realistic given their past hash rate. This constraint has ramifications for both bootstrapping and fully bonded miners (those with nb coins in the bond pool). For a fully bonded miner, we stipulate that she can change her commitment by no more than some multiple μ times her average commitment over the previous n blocks (e.g., $\mu = 2$). As a protection against Sybil attacks [10] described later, we allow up to fraction γ of the total commitment for block k_i to come from bootstrapping miners (i.e., miners who have deposited fewer than nb coins into the bond pool). If the aggregate commitments exceed fraction γ, then they are cut proportionately down to the maximum. Fraction γ, which is also tunable, should be fairly small (e.g., $\gamma = 0.05$) because new miners have much less at stake than established miners (fewer coins in deposit), and the network has observed fewer blocks from which to assess their hash rate potential. The values for both μ and γ should be set by the community at large (not just miners) and can also be updated regularly to respond to changes in miner composition.

Mining Pools. The Bonded Mining protocol is agnostic to the presence of mining pools, but the protocol does not greatly impinge on the ability to operate a pool effectively. Commitments must be made at the pool level, thus they can be changed only as often as the pool mines a block. A pool could aggregate constituent miner preferences in between blocks and update the overall commitment based on those preferences for the next block. Thus, if a pool mines a block approximately once an hour, then miners would be free to adjust their hash rate allocation to that pool with the same frequency.

3 Threat Model

Primary Threats. In the next sections, we focus our security analysis on attacks from miners who report their hash rate dishonestly. Rational miners intent on deviating from their commitment for x blocks will report dishonestly when their *preference* to deviate exceeds the greater of the value of *coins at risk* or penalty associated with honestly reporting the deviations (i.e., $x(b - f^m)$). (We discuss coins at risk in Sect. 5 and preference in Sect. 6.) In this sense, Bonded Mining can ensure miners honor their commitment only to the extent that the fiat value of bond exceeds the value of mining on a different blockchain. There are numerous ways that an attacker can falsify reports, but in this preliminary work we focus on *short-range* and *long-range* attacks. A short-range attack involves an attacker significantly deviating from his commitment for a relatively small number of sequential blocks. In this context, we are primarily concerned with the attacker committing to a large hash rate, and subsequently dramatically lowering that hash rate, which would tend to increase block times during the attack. In contrast, the long-range attacker deviates subtly from his commitment over the course of many blocks. Although not catastrophic during a short period of time, this attack can lead to systematic deviations from the target block time.

Sybil Attacks. A single miner (or a cooperating coalition of miners) can always split his hash rate to appear to be multiple, lower hash rate miners in a Sybil attack [10]. Therefore, it is important to make incentives equitable for miners of any hash rate in order to encourage honest representation of affiliation and avoid the formation of mining cabals. We tune the Bonded Mining protocol to ensure that: *(i)* after Bootstrap, the amount of bond locked up at any given time is proportional to the miner's committed hash rate; and *(ii)* the expected lockup time for bond and the expected probability of losing that bond are equal across hash rates for honest miners. We do not, however, attempt to align penalties for attacking miners with varying hash rates. That is to say, it will be possible that an attacker will suffer a lesser penalty by breaking up his hash rate among multiple identities. Note that this sort of asymmetry exists with other attacks. For example, a selfish mining attack only becomes possible for miners with a certain minimum percentage of the total hash rate.

Other Attacks. Bonded Mining's commitment validation test uses the interblock times of each miner's *own* blocks; therefore, one miner's hash rate cannot influence the test results of another miner. We do not consider out-of-band denial-of-service attacks. A selfish mining (SM) attack [12] by one miner would alter inter-block times of the other miners by increasing their orphan rate. However, bonded miners could easily adjust their commitment to account for orphaned blocks due to an SM attack. Alternatively, Eq. 3 could be modified to take into account a miner's orphaned blocks [14]. In fact, Bonded Mining makes it possible to attribute SM behavior to one or more miners who have a much lower orphan rate than other miners; we leave this analysis to future work. Doublespend (DS) attacks [24] are more challenging to carry out in Bonded Mining since the attacker must both succeed and avoid a commitment validation test failure in order to prevent loss of bond. Lastly, a miner could falsely report the timestamp in her own block headers. We do not investigate the impact of timestamp manipulation in this preliminary work, but note that the Bonded Mining protocol could impose restrictions on timestamps to greatly reduce the impact of such an attack. For example, it could stipulate that miners should synchronize clocks using NTP and that a miner should discard any block header that reports a timestamp deviating from their own clock by more than a reasonable block header propagation delay, perhaps 30 s.

4 Report Validity Test

We require a statistical test of the validity of a sequence of reports $r_1^m, \ldots, r_n^m$ from miner m. The test should have both high precision and recall in order to simultaneously prevent malicious attacks while refraining from bond slashing honest miners with high probability. It should also be capable of simultaneously detecting various types of attacks. In this section, we describe a statistical test that effectively detects both short- and long-range attacks. It requires a sample of n of the miner's most recent reports, where n varies with her hash rate, but can otherwise be treated as a black-box test within the Bonded Mining protocol.

Our approach is to use the popular one-sample Kolmogorov-Smirnov (KS) test [22] as a building block for test Valid. It is tempting to use a simpler point estimate of hash rate expected value as opposed to a goodness-of-fit (GoF) test over an entire distribution (which is provided by the KS test). There are two reasons that a distribution test is more desirable. First, the KS test is much more sensitive to systematic deviations in hash rate. Second, point estimates allow for attacks in which miners game the system by, for example, front loading all of their hash rate at the beginning of a test window; in that case the correct mean will be achieved, but the distribution will be wrong.

Consider the random sequence $\mathbf{T}^m = T_1^m, \ldots, T_n^m$, denoting the inter-arrival times of blocks $k_1^m, \ldots, k_n^m$ mined by m. By $h_1^m, \ldots, h_n^m$ we denote the actual average hash rate for miner m, with h_i^m associated with the time between blocks k_{i-1}^m and k_i^m ($h_i^m = r_i^m$ when m reports hash rate honestly). Equation 3 can be rewritten as

$$E[T_i^m] = \hat{D}_i^m / h_i^m, \tag{4}$$

where $\hat{D}_i^m$ is the average difficulty between blocks k_{i-1}^m and k_i^m; i.e. the expected time required for m to mine her ith block is equal to the average difficulty divided by her hash rate. It is well known [5,27] that each T_i^m follows the exponential distribution. That is, $T_i^m \sim \text{Expon}(\hat{D}_i^m/h_i^m)$. And because the distribution is a scale family, it is straightforward to show that $(T_i^m h_i^m/\hat{D}_i^m) \sim \text{Expon}(1)$. Now define statistic $\mathbf{X}^m = X_1^m, \ldots, X_n^m$ such that $X_i^m = T_i^m r_i^m/\hat{D}_i^m$. When m reports honestly, we have $X_i^m \sim \text{Expon}(1)$.

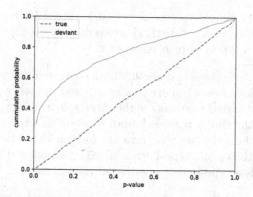

Fig. 2. Empirical cumulative distribution of p-values over 1000 trials for 500 samples drawn from the *true* distribution $\text{Expon}(1)$ (blue, dashed), and a *deviant* sequence of exponential random variables with mean drawn from a random walk about 1 with standard deviation equal to 0.01 (orange, solid). The true samples manifest p-values according to Eq. 5, in that for any given $x \in [0, 1]$, approximately x p-values fall under value x. (Color figure online)

The one-sample Kolmogorov-Smirnov (KS) test is a statistical test of the null hypothesis $\mathcal{H}_0$ that samples $\mathbf{x}^m = x_1^m, \ldots, x_n^m$, collectively an instance of $\mathbf{X}^m$, were drawn from a given distribution. In our case, the distribution is $\text{Expon}(1)$. The KS statistic is given by $\Delta_n = \sup_x |S_n(x) - F(x)|$, where $F(x)$ is the cumulative distribution function for $\text{Expon}(1)$ and $S_n(x)$ is the empirical distribution function derived from samples $\mathbf{x}^m$. The alternative hypothesis $\mathcal{H}_1$ contends that the samples were not drawn from $\text{Expon}(1)$. Now let $p(\Delta_n)$ be the p-value for Δ_n and define δ_i to be a realization of Δ_n. By definition [8],

$$P(p(\Delta_n) < \tau \mid \mathbf{X}^m \sim \text{Expon}(1)) < \tau. \tag{5}$$

Therefore, assuming that $\mathbf{x}^m$ was drawn from $\text{Expon}(1)$, we are guaranteed that the rejection region $p(\delta_n) < \tau$ will carry probability of type-I error (falsely rejecting $\mathcal{H}_0$) no greater than τ. Accordingly, large p-values provide evidence in support of $\mathcal{H}_0$ and low values provide evidence in support of $\mathcal{H}_1$.

Figure 2 shows the empirical cumulative distribution for $p(\delta_n)$ drawn from two different distributions: *true* and *deviant*. The true samples (blue, dashed curve) are drawn from distribution $\text{Expon}(1)$ while the deviant samples (orange, solid curve) are drawn from a sequence of exponential random variables with mean changing according to a random walk about 1 with standard deviation equal to 0.01. The true samples produce nearly perfect p-values in that approximately fraction x of all trials fall below p-value x, indicating an accurate estimation of type-I error. The deviant samples manifest significantly lower p-values, indicating high confidence in $\mathcal{H}_1$. Approximately 30% of deviant samples register a p-value very close to 0. We summarize a binary test based on the KS statistic below.

Definition 1. For tunable threshold $\tau \in [0,1]$, test $\text{KS}(\mathbf{r}^m; n, \tau)$ is equal to 1 when the p-value $p(\delta_n)$ of KS statistic Δ_n exceeds τ and 0 otherwise. Accordingly, the probability of a type-I error is equal to τ.

Because we are interested in detecting both short- and long-range attacks, we apply two KS tests: one over a short window n_s; and the other over a longer window n_l. Associated with these tests are the thresholds τ_s and τ_l, respectively. The window sizes and thresholds vary with both test type and the miner's committed hash rate. We combine the two tests into one to form Valid by multiplying their output. The probability of a type-I error in Valid is bounded by $\tau_s + \tau_l$ because it occurs if there is a type-I error in either of the KS tests.

Definition 2. $\text{Valid}(\mathbf{r}^m; n_s, n_l, \tau_s, \tau_l)$ is equal to 1 when both $\text{KS}(\mathbf{r}^m; n_s, \tau_s)$ and $\text{KS}(\mathbf{r}^m; n_l, \tau_l)$ are equal to 1, and it is 0 otherwise. The probability of a type-I error is bounded by $\tau_s + \tau_l$.

5   Attack Risk and Detection

In this section, we quantify the *coins at risk* for miners who report their hash rate dishonestly, which we regard as an attack. When a miner fails test Valid, her entire bond of nb coins are lost and she must proceed through the Bootstrapping state to resume mining. Thus, at a minimum, the coins at risk are the expected amount of bond lost when test Valid fails. But, any failure is also a significant step back for a miner, the γ parameter (see Sect. 2) limits her committed hash rate during bootstrap (a period of about 70 days), which amounts to an opportunity cost in terms of coinbase revenue.

We ran a simulator of PoW block mining to test the ability for a dishonest miner, with varying fraction of the total hash power, to conceal short- and long-range deviations from her commitment. We evaluated miners in three categories: *(i) honest, (ii) short-range dishonest,* and *(iii) long-range dishonest.* All miners varied their commitment each time they generated a block by performing a random walk with standard deviation equal to 1% of their originally committed hash rate. Honest miners always reported this deviation, while long-range dishonest miners reported that their commitment never changed from its original value. Short-range dishonest miners were honest about their long-range deviations in hash rate (just like honest miners), but at the end of each test window they mined at $1/5$ of their committed hash rate for approximately one week, while reporting no deviation from their commitment.

We generated block creation times by sampling from $\texttt{Expon}(\alpha T)$, where α was given by the ratio of total hash rate h_i to actual miner hash rate h_i^m. (αT is equivalent to Eq. 4 when actual hash rates are known.) Each block was randomly assigned to either the honest or dishonest groups with a probability proportional to the fraction of total hash rate possessed by the dishonest group.

Test Application. There are two separate, but related, goals for test $\texttt{Valid}$. First, given a single sequence of n consecutive blocks called a *test window*, we require that $\texttt{Valid}$ can distinguish between honest and either short- or long-range dishonest miners by means of a statistical test on sub-windows of size n_s and n_l within the those n blocks. Second, to understand the probability of ongoing attack success, we also require that $\texttt{Valid}$ continue to differentiate between honest and either short- or long-range dishonest miners during a long temporal sequence of overlapping test windows. The two goals are met by applying test $\texttt{Valid}(\mathbf{r}^m; n_s, n_l, \tau_s, \tau_l)$ with the long window $n_l = n$, short window $n_s \ll n$, short threshold τ_s, and long threshold τ_l all defined so that the probability of a type-I error remains extremely low for an entire year of mining. Our metrics for success are type-I error and attack detection rate (i.e., the rate at which $\texttt{Valid} = 0$ when applied to the attacker's test windows). The former should be sufficiently close to 0 so as to encourage honest miners. And the latter should be somewhat higher than 0 so as to discourage dishonest miners. These parameters can be chosen offline and independent of blockchain conditions.

Parameter Selection. Table 1 shows the parameter choices we made, which we applied to all experiments reported below. In general, lower hash rate miners (as a percentage of the total) require smaller windows for $\texttt{Valid}$ to produce dishonest-detection rates that would be significant to attackers. This is a desirable property because lower hash rate miners require more time to produce blocks. We chose n_s and n_l so that the expected temporal duration of each test sub-window was the same for miners across all hash rates (approximately 1.5 days for n_s and 70 days for n_l). We then chose the values for τ_s and τ_l to render the probability of type-I error close to 0, even after one year of expected mining time. These parameter choices are validated in the next sections.

Attack Detection. In Table 1 we specify short- and long-range test sub-windows for committed hash rates varying from 50% down to 1% of the network total. Given these parameters, the short-range sub-window for a miner committed to 1% of the hash rate is $n_s = 2$. To discourage Sybil attacks parameters must be chosen equitably for honest miners with any hash rate, as we argue in Sect. 3. This implies that, in its current incarnation, Bonded Mining cannot support miner commitments less than 0.5% of the total hash rate, because this would require a value for n_s that is less than 1. Note that this limitation is tied to the amount of proof of work we receive from miners; i.e., we cannot detect an attack during periods of time when the miner does not produce blocks. For BTC and BCH, respectively, 99.4% and 98.6% of the total hash rate from March

Table 1. Parameters choices for test `Valid` given committed hash rate as a percentage of the total. Short- and long-range test sub-window sizes are reported as n_s and n_l, respectively, while short- and long-range tolerances are reported as τ_s and τ_l, respectively.

Committed hash rate	Short window n_s	Long window n_l	Short threshold τ_s	Long threshold τ_l
1%	2	100	10^{-7}	10^{-7}
10%	20	1000	10^{-10}	10^{-10}
25%	50	2500	10^{-12}	10^{-12}
50%	100	5000	10^{-12}	10^{-12}

24–June 24, 2019 was contributed by miners with at least 1% of the hash rate.[2] Smaller miners do not need to be excluded from the system as they can join a larger pool. And in future work, we will explore PoW schemes such as Bobtail [5] and FruitChains [26] that have miners broadcast additional PoW information. For example, Bobtail could easily decrease Bonded Mining's minimum allowable hash rate by several orders of magnitude.

Table 2. [Attacks During Bootstrapping]. Results over 1000 trials of test `Valid` for various attacker hash rates (as a percentage of the total) and parameters selected according to Table 1. All results in this table correspond to dishonest miners who deviate from their committed hash rate during bootstrap. Long-range attackers deviate according to a random walk with standard deviation equal to 1% of their commitment. Short-range attackers drop to 1/5 of their committed hash rate for the last n_s blocks in the window, where n_s is determined by Table 1. See Fig. 3 for rates after bootstrapping.

Committed hash rate	Short-range detection rate	Long-range detection rate
1%	0.033	0.271
10%	0.108	0.653
25%	0.837	0.660
50%	1.000	0.546

5.1   Accuracy of `Valid` Over the Bootstrapping Window

Table 2 shows the probability of attack detection at the end of the bootstrapping window (block n) for short- and long-range attackers. Results are based on parameters from Table 1 for four different hash rate percentages, averaged over 1000 trials for each. We do not show the rate of type-I error when miners are honest because not a single error was encountered in the 1000 trials. However, by construction, the probability of a type-I error in any given test window is bounded by $\tau_s + \tau_l$.

---

[2] See https://btc.com/stats/pool and https://bch.btc.com/stats/pool.

The most important detection rate is for the 1% miner because a high hash rate miner can masquerade as multiple low hash rate miners. Thus, a determined attacker can avoid detection during the Bootstrapping state with 96.7% probability. From Table 1, $n = n_l = 100$; and thus there are $0.03nb = 3b$ coins at risk for the attacking miner. We show next that any initial success will likely be short-lived and the coins at risk will rise if the attacker continues as a fully bonded miner.

5.2   Attacks by Fully Bonded Miners

Once in the Fully Bonded state, a miner can remain in that state by continuing to add a new commitment for each additional block mined. Valid is tested against the sliding window of the n most-recent blocks generated by the miner. These sliding windows are highly dependent since adjacent windows share all but one block. If any single test fails, the total bond of nb is lost.

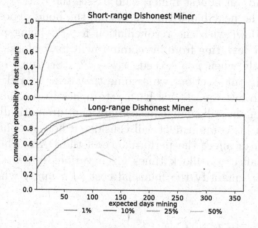

Fig. 3. [Attacks by fully bonded miners.] Probability that at least one test window fails Valid for short-range (top) and long-range (bottom) attacks spanning up to one year in duration. Short-range curves not visible are extremely close to probability 1.0 at nearly all times. (Note, for those not reading in color: in the top plot, the 50% curve appears above the 1% curve; and in the bottom plot, curves appear in the following order from top to bottom: 25%, 10%, 50%, 1%.)

We simulated one year's worth of blocks from attacking miners subject to Table 1 parameters, and we determined the mean probability (across 1000 trials) that Valid detects at least one attack during the year. Results in Fig. 3 show that approximately 30 days after bootstrapping, even attacks by 1% miners are detected with 50% probability. Thus, since again $n = n_l = 100$, the coins at risk after 30 days of attack is $0.5nb = 50b$.

Probability of Type-I Error. We ran the same experiment for honest miners who similarly varied their hash rate as in the short- and long-range attacks for multiple test windows spanning a year in duration, except that in these experiments miners reported their hash rate honestly. Over the course 1000 trials for each set of parameters selected from Table 1, test `Valid` never returned a false positive in any test window throughout the simulated year of mining across all hash rate commitments and for both short- and long-range deviations. Therefore, the probability of type-I error is no higher than 0.003 based on a 95% confidence interval from the 1000 trials [15]. Overall, miners that honestly report their commitments are very unlikely to fail test `Valid`.

6   Block Time Stability

Section 4 introduces a hypothesis test for determining when a miner reports his hash rate honestly over a window of n consecutive blocks that he has mined. If the miner fails this test, his bond of nb coins is lost; and if he misbehaves such that his probability of failure is p, then we say that he has *pnb coins at risk*. On the other hand, an honest miner, who passes the test, is eligible to submit a transaction in the next block that reconciles the bond deposited for the first block in the window, with the reconciliation fee f^m being paid according to Eq. 1. Thus, when deviating from his commitment for x blocks, a rational miner will report honestly when pnb exceeds $x(b - f^m)$. Section 5 quantified pnb for various attacks. In this section, we assume that coins at risk are high enough that the miner always reports his hash rate honestly (i.e. $pnb > x(b - f^m)$). Nevertheless, he might still change his commitment dramatically or even deviate significantly from his commitment (all considered honest behavior). Naturally, these deviations can affect the performance of the DAA. Thus, our goal is to *(i)* quantify the affect on block times given various magnitudes of hash rate deviation, and *(ii)* quantify the financial cost to a miner who performs these deviations.

6.1   Block Time Simulation

To achieve the goals of this section, we created an expected block time simulation that takes aggregate miner hash rate as input, generates blocks according to current hash rate and difficulty, and runs a DAA to update the difficulty. We compared two scenarios: one using the DAA for Bitcoin Cash (*BCH*, for brevity); and the other using the Bonded Mining protocol DAA (*BM*, for brevity). We chose to compare against Bitcoin Cash because it uses one of the newest and most dynamic DAAs currently deployed. Because the simulation reports expected block time, it is deterministic given miner preferences (Table 3). The simulation is meant to highlight DAA behavior in a variety circumstances, we leave more systematic analysis to future work.

Blocks. Both BCH and BM targeted block time $T = 600$ s. In both cases, the expected block time was calculated as $T \cdot D_i / h_i$ (a relationship that follows from Eq. 2), where h_i and D_i are the total hash rate and difficulty, respectively, during the period when block i is mined.

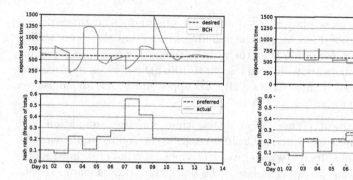

Fig. 4. Comparison of expected block times in simulation for (left) the Bitcoin Cash protocol (BCH) and (right) the Bonded Mining protocol (BM) where miner cost tolerance is $\kappa = 0.25$. Bottom facets show fluctuations in miner hash rates, while the top facets show the block time that results from the DAA response to those fluctuations. The target block time for both protocols is $T = 600\,\text{s}$. Relative to BCH, The BM DAA deviates less from desired block time in terms of amplitude and duration.

Table 3. Hash rate preferences, expressed as a percentage of total available hash rate, for miners during two weeks of simulated mining. The stated preference extends from the stated day until the next stated day or the end of the experiment.

Day 1	Day 2	Day 3	Day 4	Day 5	Day 6	Day 7	Day 8	Day 9
10%	7.5%	22.5%	11.3%	22.5%	28.1%	56.3%	42.2%	21.1%

Preference. At any given time we assumed that miners had a *preference* for what percentage of their hash rate they would like to apply to mining on the Bonded Mining blockchain. The notion of preference captures a miner's tendency to either divert her hash rate to another blockchain or stop mining altogether. Preferences could change, for example, because of changes in coin or energy prices. We do not model miner preferences but rather treat them as input.

Hash Rate. For simplicity, we adjusted all hash rates in unison every block (as though all miners acted in concert). But we set the block window to $n = 1000$ and updated commitments only every 10th block as if the network was comprised of 10 miners, each having 10% of the total hash rate. Over the course of two weeks of simulated mining time, miners developed the preference for a given hash rate at different times, expressed as a percentage of their total hash rate available (see Table 3). For BCH, miners immediately realized their hash rate preference. However, for BM, miners were restricted in two ways: *(i)* we set the commitment constraint to $\mu = 2$ (see Sect. 2) so that miners could not change their commitment by more than twice the average of their n earlier commitments; and *(ii)* miners were averse to deviating from their commitments because such deviations would result in a loss of bond. Note that BM miners were assumed to be fully bonded, so parameter γ was not relevant.

Cost Tolerance. Miners obeyed a variable *cost tolerance*, κ, expressed as a fraction of per-block bond b, which is the amount of bond that they were willing to lose per block due to deviation from their commitment. For example, if $\kappa = 0.1$ and $b = 100$ USD, miners are willing to lose up to 10 USD per block by deviating from their commitment if there exists a 10 USD opportunity cost for mining on the present chain over another.

Difficulty Adjustment. The target block time for the simulation was $T = 600$ s. BCH used the Bitcoin Cash DAA [28]: the rolling sum (over the last 144 blocks) of difficulty D (which gives the expected number of hashes performed) and block time M are calculated; the block time sum is clamped with a high and low pass filter: $M' = \mathbf{max}\{72 \cdot T, \mathbf{min}\{M, 288 \cdot T\}\}$; and the new difficulty becomes D/M'. The DAA for BM was Eq. 3: $D_i = c_i T$, which is the product of target block time and total committed hash rate.

6.2 Simulation Results

During the two weeks of simulated mining time, hash rate preferences changed according to Table 3. Figure 4 compares the resulting expected block times for each. BCH (left plot) shows several interesting features. First, we can see that expected block time deviates significantly from the target time during periods immediately following a major change in hash rate: the most dramatic drop is to 250 s, and the most dramatic rise is to nearly 1500 s. Moreover, these deviations were not corrected by the DAA for up to one day following the change in hash rate. Second, even during times when hash rate remains unchanged (for examples the period after Day 11, the expected block time oscillates about the target time. We suspect that this phenomenon is due to a feedback cycle that emerges from the DAA, which itself is essentially a feedback controller.

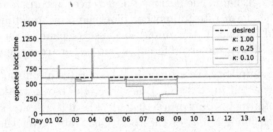

Fig. 5. Comparison of expected block times using Bonded Mining with varying cost tolerances κ. The target block time was $T = 600$ s. (Note, for those not reading in color, $\kappa = 1.0$ is the curve that varies most from desired, followed by $\kappa = 0.25$, and $\kappa = 0.1$, which has the lowest variance from desired.)

In contrast, Fig. 4 (right) shows that BM has much lower amplitude and duration of deviation from expected block time relative to BCH. And when hash rate is not changing, BM is able to maintain an expected block time that exactly matches the target time.

The Bonded Mining protocol is able to keep expected block time close to the target time by controlling the change in commitment and the tendency to deviate from that commitment. Miners are not allowed to vary their commitments by more than factor $\mu = 2$ from the average of previous commitments. Nevertheless, they may still vary their *actual* hash rate as long as they can tolerate the penalty. Figure 4 (right) shows that miners with cost tolerance $\kappa = 0.25$ will follow their hash rate preference, but only up to a point. When the preference deviates far beyond what the factor μ allows, miners cease to change their hash rate in order to avoid bond loss beyond 25% per block. Figure 5 shows how the expected block time begins to deviate more and more as the cost tolerance κ increases up to 1.0. What is not shown in the figure is that for $\kappa = 1.0$, the hash rate preference is followed perfectly, meaning that miners sacrifice as much bond as is necessary to meet their preference. For $\kappa = 1.0$, the worst-case scenario for Bonded Mining, Fig. 5 also shows that while the duration of deviation from the target block time is longer than for BCH, the maximum deviation is less.

6.3 Setting the Bond

We have intentionally not specified how the community should choose the per-block bond value b. It is a security parameter that trades off between accessibility and security of mining, and the correct tradeoff is subjective. For example, consider setting b equal to 1/10 the coinbase value. This requires a fully bonded miner (with any committed hash rate) to lockup the equivalent of approximately 10% of her mining profit for 70 days. This financial commitment is a strong deterrent for would-be attackers since failing even a single test Valid results in a total loss of bond. We do note that it could be a significant hardship for miners who operate on slim profit margins.

7 Related Work

Considerable effort has been devoted to refining the PoW difficulty adjustment algorithm (DAA) in the context of feedback control. In comparison to Bonded Mining, these past approaches do not solicit information on future hash rate from miners. They suffer from oscillations and delay in reaching the target block time because they are reactive rather than proactive.

Alternative DAAs include methods by Kraft [20], Meshkov et al. [23], Fullmer and Morse [13], and Hovland et al. [17]. Meshkov et al. note that alternating between coins has an effect on inter-block time. Grunspan and Pérez-Marco [14] recently proposed a modified DAA that includes uncle blocks in the difficulty calculation to thwart selfish mining. The recent fork of Bitcoin Cash (BCH) [1] from the core Bitcoin blockchain [2] (BTC) sparked a flurry of DAA research. The principal concern was to mitigate affects of *fickle mining* [21], where miners abruptly move their hash power from the BCH to the BTC blockchain (or vice versa). See also Aggarwal and Tan for an analysis of mining during that time [3]. Séchet devised the algorithm *cw-144* [28], which is currently used in BCH; it uses the ratio of the rolling average of chain work to block time and was based on an early model from Kyuupichan [6]. Harding's *wt-144* [16] is a similar solution, weighting block time by recency and block target. Stone [30] proposed adding *tail removal* to the algorithms above, which drives down the difficulty within a block interval in order to guarantee liveness and dampen block time oscillations.

We use a bond as collateral for a pledge to perform an offered hash rate over a specific period of time. Similarly, Proof of Stake (PoS) protocols (including hybrid PoW/PoS) [4,7,11] accept collateral as a pledge to act honestly in validating transactions over a specific period of time. We do not intend our work to be a replacement for PoS, but note that Bonded Mining similarly provides a fixed set of consensus participants that in our case are validated by PoW. For hybrids of PoW/PoS and hybrids of PoW/BFT [9,19], Bonded Mining can bolster the performance of the PoW component.

8   Conclusion

We presented Bonded Mining, a proactive DAA for PoW blockchains that sets the mining difficulty based on bonded commitments from miners. The protocol incentivizes a miner to honor his commitment by confiscating a fraction of bond that is proportional to the deviation from his commitment. We developed a statistical test that is capable of detecting short- and long-range deviations from commitments with very low probability of falsely implicating honest miners. In simulation, and under reasonable assumptions, we showed that Bonded Mining is more successful at maintaining a consistent inter-block time than is the DAA of Bitcoin Cash, one of the newest and most dynamic DAAs in production. The present work is preliminary, having the following limitations: *(i)* currently the lowest hash rate miner that it supports has 1% of the total; *(ii)* we have only demonstrated defenses against two types of attacks; and *(iii)* we only directly compare Bonded Mining to the Bitcoin Cash DAA. Future work will seek to address these limitations. We recommend these limitations be fully investigated before deployment.

A List of Symbols

Symbol	Description
m	miner
n, n_s, n_l	blocks in test window: generic, short, long
b	bond, a portion of coin held until miner passes test Valid
k_i^m	ith block mined by m
c_i	committed hash rate for entire network between blocks $i-1$ and i
c_i^m	hash rate commitment for m between blocks k_{i-1}^m and k_i^m
h_i	actual hash rate for entire network between blocks $i-1$ and i
h_i^m	actual hash rate for m between blocks k_{i-1}^m and k_i^m
r_i^m	reported hash rate for m between blocks k_{i-1}^m and k_i^m
f_i^m	reconciliation payment to m in block k_i^m
T_i^m	time between k_{i-1}^m and k_i^m
T	target inter-block time
D_i	mining difficulty (expected number of hashes per block) between blocks $i-1$ and i
$\hat{D}_i^m$	average difficulty between blocks k_{i-1}^m and k_i^m
μ	maximum multiplicative increase in commitment over previous average
γ	maximum fraction of total hash rate allowed in Bootstraping state
τ_s, τ_l	short- and long-range KS test tolerances
Δ_n	KS test statistic for window size n
$p(\Delta_n)$	p-value for Δ_n
κ	miner cost tolerance, expressed as fraction of bond b

References

1. Bitcoin Cash. https://www.bitcoincash.org/
2. Bitcoin Core. https://bitcoin.org
3. Aggarwal, V., Tan, Y.: A Structural Analysis of Bitcoin Cash's Emergency Difficulty Adjustment Algorithm (2019). https://doi.org/10.2139/ssrn.3383739
4. Bentov, I., Gabizon, A., Mizrahi, A.: Cryptocurrencies without proof of work. In: Clark, J., Meiklejohn, S., Ryan, P.Y.A., Wallach, D., Brenner, M., Rohloff, K. (eds.) FC 2016. LNCS, vol. 9604, pp. 142–157. Springer, Heidelberg (2016). https://doi.org/10.1007/978-3-662-53357-4_10
5. Bissias, G., Levine, B., Bobtail, N.: A proof-of-work target that minimizes blockchain mining variance. Technical Report arXiv:1709.08750, September 2017
6. Booth, N.: The K-1 Difficulty Adjustment Algorithm. https://github.com/kyuupichan/difficulty, (August 2017)
7. Buterin, V., Griffith, V.: Casper the friendly finality gadget. arXiv:abs/1710.09437, October 2017
8. Casella, G., Berger, R.L.: Statistical Inference. Brooks Cole, California (2002)
9. Decker, C., Seidel, J., Wattenhofer, R.: Bitcoin meets strong consistency. In: Proceedings Conference on Distributed Computing and Networking (ICDCN) (2016)
10. Douceur, J.: The sybil attack. In: Proceedings Peer-to-Peer Systems (IPTPS), March 2002

11. Duong, T., Chepurnoy, A., Fan, L., Zhou, H.-S.: TwinsCoin: a cryptocurrency via proof-of-work and proof-of-stake. In: Proc. ACM Workshop on Blockchains, Cryptocurrencies, and Contracts, pp. 1–13 (2018)
12. Eyal, I., Sirer, E.G.: Majority is not enough: Bitcoin mining is vulnerable. In: Conference on Financial Cryptography and Data Security, pp. 436–454 (2014)
13. Fullmer, D., Morse, A.S.: Analysis of difficulty control in Bitcoin and proof-of-work Blockchains. In: Proceedings of the IEEE Conference Decision and Control (CDC), December 2018
14. Grunspan, C., Perez-Marco, R.: On Profitability of Selfish Mining. arXiv:abs/1805.08281, January 2019
15. Hanley, J.A., Lippman-Hand, A.: If nothing goes wrong, is everything all right? Interpreting zero numerators. J. Am. Med. Assoc. **249**, 13 (1983)
16. Harding, T.: Implement weighted-time difficulty adjustment algorithm, October 2017. https://reviews.bitcoinabc.org/D622
17. Hovland, G., Kucera, J.: Nonlinear feedback control and stability analysis of a proof-of-work blockchain. Model. Ident. Control **38**(4), 157–168 (2017)
18. Király, T., Lomoschitz, L.: Profitability of the coin-hopping strategy, March 2018. http://web.cs.elte.hu/egres/www/qp-18-03.html
19. Kogias, E.K., et al.: Enhancing Bitcoin security and performance with strong consistency via collective signing. In: Proceedings of the USENIX Security Symposium, pp. 279–296 (2016)
20. Kraft, D.: Difficulty control for blockchain-based consensus systems. Peer-to-Peer Networking Appl. **9**(2), 397–413 (2016)
21. Kwon, Y., Kim, H., Shin, J., Kim, Y.: Bitcoin vs. Bitcoin cash: coexistence or downfall of bitcoin cash? In: Proceedings of the IEEE Symposium Security and Privacy, May 2019
22. Massey Jr., F.J.: The kolmogorov-smirnov test for goodness of fit. J. Am. Stat. Assoc. **46**(253), 68–78 (1951)
23. Meshkov, D., Chepurnoy, A., Jansen, M.: Revisiting difficulty control for blockchain systems. In: Cryptocurrencies and Blockchain Technical, pp. 429–436 (2017)
24. Nakamoto, S.: Bitcoin: A Peer-to-Peer Electronic Cash System, May 2009
25. Ozisik, A. P., Bissias, G., Levine, B. N.: Estimation of Miner Hash Rates and Consensus on Blockchains. Technical Report arXiv:1707.00082, July 2017
26. Pass, R., Shi, E.: FruitChains: a fair blockchain. In: ACM Symposium on Principles of Distributed Computing, pp. 315–324 (2017)
27. Rizun, P.: Subchains: a technique to scale bitcoin and improve the user experience. Ledger **1**, 38–52 (2016)
28. Sechet, A.: Implement simple moving average over work difficulty adjustement algorithm, October 2017. https://reviews.bitcoinabc.org/D601
29. Stone, A.: The blockchain difficulty information paradox, 17 October 2017. https://medium.com/@g.andrew.stone/the-blockchain-difficulty-information-paradox-879b0336864f
30. Stone, A.: Tail Removal Block Validation, October 2017. https://medium.com/@g.andrew.stone/tail-removal-block-validation-ae26fb436524
31. Zegers, A.: Bringing Stability to Bitcoin Cash Difficulty Adjustments, August 2017. https://medium.com/@Mengerian/bringing-stability-to-bitcoin-cash-difficulty-adjustments-eae8def0efa4

12 Angry Miners

Aryaz Eghbali[✉] and Roger Wattenhofer

ETH Zurich, Zurich, Switzerland
{aeghbali,wattenhofer}@ethz.ch
https://disco.ethz.ch/

Abstract. In this paper we investigate the behavior of miners, in terms of which type of hardware they use, based on publicly-available macro-scale data of Bitcoin. We provide a model for the market share of mining hardware, which is then used to estimate the energy consumption, the distribution of electricity price among Bitcoin miners, and the total investment in the backbone of the Bitcoin network.

Keywords: Bitcoin · Data analysis · Mining hardware

1 Introduction

Raw materials such as minerals, metals, coal, oil, gas, or gems are the foundation of the world's economy. Mining these materials is big business, with major investments. The raw material mining companies are somewhat secretive about their mining equipment and investments.

Cryptocurrency mining shares some of this secrecy. Cryptominers do not publish how much they invest in mining equipment or how much they actually mine. Mining pools advertise their hashing power to attract more miners, but their self-announced numbers should be handled with care, as there are various incentives to claim both higher or lower hash rates.

However, in contrast to raw materials, a lot of cryptomining information is available in the public cryptocurrency blockchain. In this paper, we ask to what degree we can understand the cryptomining business by analyzing public information. What mix of mining equipment is being used? How much energy is spent, and at what cost? When is it no longer profitable to mine with outdated equipment, and when does it become profitable again? In our paper, we focus on the largest cryptocurrency, Bitcoin.

Bitcoin mining became a serious business already in the early years of Bitcoin's inception. It also created businesses like ASIC manufacturing, mining pools, and cloud mining around it. Although the hashing power and mining hardware are integral parts of this ecosystem, since not much data has been publicly released by miners, hardware manufacturers, or mining pools, data-driven studies focusing on the hardware market have been difficult and rare.

On the other hand, the money invested in any cryptocurrency shows how interested people are in its success. Many have argued that people who have

© Springer Nature Switzerland AG 2019
C. Pérez-Solà et al. (Eds.): DPM 2019/CBT 2019, LNCS 11737, pp. 391–398, 2019.
https://doi.org/10.1007/978-3-030-31500-9_25

stake in Bitcoin have the incentive to keep it alive. The market capitalization shows a part of the total stakes, however, another important, if not the most important, group of stakeholders are the miners. Yet, we do not know how much stake they have in this business.

2   Related Work

Analyzing Bitcoin's or other altcoins' blockchain data has been the focus of various recent work. In [7] Decker et al. studied the propagation of messages and their delays on the Bitcoin network. The graph of transactions in Bitcoin was extracted by Ron et al. in [10]. Reid et al. showed that Bitcoin does not preserve anonymity [9]. Anderson et al. [3] did some empirical analysis on Ethereum, Namecoin, and Peercoin. Bartoletti et al. [4] analyzed the Ethereum blockchain to detect Ponzi schemes. Cong et al. [6] investigated the mining pools. Recently, Ma, Gans, and Tourky [8] analyzed the equilibria of the Bitcoin mining game.

Two simple approaches have been used to estimate the energy consumption of Bitcoin. These estimations were popular when the Bitcoin price exploded in 2017, with a lot of press coverage. One approach is to divide the total hashing power by the hash rate of the most efficient (or average) mining hardware. Another approach is to take the value of mined Bitcoins in a period of time, and divide it by either the cheapest (or average) electricity cost of world region where most of the mining actually happens. These approaches showed that the mining energy consumption of Bitcoin is significant ("in the order of a medium European country"). Since our approach models the share of hashing power by each device, we can have more accurate estimations of the energy used.

In a series of reports, Bendiksen et al. [5] investigated the electricity consumption of Bitcoin and its distribution and types of sources that it come from. Their work presents good insight into the energy aspect of Bitcoin, using a journalistic approach. They gather data from many different sources and verify them, while using expert knowledge when the data is not credible enough. The results that we present in this paper can be used in similar reports as another source of information about another aspect of this market.

3   Data

Our Bitcoin blockchain data is fetched from blockchain.com [1]. We use the price, hash rate, transaction fees, and difficulty, which are reported daily. The hash rate h is calculated from the number of blocks found, n, in a day by the following formula:

$$h = \frac{n}{144} \cdot d \cdot \frac{2^{32}}{600},$$

where d is the difficulty. The process of mining blocks is a Poisson random process and hence the number of blocks mined in a day might change even if the hashing power has remained constant. So, it is important to distinguish the perceived hash rate from the actual hash rate. We use the smoothed hash rate as the actual hash rate of the network.

The data for mining hardware was extracted by a manual process. We consider the S series (S1–S11) of AntMiner by Bitmain, which became the dominating company to produce ASIC based mining hardware. We also consider one sample hardware from each generation of the pre-ASIC era, i.e. a CPU (Intel Core i7 920), a GPU (Nvidia GTX 460), and an FPGA (BitForce SHA256 Single). The specifications of devices used can be found in Table 1. The hashing power, and electricity usage of each device comes from what the manufacturer advertised, and the prices and release dates are either from the manufacturer or the main seller. The release date of the GPU is chosen as the first date which there is evidence of GPU mining happening in Bitcoin. We also added 30 days to the release dates of the Antminers before they can be used to account for the delivery times.

Other mining hardware, like Avalon miners, DragonMint miners, and even other CPUs and GPUs, have also been available on the market and used in mining. Nonetheless, we wanted to select a sample of devices that can be representative of all mining equipment, and avoid the factors of marketing and sales that affect the use of similar devices. So wherever our data refers to a particular mining device, the reader can imagine any device with similar specifications.

Table 1. Mining hardware that we consider and their specifications.

Device	Release date	Hash rate (TH/s)	Power consumption (KJ/TH)	Price (USD)
CPU	2008-11-01	0.0000192	4166.67	305.0
GPU	2010-07-18	0.00006831	2341.92	229.0
FPGA	2011-06-01	0.000832	96.1538	599.0
S1	2013-09-21	0.18	2.0	299.0
S2	2014-04-01	1.0	1.1	2259.0
S3	2014-07-10	0.4	0.77	382.0
S4	2014-10-16	2.0	0.7	1400.0
S5	2014-12-27	1.155	0.51	370.0
S5+	2015-08-14	7.722	0.44	2307.0
S7	2015-09-30	4.73	0.25	479.95
S9	2016-06-12	13.5	0.098	1987.95
S11	2018-11-01	20.5	0.075	1173.38

4 Model

Cryptomining is a serious industry, and almost all miners are participating because of monetary incentives, namely block rewards and transaction fees. Since buying and running mining hardware is expensive, people that mine because of other reasons are few, without considerable hashing power. So we assume all miners to be rational agents, i.e. they want to maximize their profit. To gain more profit, a miner might want to increase or decrease its hash rate based on Bitcoin's current or future price, its electricity costs, in relation to the total hash rate of the network. The actions an agent can take are to either turn off some running hardware, buy hardware, or turn on some hardware that is currently off. We will now explain how the model works in detail.

Starting from the first day of Bitcoin we calculate the hash rate share of each device on a daily basis from the total hash rate. Each day that the hash rate has increased from the previous day, some miners have either turned on some unused hardware, or they have bought new mining equipment. If the hash rate has decreased, then they must have turned off some of their hardware.

Case 1: Hash rate has decreased. From all currently running devices we start with the lowest performing one (in terms of both the power consumption and hash rate), which in our list is the oldest one. We decrease its hash rate share until it reaches zero hash rate and then move to the next lowest performing device, or the amount of decrease is smaller than the hash rate of that device, where in this case we move to the next day after decreasing its hash rate, such that the sum of hash rates matches the hash rate of that day. As long as a device is making profit, the rational decision is to keep it running. So when a mining hardware is turned off, it means that its profitability just dropped below zero. Knowing the profit is zero and the revenue of mining from the price and hash rate data, we can calculate the running cost in this situation, which is almost completely the electricity cost. Extracting the electricity price from this model is a result of the novelty of our approach. The electricity price is calculated by dividing the revenue per hash rate of that specific device by the hash rate deducted from that device. We have the revenue per hash rate for each day in the history of Bitcoin by dividing the total value of Bitcoins mined in that day by the hash rate in the same day. The reduction in hash rate is saved as "turned off" to be used in the next case. We also save the electricity price for each hardware that has been turned off.

Case 2: Hash rate has increased. We start by looking at the currently turned off hardware ordered from highest to lowest performing. We perform the following routine until the increase in total hash rate is justified in our model. If buying the newest available hardware on the market breaks even on electricity costs alone in less than three months, then we assume the miner buys the new device (when considering the break-even time based on the electricity costs alone for various mining hardware there is a large gap, which justifies this assumption). Otherwise, if the new hardware breaks even in a longer period of time, the miner turns on the highest performing devices that she has. Calculating the break-even point requires the knowledge of electricity price of the miner. We know the electricity prices by saving the average electricity cost of miners that use this particular device when they are turned off in the previous case. If the increase in hash rate is still higher than what we already considered, then we assume new hardware was bought. Similar to the previous case, the rational reason to turn on or buy hardware is that it is profitable. So we can again compute the electricity prices. However, these electricity prices that we calculate are not exact but bounds. So when we calculate electricity prices in Case 1, they are the lower bounds for the electricity prices of those miners, and in Case 2 they are the upper bounds. The reason that these are not exact prices is that in assuming agents to be rational we only consider current profits but not price and market speculations.

The biggest limitation of our model is that since the granularity of our data is limited to a day, we cannot capture changes that happen with higher frequency. However, as the nature of Bitcoin mining has randomness (the number of blocks mined in a day follows a Poisson distribution), shorter time frames cannot provide data with good accuracy. Even the daily data has fluctuations in hash rate caused by randomness. To avoid the error caused by these random fluctuations, we smooth the hash rate by taking a weighted average of neighboring points.

Using the total hash rate we cannot detect some changes that happen at the same time, such as when some mining device breaks down and somewhere else a device is turned on. These two events would cancel out each other and not appear in our analysis.

Our model inherently considers cases that are not mentioned above. For example, in recent years there have been some other cryptocurrencies that could be mined using the same hardware as Bitcoin's and some miners "coin hop", which is mining the most profitable coin and frequently switching between them. We incorporate this behaviour in our model by taking the Bitcoin Cash (BCH) data (total hash rate and price) and assuming when it is more profitable to mine Bitcoin Cash, the hash rate on Bitcoin decreases, and the hash rate on Bitcoin Cash increases, miners are moving from Bitcoin to Bitcoin Cash, and the other way around. The reason that we chose Bitcoin Cash is that it is the most popular cryptocurrency that uses the same proof-of-work as Bitcoin. Moreover, other altcoins with the same proof-of-work do not have noticeable hash rates and we do not have as much data as we need for them. Another case is when a wealthy miner which already has a mining farm running on miner x decides to buy miner y because it is more profitable. This miner might want to sell his x miners and buy y miners. Note that since xs are still profitable, not using them is not rational. Hence, this scenario is also captured by our model, as those x miners are considered to still be running (by another miner).

5   Analysis

Figure 1 shows our main result, which is the market share of each mining hardware through time, in logarithmic scale.

From the hash rate share we immediately get the power consumption by device, which is shown in Fig. 2 and compared to [2]. Although both estimates follow the same trend, the difference is the methods used. We use the hashing power of running devices to calculate how much electricity they use. In [2] the price of Bitcoin is used to estimate the energy costs, and by assuming an average of 5 cents per KWh, the total energy consumption is achieved. It is important

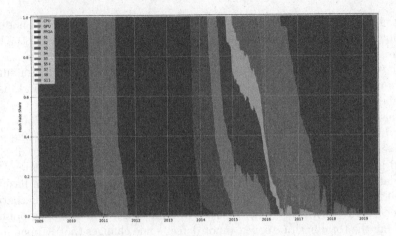

Fig. 1. Market share for each mining hardware.

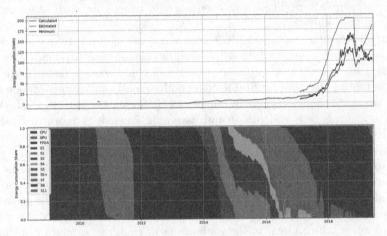

Fig. 2. The top plot shows the daily total energy consumption of Bitcoin mining in comparison to [2]. The "Calculated" line is our result, and the other two are from [2]. The bottom plot shows the daily electricity consumption share by mining device.

to note that the hash rate and price of Bitcoin are not always synchronized. There have been times where the hash rate lagged behind price, and there have been times where the price lagged behind the hash rate. We refer the reader to Bitcoin's charts on [1].

Since we have the price of Bitcoin and the total hash rate, we can compute how much revenue each unit of hashing power makes in each day. As mentioned in Sect. 4, turning on a device or buying new hardware shows that it just became profitable to mine using that device or hardware, and turning a device off shows that it just became unprofitable to mine. Hence, we can calculate the points in time where the profitability changed, and what the electricity price was for those points. The distribution of electricity prices for miners who turned off their devices can be seen in Fig. 3.

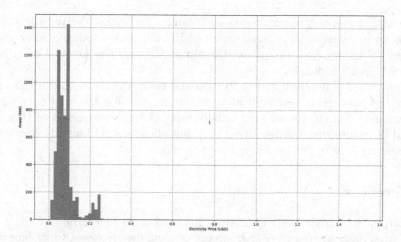

Fig. 3. The distribution of electricity prices for miners who turned off their device at some point. The distribution is weighted by the amount of energy used.

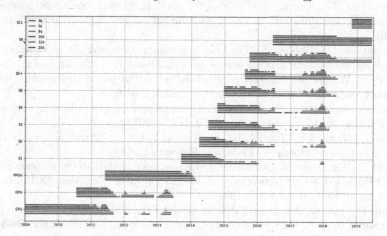

Fig. 4. The profitable period for each mining hardware is shown by colored lines (each cluster of lines represents one mining device). Different colors show different electricity prices. (Color figure online)

We can see when each of the 12 miners were profitable and when they became unprofitable in Fig. 4.

6 Conclusion

In this paper we present an empirical model for the market share of mining hardware. Using this model we inferred information about the energy consumption, which we found is 0.14% of the global electricity usage, the profitable times for each mining equipment, and the distribution of electricity prices for Bitcoin miners, that matches many reports and claims. We further derived the amount of money invested in the mining sector of Bitcoin, which today is more than 7 billion US dollars, and throughout these past 10 years has summed up to more than 9 billion US dollars.

Our model is implemented in Python and open sourced on Github[1], so that interested researchers can replicate our results, and use them to gain further insight into this market and potentially others. A nice extension of this work would be to implement a web-based tool that allows users to modify parameters and assumptions of our model and see the effects.

References

1. https://www.blockchain.com/charts/. Accessed on 20 June 2019
2. https://digiconomist.net/bitcoin-energy-consumption. Accessed on 20 June 2019
3. Anderson, L., Holz, R., Ponomarev, A., Rimba, P., Weber, I.: New kids on the block: an analysis of modern blockchains. arXiv preprint arXiv:1606.06530 (2016)
4. Bartoletti, M., Carta, S., Cimoli, T., Saia, R.: Dissecting ponzi schemes on ethereum: identification, analysis, and impact. arXiv preprint arXiv:1703.03779 (2017)
5. Bendiksen, C., Gibbons, S.: The bitcoin mining network-trends, marginal creation cost, electricity consumption & sources. CoinShares Research **21** (2018)
6. Cong, L.W., He, Z., Li, J.: Decentralized mining in centralized pools. Technical report, National Bureau of Economic Research (2019)
7. Decker, C., Wattenhofer, R.: Information propagation in the bitcoin network. In: IEEE P2P 2013 Proceedings, pp. 1–10. IEEE (2013)
8. Ma, J., Gans, J.S., Tourky, R.: Market structure in bitcoin mining. Technical report, National Bureau of Economic Research (2018)
9. Reid, F., Harrigan, M.: An analysis of anonymity in the bitcoin system. Security and Privacy in Social Networks, pp. 197–223. Springer, New York (2013). https://doi.org/10.1007/978-1-4614-4139-7_10
10. Ron, D., Shamir, A.: Quantitative analysis of the full bitcoin transaction graph. In: Sadeghi, A.-R. (ed.) FC 2013. LNCS, vol. 7859, pp. 6–24. Springer, Heidelberg (2013). https://doi.org/10.1007/978-3-642-39884-1_2

---

[1] https://github.com/EAAL/Bitcoin.

Author Index

Printed in the United States
By Bookmasters